COMMUNICATIONS SATELLITES

Global Change Agents

TELECOMMUNICATIONS
A Series of Volumes Edited
by Christopher H. Sterling

COMMUNICATIONS SATELLITES

Global Change Agents

Edited by

Joseph N. Pelton
Robert J. Oslund
The George Washington University

Peter Marshall
Communications Consultant

Routledge
Taylor & Francis Group

NEW YORK AND LONDON

First Published by
Lawrence Erlbaum Associates, Inc., Publishers
10 Industrial Avenue
Mahwah, New Jersey 07430

This edition published 2011 by Routledge
711 Third Avenue, New York, NY 10017
2 Park Square, Milton Park, Abingdon, Oxon, OX14 4RN

Cover design by Kathryn Houghtaling Lacey

Library of Congress Cataloging-in-Publication Data

Communications satellites : global change agents / edited by Joseph N. Pelton,
Robert J. Oslund, Peter Marshall.
      p.   cm. — (Telecommunications)
   Includes bibliographical references and index.
ISBN 0-8058-4961-0 (alk. paper)
ISBN 0-8058-4962-9 (pbk. : alk. paper)
   1. Artificial satellites in telecommunication.   2. Artificial satellites in
telecommunication—Social, technology, political, and economic aspects.
I. Pelton, Joseph N.   II. Oslund, Robert J.   III. Marshall, Peter (Peter P.)
IV. Telecommunications.

TK5104.C625   2004
384.5'1—dc22   2003064927
                                                             CIP

Publisher's Note
The publisher has gone to great lengths to ensure the quality of this reprint
but points out that some imperfections in the original may be apparent.

Contents

I. INTRODUCTION

II. TECHNOLOGY

VI. FUTURE TRENDS

VII. SYNTHESIS AND CONCLUSIONS

Foreword

Christopher H. Sterling, Series Editor
George Washington University

REMEMBERING THE EXCITEMENT . . .

In October 1957, I was one of thousands across the country watching the early evening skies for signs of something new in space—the Russian *Sputnik*, the world's first artificial satellite, launched earlier that month. We could make out the tiny moving satellite orbiting just high enough to reflect the setting sun. It would streak across the visible sky in just a few moments on its 90-minute orbits around the earth, but at least we could see this latest step in heading for the stars. Then just entering high school, I was too young to understand how the Russian's surprise launch of this pioneer struck the Washington, DC, military and policy communities. Nor did I realize then that Arthur C. Clarke had predicted orbiting "rocket stations" (communication satellites) in a British journal just a dozen years earlier, although he thought decades might pass before the first appeared. Years later, I saw a replica of the basketball-sized *Sputnik* at the United Nations, a gift of the proud nation that launched it.

Space travel and communication were hugely exciting things in the 1950s. Many of us had grown up with the other-worldly paintings of space travel by Chesley Bonestell illustrating the pages of weekly magazines and a subsequent series of popular books, including *The Conquest of Space* (1949), *Across the Space Frontier* (1952), and *Conquest of the Moon* (1953). Science fiction stories and novels, articles predicting space travel, and movies were all the rage. In an era of relatively slow propeller-driven airliners and jet fighter aircraft, the idea of *space* satellites or traveling was positively energizing.

ix

Not long after *Sputnik* (and a later companion that took a dog on a one-way trip into orbit—and sent animal rights people nearly into orbit as well), I got an early introduction to how interdisciplinary the field of space already was. After school, I would take a bus downtown to sit in the back of a large University of Wisconsin lecture hall and listen to a series of slide-illustrated lectures on developments in satellite and space technology. A lot of this was way over my head, especially the engineers and their fascination with equations, but others were gripping in their projections of what was coming. One of the speakers was Werner von Braun, well before he was widely known in the United States. Those lectures were packed, and the talks were followed by avid question-and-answer sessions. My own fascination did not, however, help boost my math or science grades any, and thus it quickly became clear I was not going to be *doing* anything in this field. At least I could read about the latest developments and watch them on TV. Thus, I saw the attempt to launch the pencil-thin U.S. Navy *Vanguard* missile (from a strange place called Cape Canaveral), only to see it come crashing down on the launch pad when its rocket engines malfunctioned. Even on black-and-white TV, that explosion was pretty impressive. The failed launch made clear just how close to the edge of technology satellite engineers were working.

Most exciting, of course, were the steps to place *man* in space. The selection of the first seven astronauts was headline news, and they quickly became "skygods." President Kennedy electrified us with his promise to place a man on the moon by the end of the 1960s. Soon we all watched TV network coverage of Scott Carpenter and then John Glenn in early *Mercury* one-person orbital missions. These were followed by the first "space walks" in the mid-1960s from *Gemini* two-man capsules. Yet the risks were high, the terrible space capsule fire that killed the three *Apollo 1* astronauts on the launch pad in 1967 being but one tragic example.

By this time, of course, satellites and space travel were of interest to more than star-struck kids and the military. When Communications Satellite Corporation (Comsat) was formed in 1962, its stock was quickly grabbed up by a technology-happy investment community. We watched in amazement as AT&T's *Telstar* made possible the first experimental two-way TV links from Europe to the United States only a few years after undersea telephone cables had entered service across the Atlantic. That satellites might have continuing practical applications was dawning on a growing number of people.

Flash ahead to Christmas Eve 1968, as millions of us watched network TV relay pictures from the formerly unknown far side of the moon. Astronaut Frank Borman aboard an *Apollo* vehicle orbiting the moon read from the Bible as we saw something no human had ever observed before—the earth rise over the surface of the moon. It was a magical moment, and memories of it still give me chills. Six months later, the world was watching again as astronauts Neil Armstrong and Buzz Aldrin became the first two men to walk on the surface of the moon. Then completing graduate school, several fellow students and I jury-rigged three TV sets side by side so we could monitor what each of the networks was showing on that landmark night in July 1969. Walter Cronkite of CBS had become the acknowledged network guru of space flight, always there for the critical launches and recoveries. As you learn in the pages that follow, the first moon landing drove Intelsat to complete full global coverage of its satellite network so that the entire world could see this event "live via satellite." Several other moon missions followed in the early 1970s, including one using a motorized "Rover" vehicle to cover greater distances. Landings on Mars seemed well within reach.

Yet we were continually reminded of how technically difficult and sometimes dangerous all this was. In 1986, network reporters broke into scheduled daytime programs to tell us the Space Shuttle *Challenger* had blown up shortly after launch, killing the crew of seven aboard. There had been almost no live coverage of the launch as we had grown accustomed to successful launches being the norm. The same shuttle system had another jarring setback early in 2003 when the *Columbia* came apart on reentry—and we watched contrails of multiple pieces when only one should have been moving across the sky. Again a crew was lost, and again scientists and engineers began piecing together information to determine what happened. Going into space was not without its costs.

The space industry and its satellites are, of course, also affected by what happens back on earth. The telecommunications industry's dramatic meltdown that began in 2001 sharply cut investment and thus the number of satellites once scheduled to be launched. Most mobile satellite projects—once the cutting edge of the industry—have gone bankrupt or been closed down altogether. As older satellites reach the end of their useful life, we increasingly lack the valuable redundancy that ensured continuing service. Military leaders are concerned about the decreasing availability of satellite bandwidth to "deliver the goods" where required and when. Commercial users of satellites worry when and where the needed new "birds" are going to originate. Things may be turning around—the New Iridium and Globalstar/ICO organizations seem to be on the rebound. Direct broadcast radio satellite systems (Sirius and XM Radio) are expanding, while direct to home (DTH) satellite TV systems continue to add millions of subscribers.

As we approach a half century since the first human-made satellite was placed into orbit, my university students cannot begin to comprehend how different their world is, thanks in considerable part to the role satellites play. They take satellite-delivered news on a 24/7 basis as a given. Live reports from Baghdad as a war begins are nothing new to them. Time zones have been erased. Distance does not matter. Few have any idea that satellites make possible their many evening entertainment choices, so much wider than those we knew a half century ago. They *assume* that hundreds of video entertainment channels, delivered nationally and sometimes worldwide, are a given. Satellites have become such an integral part of their everyday lives that life without their service is not imaginable. The technology has become essential.

The chapters that follow, all written by authorities with extensive experience, help clarify how essential satellites have become and how integrated they are within our global economy and culture. Both the positive and negative sides of satellite systems are explored. Thoughtful attention is given to the many dilemmas that satellites and information technology now pose to our lives. Drawing on the expertise derived from a menu of disciplines, this volume forms a valuable assessment of the many ways satellites have impacted human life—both for better and for worse. The authors review how we got here, where we are, and what is likely to happen. The breadth of coverage here underlines just how vital satellite communication has become. There are few aspects of daily life—whether commercial, military, or diplomatic—that are not in some way impacted by the ability to instantly communicate almost anywhere.

Getting to this point in satellite communications has been neither easy nor cheap. The result, however, has been well worth the effort. The authors and editors of this book make that abundantly clear.

Preface

The field of satellite communications is overcrowded. We are not talking about the problem of a crowded spectrum or crammed geosynchronous orbit, although this is now packed with a gaggle of high-tech machinery. No, the crowding of the satellite field we address here refers to the presence of so many disciplines.

Satellite systems actually involve many advanced technologies (from rocket science and orbital mechanics to transmission and multiplexing systems as well as material sciences and power systems), and this is just the start. Satellite operations include economics, marketing and business, news and entertainment, international law and regulation, international politics, military applications, education, health, sociology, and culture. Any serious attempt to explain and interpret the past, present, and future of satellite communications should recognize the interdisciplinary nature of the task.

Far more than technical knowledge is needed to understand the "why," "how," and "wherefore" of space communications. In fact over the past 40 years, satellites have created or stimulated a multitude of changes and innovations in our world. These impacts or breakthroughs have come in diverse arenas such as tele-education in China, Indonesia, and India; and health care in the Caribbean and East Africa. Then there are the political and cultural changes stimulated by international treaties and executive agreements as well as satellite TV news, entertainment, and intelligence gathering all over the world. E-businesses driven by satellite links to emerging markets have changed patterns of trade and brought innovations in international law and business. In one sense, it could be said that satellites have had a major role in redefining the nation-state. Satellites, by creating an immediacy of contact around the world, have altered the nature of international relations and even our understanding of war and peace.

In short, satellite networks have engendered technical, political, legal, and economic impacts across the broad reaches of our planet and even into outer space. These issues and how satellites have reshaped (and are reshaping) our world are ones addressed throughout this book, with an attempt at a broad overview in the first chapter. Thus, we have organized the book into seven parts that reflect the many disciplines in the world of satellite communications.

Satellites represent the key new media that have brought us "live" to the Olympic games, the first moon landing, and a bird's eye view of wars, skirmishes, and border disputes. Via satellite we now see the horrors of AIDS, genocide, and mass starvation. Yet this is also how we see beauty pageants, global rock concerts, and Pamela Anderson's Stripperella. Satellites are the key to Internet connections to over half of the countries of the world. It is space technology that connects the most remote parts of the world, as well as brings us weddings, beauty contests, and the most trivial of movie star news. Cable channels such as CNN or E! could not exist without satellite networking. To offset some of their "electronic excesses," satellites have also brought new health and educational links to millions of people otherwise denied such service. Satellites help to create a new vision of a "World without Walls." Satellite access and related electronic technologies are creating a new world that might well be called the *e-Sphere*.

It is not easy to document accurately and fairly all the major changes that satellite communications have brought to our world in four decades, but the results are undeniably significant and manyfold. Some of the most important developments include:

• Creating the ability for billions of people to witness global events such as coronations, political elections, the Olympics, and other key historical moments in real time—on all the continents of the world. Yes, even Antarctica now has satellite service.

• Allowing a truly global economy to exist via such means as electronic fund transfer (EFT) networks so that satellites, along with fiber and other terrestrial telecommunications systems, can now process nearly $400 trillion (U.S.) in "instant money" transactions around the world.

• Making possible the true globalization of the Internet and extending e-commerce to South and Central America, the Middle East, the Asia-Pacific, and Africa. (By the end of 2005, there may be as much as $750 billion [U.S.] in e-business, and satellites will provide the key link to allow such Internet business to connect to many developing countries.)

• Enabling electronic diplomacy so that heads of state, through the intervention of TV network commentators, can discuss the crisis of the day live via satellite. (This is a process that may ultimately undercut the power of presidents and lessen the role and influence of nation-states as we begin a new millennium.)

• Creating a global capability to transmit over 12,000 high-quality TV channels on a simultaneous basis throughout the world, distributing drama, concerts, sports, news, sit-coms, and movies to even the most remote locations.

• Contributing to the dramatic spread of scientific knowledge and information anywhere, anytime to virtually anybody and spurring and nurturing the growth of new technology at unprecedented speeds.

• Reaching, connecting, and delivering health care and education to even the most remote and developing countries (with over 7 million in China and India now relying on satellites as an educational lifeline).

• Building and frenetically interconnecting an ever-changing and dynamically live global network. The Internet is an electronic network that is redefined and updated in tiny increments of a second. It creates planetary linkages, but even so global satellite networks have not solved the problem of the digital divide. Those in the satellite world remain greatly concerned about societies that are being left behind economically, educationally, and socially.

There has been a surfeit of good, scholarly, and useful books about satellites over the years, although some have been as dull as thick molasses dripping on a pancake. Few have looked at the synoptic impact of satellites on our world and how the various elements of the satellite world interact and change global markets, world news and entertainment, telecommunications technologies, education, health care, politics, and ultimately our future potential. Show us an exciting and stimulating book about science and technology and its policy impacts, and we will show you an exception to the rule.

Some of these satellite books have described the vivid history and pioneering thinkers and innovators who were involved in the development of satellite networks. Other publications have addressed the policies and legal/regulatory issues of the satellite world as they have unfolded over time. Some treatises have explored the fascinating and difficult technology required to design larger, more capable, and cheaper satellite systems. Yet other analyses have addressed or sought to address the business, financial, cultural, social, and educational aspects of satellite communications.

This book has a different mission. Its goal is to approach the subject of communications satellites in an interdisciplinary way. We have tried to present all these different and sometimes conflicting facets of satellite communications together in a synoptic way. We have tried to capture, integrate, and analyze the broader fabric of change and innovation that satellite communications have brought to our planet over the last four decades and then interpret what changes the satellite systems of the 21st century will bring tomorrow.

We believe that such a comprehensive and interdisciplinary overview is needed to understand how all the pieces fit together. Thus, we have invited some of the leading people in the field from various disciplines to interpret what has happened. These are people with interesting backgrounds. Their short biographies will give you a feel for their expertise and special talents.

We seek to interpret the whole of the change and innovation that has occurred. In the pages that follow, you will learn not only how satellite systems have changed, but how these systems have changed the world. Through this process, we think we can also make sense of what the future trends might be. We believe that satellite networks, along with other new and evolving technology, will continue to change and integrate our world in remarkable and profound ways.

From the historical perspective of how technology changes society, in *Understanding Media*, McLuhan (1966) gave us many useful insights. None of these, in our opinion, was more important than his view that the printing and mass distribution of books in the 16th century changed our world and created individualism. If this is true, one may also speculate that a key offshoot of individualism is the modern concept we associate today with nationalism within a democratic state where there is freedom of choice.

We believe and hope to demonstrate that satellites—together with their companion technologies of computer networking, fiber optic, and wireless systems—are creating new and comparable waves of change in modern society. Satellites can be powerful tools of empowerment, avenues to knowledge, and a means to ensure human rights and liberty in remote areas that are isolated by the so-called *digital divide*. This is a young technology, and its impact on the world is still unfolding.

Periods of dramatic change are often hardest to understand and comprehend until the transition is complete. The full cycle of change may still be decades away. Only in time will we fully understand the full meaning of *globalism*. This concept is being strongly driven by satellites, networking, and the Internet. Likewise it will be some time before the full implications of the national tragedy known as 9/11 are understood and how satellites can help prevent or ameliorate such attacks.

New concepts such as *electronic decentralization* and *supranational processes* are closely related to modern satellite networking. A new world is being created by satellite-fed global TV news networks as well as by satellite-linked computer access systems that allow information, as well as music and videos, to be acquired and shared in totally new ways by millions of people around the world. It is satellite networks as much as personal computers that allow people the world over to *share* their CDs, videocassettes, and DVDs. Satellite networks cross national boundaries effortlessly and make censorship and political oppression much more difficult. Satellites have the potential to transcend religious, cultural, social, ethnic, and national barriers.

Yet these electronic gateways in the sky also offer the opportunity to spew forth propaganda, prejudice, and misinformation from anywhere to anywhere in an instant. New satellite systems, such as Worldspace and Wild Blue, have the potential to redefine our world in the blink of an eye. Before regulators know it, innovations in satellite technology can quickly make national boundaries amazingly permeable and potentially obsolete.

Only in time will we truly see the full scale and dimensions of the new information systems that modern telecommunications and computer machinery are furiously creating at increasing and unparalleled rates of acceleration. Even if we allow that technologically fueled change is far from finished, we see clear evidence that space telecommunications systems and their "kissing cousins," the other electronic technologies, are redefining how we live, where we work, and how we interact with people at home and abroad. Today we have over 1 million electronic immigrants who live in one country and work in another.

The first to clearly recognize the big picture of how world economic, scientific, educational, and other systems transcended national systems of knowledge and communications and interpret these trends for us in a truly coherent way was probably Tielhard de Chardin. In his writings in the early 20th century, Chardin described the evolution that has been unfolding since the High Renaissance. He called this process the *Noosphere*—the world intellectual climate or network that allowed humans to share their thoughts and discoveries in largely unfettered ways. Clearly the interventions of two world wars and now acts of terrorism have served to close, or at least shrink, the channels of human intellectual intercourse. Nevertheless, the last half of the 20th century fostered transnational or supranational invention, encouraged global business and financial systems, and spurred entertainment and travel to escalate to new heights.

We may still move beyond globalism, the Noosphere, or McLuhan's Global Village to experience new paradigms via electronic means. The wonder of it all is that, although satellites and fiber optics are real and tangible, the future they will enable and stimulate is actually quite unknown. Within the 21st century, we will encounter a new form of "planetary intellectual, commercial, and cultural ether" that far exceeds the vision that Chardin imagined at the start of the 20th century. These new visions have been previously and certainly inadequately described in such terms as the *worldwide mind*, the *global brain*, or *E-Sphere*, but we know that a new system of thought awaits us and that our electronic tools are accelerating our journey to that future.

In the following chapters, we explore how satellite communications—especially global satellite linkages—came to be, technically, economically, and institutionally. It is a vivid and interesting history. It is filled with people of imagination, genius, vision, intrigue, and sometimes avarice. All in all, it is an interesting and absorbing story.

We also explore how satellites, more than any other technology, gave rise to global TV, worldwide entertainment networks, and the truly worldwide Internet—something that fiber cannot and will not accomplish even by the end of the 21st century.

This is not *just* a story about technology or technological competition, but rather an interdisciplinary investigation of how satellites have changed the world. We explore as much as we can from what might be called the *bigger picture*, or at least the interdisciplinary one. We seek to look at how business, entertainment, and social and moral changes in the family and workplace will never be the same. We examine how worldwide news, new ways of waging war with GPS-guided "smart bombs," and unsuspected cultural and technological forces have been affected by the world of satellites.

We look into how satellites have changed just about everything in terms of how we live, how we are entertained, and how world trade and global finance work. In the process, we address some basic questions as to why the world is moving more and more quickly. We look into why economic cycles are increasingly more global than national in scope. Likewise we examine why global businesses in terms of stock markets, travel agencies, banks, insurance companies, and even manufacturers are increasingly migrating to a 24/7 work schedule. As humans spend less time at the office or plant, machines put in longer and longer hours.

Ultimately, this is a story about how satellites—together with fiber, computers, and other technologies—have changed global economic, social, and cultural systems in new and unsuspected ways. It is a story about how technology and fundamental values remain in conflict. It is a story of how the 20th century brought us skyscrapers, megacities, mass transportation, superpowers, and centralization, and how this will now change in response to terrorism and new technology. In many ways, only modern communications could enable all of these things to transpire, but this is only half the story. The other parts of this book also explore future trends and the shifts toward decentralization that the new millennium will bring. This is, in part, a story of how satellites and other information technology (IT) capabilities may bring us long-term security. The desire to cope with pollution, soaring energy costs, and security risks will also change our world. These forces could move us toward decentralization, telecities, telework, and possibly new forms of intercultural understanding.

In this book, our goals are to (a) explore how satellite communications have impacted and shaped the world, and (b) analyze how they might dictate events in the 21st century. To this end, we have asked over a dozen knowledgeable people, who understand satellite systems from a variety of backgrounds, to contribute to this undertaking. These contributors participate from the perspectives of their various disciplines. They have a breadth of understanding of the field that ranges from technology; to business and economics; to historical, legal, and political systems; to society and culture. We hope that the collective result is a better understanding of the broad patterns of global change that satellites have caused and will cause in the future. Specifically, this collective process has resulted in the identification of the top 10 conclusions we jointly reached about past and future trends that satellites have already or will soon generate for our increasingly small and interconnected planet.

Our mission has been to recount, probe, explore, analyze, and interpret the past, present, and future of satellite communications. We have embarked on this venture with a dollop of historical zeal, a measure of technical insight, and a smattering of social and political insight. On occasion there have been attempts at touches of whimsy, irony, and even old-fashioned humor. To make this book as readable as possible, we have avoided footnoting. Instead at the end of each chapter, we provide a listing of references for those who wish to pursue their studies of the field. When particularly needed, brief authors' notes are provided along with the text.

We hope the reader finds this analytical and interpretative story of satellite communications—past, present, and future—to be interesting and insightful.

REFERENCE

McLuhan, Marshall, (1966), *Understanding Media: The Extensions of Man*, New York: Signet Books.

—Joseph N. Pelton, The George Washington University
Robert J. Oslund, The George Washington University
Peter Marshall, Communications Consultant

DEDICATION

This book is dedicated to the thousands of men and women who make the global satellite communications systems of the world possible. These people support nonstop operation of hundreds of satellites to provide so many vital functions in virtually every country and territory, on the open seas, and even in outer space. This process started many years ago with the intellectual giants who conceived of such a possibility, such as Sir Isaac Newton and Sir Arthur Clarke. This great enterprise continues with scientists, engineers, broadcasters, telecommunications network providers, manufacturers, farmers, financiers, social scientists, doctors, educators, and the many others who make "live via satellite" a reality 24/7. We salute them.

PART ONE

INTRODUCTION

CHAPTER ONE

Satellites as Worldwide Change Agents

Joseph N. Pelton
The George Washington University and Arthur C. Clarke Institute

> *The medium is the message means . . . that a totally new environment has been created.*
>
> *We have extended our central nervous system itself in a global embrace, abolishing both space and time as far as our planet is concerned.*
> —Marshall McLuhan (1966, pp. ix, 19)

Communications satellites have redefined our world. Satellites and other modern telecommunications networks, together with TV, have now altered the patterns and even many of the goals of modern society. Satellites, for better or worse, have made our world global, interconnected, and interdependent. Worldwide access to rapid telecommunications networks via satellites and cables now creates widespread Internet links, enables instantaneous news coverage, facilitates global culture and conflict, and stimulates the formation of true planetary markets. Satellites change our world and affect our lives. This book is an exploration of how satellites are a global agent of change in an incredible range of ways. These include the spanning and spread of new technology, news, culture, sports, entertainment, economic markets, global politics, and much more.

The editors' prior introduction is intended to be a roadmap guide to our efforts to explain the many sides of the satellite world—past, present, and future. This chapter seeks to introduce in a general way the wide dimensions of that world and how it impacts our planetary society in a myriad of ways. Some of these are now so familiar that they are almost hidden from public view. For instance, most people know that direct broadcast TV comes from satellites, but are unaware that cable TV (from HBO to CNN) needs satellites to obtain most of its programming. A U.S. congressman is purported to have said: "Why do we need weather satellites when I can turn on my television and see the latest report on my local channel?"

Satellite technology and systems have evolved tremendously in the past 40 years, but as is seen in the technology sections of this book, this development continues apace.

SATELLITES AND THE NATION STATE

Ever since the Treaty of Westphalia in 1648, the *world* has been defined by what is called the *nation state*. Prior to that time, conflicts and wars often revolved around the borders of city states and communities of interest, not nations. Prior to the modern national state, wars were fought over disputed territory, stolen property, kidnapped damsels, religious conflicts, and ethnicity, not on the basis of the national interests of independent countries. For the last 300 years, however, national sovereignty has been king. Under the Westphalian system, as it has become known, the established authority of a state was responsible for internal laws, national subjects, national punishment, and even defined national religious beliefs. "No man is an island," but under international law countries were supposed to be. The international system, as it evolved based on the sovereignty of the nation state, had weaknesses, but it did create an international law of minimum order. As Fernand Braudel has explained in his massive trilogy on life up through the 18th century, institutional structure is key to the advance of civilization (Braudel, 1979a, 1979b, 1979c).

Until the 1950s, nations or national empires such as those of Britain and France were amazingly self-contained. Admiral Perry may have opened up Japan, and Europe may be becoming a community of nations, but prior to 1965—the date when Early Bird, or Intelsat 1, became operational—overseas communications were incredibly limited and expensive. Most communications (over 99%) were within national boundaries. Likewise international commerce was quite sparse, and multinational corporations were only just beginning to emerge. In the age of satellites and fiber optic cables, it is now possible to operate global corporations, carry out free trade on a worldwide basis, and instantaneously control systems from halfway around the planet. Millions of telecommuters, what might be called *electronic immigrants*, now work daily in another country via modern telecommunications connections. In many ways, satellites have made such an international world possible. It has enabled a world that operates beyond the nation state technically and economically viable.

Analysts such as Robert B. Reich (1991), author of *The Work of Nations: Preparing Ourselves for 21st Century Capitalism*, believe that the nation state is becoming passé or only marginally relevant. Such analysts, often associated with a group known as the Tri-Lateralists, perceive the growing influence of international corporations, global financial transactions, and science and technology, and suggest that the nation state is losing influence and control in our modern world. Each year those who make their way to the Global Economic Conference in Davos, Switzerland, seem to increasingly define those individuals and entities with true power in a 21st-century world. At the opposite end of the spectrum, religious mullahs are also attacking the nation state on the basis of cultural and ethnic grounds.

The world has certainly experienced more change in its economic, cultural, social, scientific, and political systems in the last 40 years than at any time in history. Further, the rate of global change due to satellites, telecommunications, information technology, electronic funds exchange, and other forms of advanced science and technology will, if anything, continue to accelerate ever more rapidly in the years ahead. When one considers that the value of global electronic fund transfers in 2002 were about $300 trillion and ran at over $1 trillion per day for 2003, it is hard to deny that the world today is dramatically different than only two generations ago.

In short, this is not an easy time for a world shrunk by technology and yet fractured by fundamental religious beliefs and terrorism. One part of our modern world believes that the structure of our world can be defined in dollars, patents, and gigabytes of information, but another part, steeped in radical beliefs and concepts such as Jihad, define our world much differently. The question is whether the 300-year-old concept of the nation state can cope effectively with this ever-speeding swirl of change and conflict. As satellites have made our world smaller and smaller, the elements of conflict and cultural difference have begun to emerge as increasingly important, and the nation state appears to be in need of redefinition. Some would say that the terrorists' attacks on September 11, 2001, and the U.S. response in terms of a war against terrorism mark the start of the end of the Westphalian concept of the world politic. Yet the rise of the world corporation, the emergence of global entertainment and news, and the pervasive reach of modern telecommunications and information networks gave rise to fundamental shifts in the world community decades before the Al Qaida attacks on the World Trade Center buildings and the Pentagon.

Thus, we seem to have advanced technology in conflict with the traditional nation state as well as with traditional religious beliefs and cultural values. At the start of the 21st century, a third axis of technological conflict with yet another value has increasingly emerged. There is increasing concern about technology and development and its potential conflict with our environment, global warming, and the need for sustainable patterns of living over the longer term. Some fear that population explosion, unchecked consumption of petrochemical fuels, and other destructive environmental practices threaten our future. Runaway development, consumption, and unchecked technology are seen by many as a challenge to the sustainability of the human race on Earth.

Recent projections showing that the population growth could ebb by the mid-21st century and the world population could actually shrink in coming decades are seen as hopeful signs. Other new environmental, remote sensing, and meteorological technologies also give new hope for a "new greening" of the planet another generation or so hence. Yet environmentalists from the Club of Rome, the Sierra Club, and other groups are not equally optimistic about the future. On both sides of the environmental debate, satellites are peculiarly positioned to play a powerful role. As one reads further into this book, one can find satellites represent a powerful technology whose force can be channeled in many directions.

For better or worse, satellite systems will continue to play a key part in redefining the political, scientific, cultural, and economic systems of the 21st century. Here are some of the ways that satellites will define our future. These themes are replayed and amplified in the chapters that follow.

SATELLITES AND GLOBALISM

Over the past decade, space activities have contributed nearly $1 trillion to the global economy. Satellite communications represent about a third of that figure. During 2003, satellite communications directly represented $40 billion in worldwide revenues and nearly $75 billion in directly related economic activities. If one looks beyond the role of satellites (in narrow terms of telecommunications) and begins exploring the multiplier impact on the global economy, the overall impact is vastly more important on our planet as a whole. For instance, a good percentage of the world's hundreds of trillions of dollars in electronic funds transfer (EFT), as reported by the World Bank for 2003, went via satellite. This stunning amount represents a figure that is more than four times the global Gross National Product (GNP) for all countries of the world—nearly a $100 trillion. Money in a service economy circulates faster and faster and must do so to create an actual payoff that can be measured as national economic productivity.

Satellites now beam down to us 24 hours a day, 7 days a week—a Niagara of information, entertainment, and business transactions. This massive dump of data is now accomplished by means of over 12,000 satellite video channels, produced not only in Hollywood, but in Bollywood, India; Cairo, Egypt; and other film production centers as well. Connectivity and electronic access around the world have reshaped just about everything that defines business, cultural, and even religious life in the Third Millennium. An evangelist or a pornographer can reach a billion people "live via satellite."

The operation of communications satellites in today's society is now so pervasive that they have become invisible like electric motors or indoor plumbing, but occasionally an event occurs that signals both their presence and importance. A few years ago, a Panamsat Galaxy satellite (then owned by the Hughes Communications Company) covering much of the United States failed. That satellite was carrying the video distribution for CNN News and the CNN Airport channel, and thus many people were aware of the failure. Far more important, this satellite was also providing the paging service for a large percentage of the U.S. public, including a high percentage of doctors. This disrupted the medical system of the entire United States, and thousands of operations and the provision of emergency care at hospitals all over America were thrown into a great tangle. On the Evening News Hour with Jim Lehrer, the executive director of the Satellite Industry Association and I were invited to explain how not only the United States but the entire world was now dependent on hundreds of communications satellites that provide communications for business, education, health care, and news. We made clear that improved restoration planning was critical to avoid a repeat of the "satellite paging crisis" of the day.

On a more positive note, Japan has turned to satellites in the wake of the great Kobe earthquake of the early 1990s. When the earthquake hit, all terrestrial communications via fiber optics and coax cable were instantly destroyed. Only the wireless satellite circuits for national and international connectivity survived. Today, throughout Japan, there is a Local Government Wide Area Network (LGWAN) that consists of over 5,000 satellite terminals that have been installed against the possibility of another major earthquake disaster, and soon this network will include 10,000 very small aperture antennas (VSATs). Around the globe, satellites are increasingly seen as a way to maintain communications in case of a natural or man-made disaster or terrorist attack. No other technology interconnects the world so incredibly well and reaches so many remote locations. Some 220 countries and territories are linked together via satellite. Fiber optic systems connect less than half that number of countries. The smallest, most isolated, and most land-locked countries and territories in the world are linked via satellite. As you see later, this has changed things forever.

One can argue that the impact of satellite communications around the world on business and the citizenry has been good, bad, and occasionally even indifferent. In the chapters that follow, we see evidence of all three results.

SATELLITES HAVE CHANGED THE WORLD AS MUCH AS TV OR THE TELEPHONE

Change spawned by electronic connectivity is everywhere—from news to terrorism, from education and health care to warfare. Marshall McLuhan, the Father of the "Global Village," was wrong—or at least only partially right—when he said that the new "cool" and "mentally involving" media of TV was the message! As we explore our electronic world, we find it is actually much more complex than even McLuhan sought to describe.

However, the futurist James Naisbitt (1980) was probably right when he suggested that global connectivity was perhaps of equal or greater importance to media content in shaping today's world. Naisbitt thus challenged McLuhan when he suggested that instantaneous distribution of information globally was the key to the future. In his book *Megatrends*, he suggested that satellite communications, by making modern electronic media "omnipresent" and "instantaneously accessible," was creating the "ultimate force of change" in modern times. In short, Naisbitt argued that satellite systems were creating the global village—not TV by itself.

Instantaneous information also has other unintended consequences:

1. An abundance of rapid information overloads our sensory systems and competes for attention. Thus, information in the modern context must be as contemporary as possible. As a result, information today is less enduring.
2. New constraints on decision makers emerge as news is contemporaneous with policymaking.
3. Policymakers try to avoid this amplification of news by ever-increasing forms and processes of secrecy and intelligence.
4. The media can also be fed false and misleading information, and the pace with which it flows is so short and fast that facts cannot be checked adequately—the major problem is to determine what is really real.

Indeed, it was satellites that brought TV and global telecommunications rather quickly and dramatically to an increasingly small planet in the 1960s. The 1968 Olympics in Mexico City was the first true international electronic media event, but satellite TV service was still limited at that time. The countries of the Indian Ocean region were not yet included in the world of instant messaging in 1968. That was to change forever 1 year later.

The Apollo Moon Landing in the middle of 1969 became the world's first truly global event. Satellite service to the Indian Ocean region via an Intelsat III satellite was achieved just a little over 1 week before this huge media and scientific event changed our view of our small planet in a significant way. Over a half-billion people watched "live" what had seemed pure science fiction only a decade before. By the time of the Montreal Summer Olympics in 1972, satellite technology had made possible a global audience of over 1 billion. In less than a decade, the earth had been "dramatically shrunk" by satellite networking and broadcasting. Global TV programs evolved rapidly over the decade that followed. At the start of the decade, they were revolutionary events. By the end of the 1970s, they were commonplace occurrences, and the phrase "live via satellite," which once inspired a sense of awe, began to disappear from TV screens as irrelevant information.

Satellites, as much as any other technology, thus served to create what can legitimately be called *globalism*. Communications satellites reshaped our vision of humanity when we first saw "Earth Rise" from the moon in the last gasp of the 1960s.

The 21st-century world is different in many ways, but global interconnectivity is certainly one of the most fundamental shifts that now separate us from our past history. Just as mass access to books and newspapers created a new world for us centuries earlier, global media separates us from the near-term past.

Before satellites (i.e., as recently as 1964) there were only a few hundred poor-quality telephone circuits to interconnect the entire world. These submarine cable voice links often resulted in "clipped-off" conversations due to an antiquated technology called Time Assigned Speech Interpolation (TASI). This was a poorly chosen acronym, but perhaps well suited to an even poorer technology. Today, Sri Lanka, Kenya, and Uruguay each operate more overseas voice circuits than were in worldwide use 40 years ago. Well over 200 countries have ready access to international radio and TV broadcasts via satellite, and this is viewed as routine.

Before the end of the 21st century, we will be linked via space communications not only to everywhere on our planet, but to the moon and maybe other parts of our solar system. The national, regional, and global linkages that satellites provide are comparable in impact on our planet to that of telephones, TV, the jet airplane, or nearly any other modern technology.

THE GLOBAL INFORMATION REVOLUTION AND INFORMATION OVERLOAD

Communication satellites, more than any other of the advanced electronic media during the short period since the 1960s, have stimulated a true global information revolution. Submarine cables and fiber optic networks have served to integrate the most economically advanced countries together, but they have not been at the core of creating true global interconnectivity. As described in more detail in a later chapter, in the 1980s and early 1990s, Sidney Pike of CNN almost single handedly brought satellite TV to scores of countries that had never seen international news transmissions before.

Today, satellite systems such as Intelsat, Panamsat, SES Astra, Eutelsat, Asiasat, and nearly 100 other space networks provide literally thousands of international connecting links between the countries of the world, in comparison with fiber systems that are typically providing only a few score of point-to-point link-ups. Global TV is still largely a satellite-based phenomenon and will remain so for some time to come.

Global society—at least as measured in simple quantitative terms such as human population, information available to global society, cultural linkages, medical science, and longevity—has changed more since the 1960s than the patterns of human life changed during 1 million years of cave man civilization.

This startling rate of change in the state and nature of human society is not hyperbole, but simple fact. Nor should it necessarily be considered a positive accomplishment. Actually such "superspeed" change should be comprehended as both a source of wonder and even more so of concern. The last 35 to 40 years—and not coincidentally also the era of satellite communications—represents a sea change for world society. We have witnessed much more change (at least in tangible things that we can measure) than ever occurred in vast spans of time in the early days of human history.

World population has expanded by over 3 billion people in the last 40 years. Further, according to UN projections, the people to be educated in the next 40 years outnumber all those people educated since the dawn of human society. Satellites, in terms of expanding international business, broadening international scientific interchange, increasing media distribution, and extending Internet to the world community, have often driven this curve of change.

In fact if we look at the 5-million-year history of humankind and proto-humans as being represented by a 10,000-story building that is 20 miles high, we find that the last 40 years (the time that we have had broadband global communications capability via satellite) represent only a matter of some inches of space on the top floor of this imaginary megastructure (see Fig. 1.1).

Yet if we were to envision a building representing not human history, but the accumulation of global information (Fig. 1.2), we see a dramatically different building from this perspective. The top 8,500 stories of the building of global information would have been built since the launch of the first operational satellite, and some 6,000 floors would have been added since the advent of the Internet. In the next 15 years, if we maintain the same index scale of "progress," the currently projected amount of information in our building of global information would soar outward at an even more astonishing rate. In fact the information in our global database is now doubling perhaps every 3 years (and some would argue it is doubling in only 9 months time). Thus, the "building of global information" could mushroom to 160,000 stories (320 miles into space) before we reach 2020, and of this information only 2,500 stories would predate the age of satellites. This building, which represents our global database, would represent the ultimate Tower of Babel and truly reach into outer space.

This is to say that in about 15 years the totality of global information systems could increase some 16 times in size. Just the information represented by environmental data captured by NASA's Mission to Planet Earth, called the Earth Observation System, will have increased to nearly 3,000 terabytes of information or the equivalent of 1,000 Libraries of Congress. If we sought to organize, store, and retrieve this remote sensing information on a fully interactive basis on a national and international scale, this would create problems of communications almost unimaginable a few years ago. (Some people even fear that without some new standards of information management, the Internet as presently organized could reel out of control due to simple "information overload.")

BEYOND THE GLOBAL VILLAGE INTO THE AGE OF THE WORLDWIDE MIND

The satellite, other wireless systems, and fiber communications networks will not only assist in creating the oft-quoted "Global Village," but will soon start to fashion what might be called a "Worldwide Mind." By the end of the 21st century,

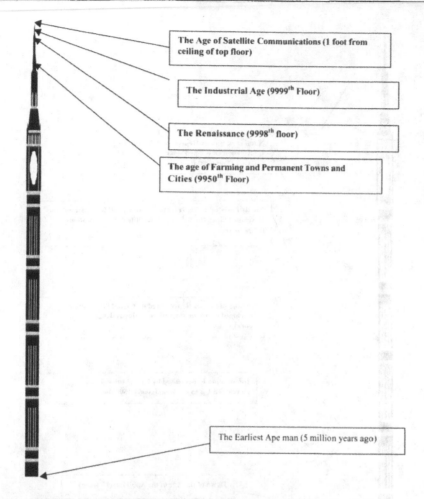

The Age of Satellite Communications (1 foot from ceiling of top floor)

The Industrrial Age (9999th Floor)

The Renaissance (9998th floor)

The age of Farming and Permanent Towns and Cities (9950th Floor)

The Earliest Ape man (5 million years ago)

FIG. 1.1.   The building of human history (20 miles and 10,000 stories high).

broadband systems will link humanity together in almost unimaginable ways (see Pelton, 1999). In another century, this "intellectual network" can be expected to reach not only to the most remote reaches of our earth and ocean, but also out across the broad reaches of our solar system. Some day we will turn on our TV and check the weather for the day not only on our planet, but others as well. Since the events of 9/11, however, we understand that quite different scenarios are possible. Technology does not always promise progress and better horizons.

Technological innovation and broader band instantaneous communications can no longer be seen as "guaranteed progress" for the human species. There are clearly problems ahead. Serious issues arise from the fact that some so-called *developing* economies and societies live virtually completely outside this rapidly expanding "E-Sphere" of accelerating scientific knowledge and electronic data sys-

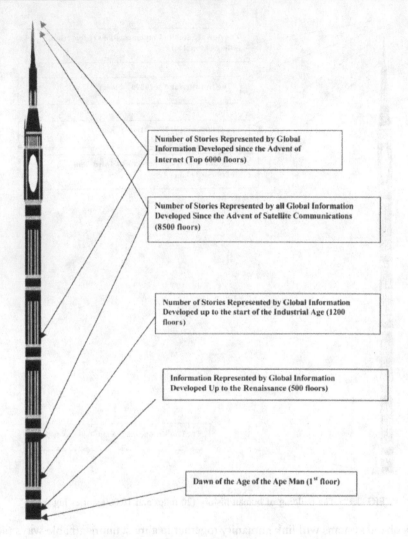

Number of Stories Represented by Global
Information Developed since the Advent of
Internet (Top 6000 floors)

Number of Stories Represented by all Global Information
Developed Since the Advent of Satellite Communications
(8500 floors)

Number of Stories Represented by Global Information
Developed up to the start of the Industrial Age (1200
floors)

Information Represented by Global Information
Developed Up to the Renaissance (500 floors)

Dawn of the Age of the Ape Man (1st floor)

FIG. 1.2.   The building of global human information (20 miles and 10,000 stories
high).

tems. These problems will be heightened if the G-8 and other economically
advanced countries continue to be thoroughly integrated into this global web of
information while other nations and societies, for technical, economic, or cultural
reasons, are excluded.

In only a few decades, the most developed countries of the Organization of
Economic Cooperation and Development (OECD) can extend their knowledge
base and economic wealth well beyond limits that we can now only imagine. Yet
what does this portend for our increasingly small planet if the process leaves the

poorest and least educated societies behind? Today we talk of the issue of the "Digital Divide," but the potential scale of the problem is not well understood. The prospects of staggering gaps of information on one small planet are fundamental and sobering to contemplate.

This issue of the Digital Divide could easily escalate rapidly in only 20 years. The United Nations, national aid agencies, and nongovernmental organizations that provide international aid and relief services have now identified the Digital Divide as one of the world's most important issues and problems. Satellite networks can serve to extend this problem by providing advanced electronic systems to promote the growth and expansion of multinational business enterprises or minimize this problem by the provision of teleeducation and telehealth services. It is likely that advanced satellite networks will serve both causes.

Yet another way to consider the issue of satellites and their role in the future life of our evolving "E-Sphere" or "Worldwide Mind" is that global information is being collected and stored at an astonishing rate. In mathematical terms, the growth of information globally is not just accelerating, but growing in accord with a third-order exponential. This is to say that global information systems are not merely accelerating—the rate of acceleration is increasing. This is what physicists call *jerk* for obvious reasons.

More simply we can say that the generation of global information is growing at least some 200,000 times faster than our global population. (This, in mathematical terms, is a swift turtle trying to catch the Space Shuttle.) The problems of surviving and adapting to such a "Super Speed" world are not small. In fact we are now looking to future satellites and computers that can operate hundreds of times (if not thousands of times) faster than today's impressive machines. These machines of tomorrow, most likely powered by concepts known as quantum communications and quantum computing, may possibly allow us to keep up with the flow of new information that will flow across our increasingly small "intellectual telecity," which I call the "E-Sphere" or the "Worldwide Mind."

Intel has already indicated that by 2005 or so it will have desktop computers that process at speeds of 10 GHz from a single chip. This means that easily accessible computers for the mass consumer market can process the equivalent of a human lifetime of reading and speaking in a single second. Satellite links in another decade could be operating at the hundreds of Gigabit level and fiber optic connections providing connections at the terabit/second level. (This means moving the equivalent of several Libraries of Congress per minute.)

The continued expansion of digital services in the areas of video, multimedia, broadband Internet, and IP services—barring some catastrophic disaster—will thus only continue to mushroom in the coming years. This increased speed of processing and broadband communications will give rise to a staggering array of new services.

The speed with which this is projected to occur over the decade ahead is shown in Fig. 1.3, with the different shadings identifying existing, new, and projected

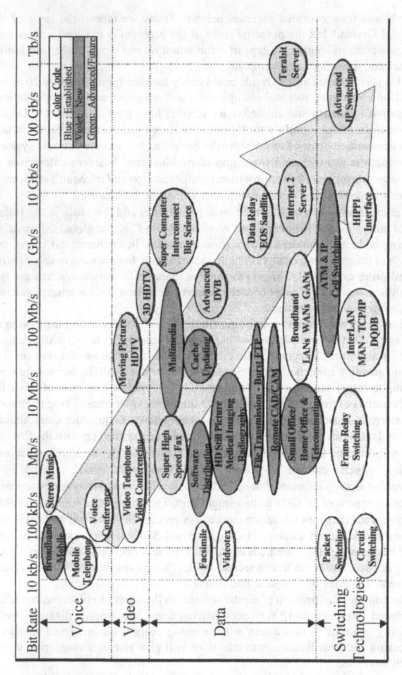

FIG. 1.3. The continued growth of broadband services 2002–2015.

14

services to occur within the next 10 years. The driving force will be cheap processing power. The challenge will be for fiber optic and satellite networks to keep pace—not only by providing faster and faster transmission speed, but by furnishing them at low cost and with high reliability and security, especially in wireless and mobile systems. The biggest challenge of all may be filtering out unwanted information.

NEW PATTERNS OF GLOBAL TRADE, POLITICAL PROCESS, AND SOCIAL CHANGE

Although a good case might be made for other instruments of change, this book presents the case for satellites and how they will represent the prime key to globalization. The rise of globalism will undoubtedly bring changes to education, health care, entertainment, and business. Low-cost services and user terminals that support easy communications by satellite are likely to complicate international trade in products, but especially so in relation to services. This will seem particularly true as more and more electronic immigrants work across international boundaries. Today more than 1 million people are what I call *electronic immigrants* or *international teleworkers*. This special class of workers might live in Barbados, Jamaica, India, Pakistan, or Russia, but work in the United States, Japan, or Europe. These international teleworkers actually have been around for decades, and many now accept such a way of life as normal.

Within a decade, it could well be 10 or even 20 million people who are professional business nomads and who will find their way to work on electronic beams of digital information, sometimes halfway around the world. Such trends, and the parallel efforts to facilitate global communications, will make national and international satellite links just as common and convenient as cell phones or global TV shows are today. In 20 years, the cost of communicating across town, across a country, or around the world could well be much the same—just as Arthur C. Clarke predicted decades ago. What is not known is whether this will generate supranational relationships and institutions that could well confound the working of today's national states or of regional, ethnic, or religious cultures.

The nation state, a modern invention, is only a few centuries old. Countries are actually the manifestation of what is fundamentally a 16th-century technology (i.e., low-cost printed materials in mass distribution) and then of the 18th-, 19th-, and 20th-century industrial revolutions that followed. The role of the nation state in a global service economy has yet to be clearly defined.

Marshall McLuhan, Fernand Braudel, and others taught us to analyze the relationship among communications, technology, and political history. They and others such as Georg Hegel, Jacques Ellul, Lewis Mumford, Victor Ferkiss, and Norbert Weiner, from widely different perspectives, examined how technology has reshaped modern life. We certainly know that the rise of mass printing and the dissemination of knowledge on a broad scale in the 16th century (a lag time of about 100 years after Guttenberg's famous press appeared on the scene in Europe) changed the world of the Renaissance. The free exchange of ideas and the resulting intellectual intercourse gave rise to the concepts of both individualism and political systems, which were the precursor to the modern nation state. Political and technological philosophers as diverse as Siegfried Gideon in *Mechanization Takes Command* and Jacques Ellul in *Technological Man*, on the liberal side of the argument, and Marshall McLuhan, Norbert Weiner, and Buckminster Fuller, on the "scientific side" of the spectrum, have all suggested that these intellectual changes reshaped human society in fundamental ways.

Essentially philosophers such as Henry David Thoreau, Jacques Ellul, and Lewis Mumford have seen technology as adversely affecting and "dehumanizing" society, whereas others such as Norbert Weiner and Buckminster Fuller have seen technology not only as positive, but almost as an inevitable outcome of humanity's intellectual development. Thus, the technological positivists see technology as the ineluctable quest of human civilization.

Regardless of which interpretation we choose, the nation state is showing its age as we start the 21st century. All of these writers and more suggest that political institutions may need to be reinvented. Exactly how national political systems adapt to an electronic world whose business and economic systems work on the basis of split-second communications across the world is clearly a challenge. Political systems work within relatively small, intimate, and low-moving communities and cultures, and thus are often ill equipped to cope with their faster (and other more wealthy) protagonists. This is just one of the more interesting puzzles of our technological age. Certainly the institutions that provide global communications, including satellites, are currently in a hubbub of change and transformation. One thing we know: "Faster is not necessarily better."

Today with the rise of global information systems, worldwide e-commerce and media, and planetary science and engineering, the dawning of a new age seems to be occurring. Yet we see another landscape: We see fundamentalist and closed cultures with absolutist political systems. These societies are many centuries old, and in many cases the political and religious leadership of these entities actually despises the results of global connectivity and modernity. In many cases, they despise Western culture, human and political freedoms, and civil rights, and they have contempt for the "loose morals" and "materialism" of what seems to them a depraved society.

We are certainly living in a time of turmoil and cultural striving. These "gaps" in philosophy and culture impact not only the "West's relationship" with Afghanistan, Iraq, Pakistan, Borneo, or the rain forests of the Congo. These gaps also stretch across the hopes and values that divide urban and rural America and separate suburban communities and the inner city. Similar conflicts are present throughout other advanced nations as well.

The point to note here is that satellite communications is neither the cause nor the focus of change. Yet it is these systems, along with fiber optics, cybernetics, and computer networks, that speed up the process of conflicting knowledge and values both within countries and across the broad reaches of our globe. In some ways, we have been warped by a new kind of electronic reality. This has been called *telegeography*. Within this concept, distance (at least in the modern commercial world) no longer has much meaning, and it is culture, intellect, and speed of electronic connection that now create dysfunctional gaps in our planetary system, not physical separation.

In the age of broadband networking, New York, Tokyo, Paris, Singapore, Berlin, Moscow, and London are closer together than Okmulgee, Oklahoma is to Arkadelphia, Arkansas or Zamengoe, Cameroon is to Gaborone, Gabon.

Digital communications machines, for better or worse, now integrate our global business systems and many national and ethnic societies. Our versatile space communications systems link together our TV sets, telephones, GPS receivers, computers, terabyte databases, and portable Internet phones. This process has made our world smaller and more intimate in new and different ways. We have now seen the advent of new and more powerful satellite systems. Truly anybody can stay in touch anytime and anywhere—if they want to do so. Via the Worldspace satellite system, it is possible to receive hundreds of radio channels in the world's most remote rain forest or arid desert. Via mobile satellite systems such as Inmarsat, Thuraya, New ICO, or New Iridium (to name only a few), one can use a specially equipped cell phone or lap top transceiver to communicate to the world or link to Internet from ships at sea, from aircraft in the stratosphere, or from the most remote island—and for an increasingly low cost.

Furthermore, billions of people can choose what they wish to watch on any one of hundreds of satellite TV channels. They can do this by direct satellite, satellite-fed cable TV, or simply by going to the Internet to watch their favorite show via "video streaming" or listen to their favorite music by CD downloading. Ultimately one will be able to watch virtually any movie ever made via the right connection on the Web. Today, people in developed countries and a growing number of those in developing countries can quickly connect to tens of thousands of "information networks" to learn about virtually any conceivable subject or watch any form of entertainment or amusement. Censorship will have become obsolete.

Soon we may be paying more for screening out unwanted information than for connecting to the global network. Yet those without access to the networks or those unable to pay to access information may be pushed further away from modernity. These isolated communities may not rue being isolated from global entertainment. (Many would say it is a plus.) Yet these individuals are also being limited in their access to education, health care, and unfiltered news. At the most fundamental level, there is also a question that goes beyond how to access global networks and whether it will be affordable. This is whether one wishes to have access to and intercourse with Western technological society at all. Is such access good or bad? Right or wrong? Beneficial or destructive?

AN INCREASINGLY COMPLEX
AND INTERDISCIPLINARY WORLD

This book is focused on satellites. The information here is thus largely based on how satellite technology came to be, what is or is not available from satellite systems, and what impact they have made on our world—past, present, or future. That is not to say that satellites are the source of all modern telecommunications or networking systems. Clearly they are part of a "telecommunications gestalt" that now enables planetary communications to work with great efficiency at an increasingly low cost.

Satellites have made an enormous impact, as the following chapters clearly explain. Nevertheless, companion systems and technologies are also a key part of the digital revolution that is bringing us such change at a constantly increasing pace. If we look around, there are other candidates that might be labeled the "prime movers" of modern change. Thus, we now find an array of "kissing cousin" technologies that create electronic intimacy and information formation and exchange. The list of prime change agents for Western technological society could certainly include computers, biotechnology, artificial intelligence, space exploration, telecommunications devices, the Internet, missiles, atomic weapons, lasers, fiber optics, or TV.

We do not dispute that these other technologies have also had a profound effect on the world and have also reshaped our lives. In fact we try to show in this interdisciplinary investigation of the field of satellite communications just how communications and information technologies fit together. It takes a good deal of engineering and standards making to create seamless networks that enable global media, global business, global science, global education, global health care, and global news to all work together. That having been said, the impact on satellite technology and modern planetary systems simply cannot be denied.

The following chapters describe how satellites can and will fundamentally alter the state and nature of human existence. Satellite news and entertainment have made us more globally aware and interconnected. Today, telephone connections and satellite-based Internet penetrate to the core of urban ghettoes, the battlegrounds of the Middle East and Afghanistan, and the remote reaches of schools and clinics in the midst of jungles and deserts. Satellites reach from university laboratories and mountain top telescopes to the living rooms of houses—all the way from the United States to Borneo.

By a combination of history, political analysis, business, and trade information, a diverse mix of people who have lived through the satellite revolution will attempt to explain, probe, and analyze how satellite communications have swept forward the wild fire of change, innovation, and globalism.

Today the six sextillion-ton mass of dirt, stone, water, and air that orbits the sun as the third planet in the solar system is much the same in its chemical and physical composition as it was thousands or even millions of years ago. Yet in terms of the organization of the human bio and political systems, life is fundamentally different today than it was at the dawn of the space age less than a half century ago.

SATELLITES AND NEW PATTERNS OF WORLD LIVING

Human civilization has changed in significant ways from the days when we attempted to launch the first tiny artificial communications satellites in the early 1960s. The first steps came quickly with the launch of Score (1958), Courier (1960), Telstar (1962), Relay (1962), Syncom (1963), and the world's first commercial communications satellite (spring 1965). Today's largest and most powerful satellites are some 10,000 times more capable than the first small satellites of the mid-1960s in terms of performance and lifetime. The first operational satellite resembled a largish 80-pound coffee can with a pipe sticking out the top. It showed little commonality with today's monster satellites, with a wingspan of some 100 feet (about 30 meters). Contemporary satellites include huge, high-gain antennas that are some 40 feet in diameter (i.e., 12 meters across). The history of this technology and how it happened is told in greater detail in chapter 2.

The most important thing to know is that the story of communications satellites and the development of new and exciting technology is far from over. The Japanese experimental satellite—the ETS VIII—will soon deploy a huge antenna that is 17 meters by 19 meters in size. The next three decades will likely produce satellite designs that are radically different from the golden boxes from which protrude solar panel wings, considered the latest in technology today. In later chapters, we discuss and show concepts of the possible designs of the future. The most impressive designs for the future, however, may not be the satellites in the skies, but the tiny broadband communications, computing, and navigational devices that we can wear on the wrist or perhaps even implant in our bodies.

When it comes to identifying global change agents, satellites and other related media have much to brag about. Yet satellite system designers and operators from DirecTV to Astra and from Intelsat to Panamsat have a good deal to be concerned about as well. The technology that brought us the potential for worldwide education and health care, as well as electronic diplomacy and unprecedented economic growth, also spawned global MTV, Cinemax, the "Gong Show," "Baywatch," and nonstop game shows and commercials. Communications and navigational satellites can also be powerful instruments of war that ensure that "smart bombs" are delivered to their targets with deadly accuracy. In short, the communications satellite can be seen as a mixed blessing; satellite systems can instantaneously bring us the ravages of war as easily as nudity and violence. These capabilities, the result of a free and open society, are a concern to an astonished and often outraged global audience. Satellites have certainly forced us to view our expanding space capability with everything from glee to horror.

Television first occurred within national boundaries. Programming was sent out over localized radio waves within neighborhoods, cities, and sometimes across a nation. However, this was extremely difficult to accomplish in the 1950s with the terrestrial transmitters and coaxial cable systems that existed then. The early days of TV spearheaded change, but not global revolution. It was actually only when satellites came on the scene in the mid-1960s that worldwide change began to occur.

In the 40 years that have followed, satellites have brought us change in every possible way—in terms of culture, business, and even warfare. It is satellite systems that have made Internet a truly global experience. Without satellites, the Internet would not be a force in countries as diverse as Mongolia, Bolivia, Zaire, and Vanuatu. Today, fiber optic networks are a key part of the global network, but fiber only interconnects about half of the world's nations. If one wants to reach the more than 200 countries of the world—places like Lesotho, Truk, or Tuvalu— satellites are the medium of choice.

In his book of essays entitled *Utopia or Oblivion?*, Fuller (1971) singled out the unique aspects of satellite technology. Fuller explained nearly three decades ago why geosynchronous satellite systems represented a "breakthrough technology" that allowed us to accomplished much more with much less. He called this phenomenon *ephemeralization*; Buckminster suggested that this breakthrough in intelligence is critical to human evolution and to achieve his ultimate dream—that "intelligence in the universe" might ultimately overcome entropy.

He argued that the satellite created a virtual sea change in human existence (akin in its force and range of impact to the printed word or electrical power). In 1973, at the White House World Communications Year observance that I luckily got to attend, Fuller, 2 weeks prior to his death, argued that the uniqueness of human intelligence is in the ability to do things in totally new ways. He suggested that ephemeralization is not only essential, but forms an ineluctable part of human development. Thus, ephemeralization is the ability to achieve tasks or activities of the past in ways that are orders of magnitude more efficient in terms of time, energy, consumption of resources, and reduced cost. In "Bucky's" view, it represents the best long-term hope for human survival.

THE AMAZING NEW SATELLITE TECHNOLOGIES YET TO COME

Advances such as books, electricity and electric motors, artificial intelligence, solar energy, the telephone, TV, the Internet, computers, and satellites have unlocked a new future for humanity. Fuller's view of the need for humans to search for new pathways to the future takes on almost religious implications in his writings. These breakthroughs or fundamental discoveries are needed to divert us from conventional pathways to profound societal and physical change.

To Fuller, the satellite, electronic computers, the transistor, and parallel inventions are destiny.

Such basic knowledge, he argued, can unlock capabilities and possibilities that are a basic departure from simple extrapolations that represent the trajectories of past civilizations. Figure 1.4 shows how satellite systems are continuously able to do more with proportionately fewer resources. As noted earlier, the satellites of the near-term future, although they require only 10 to 20 times more resources to deploy, are as much as 10,000 times more capable than the earliest satellites. The marvel of these new satellites is that they can work with user terminals that are ever smaller in size and cheaper in cost. Thus, these new microterminals can soon become almost universally available anywhere at anytime. This type of dramatic technical innovation helps to generate and sustain a global revolution in worldwide TV, communications services, and expanded access to the Internet.

THE NEW SATELLITE APPLICATIONS ON THE HORIZON

What do Buckminster Fuller, James Naisbitt, and others mean when they refer to satellites and other telecommunications technologies as giving rise to a global revolution? Ultimately they are referring to a change that impacts us all and can be seen in just about everything. The implications reverberate everywhere—to business, education, health care, entertainment, public safety, and even armed conflict. There is today global trade at the level of tens of trillions of dollars (U.S.) that only 30 years ago was measured in billions of dollars. Even discounting inflation, this is a huge increase.

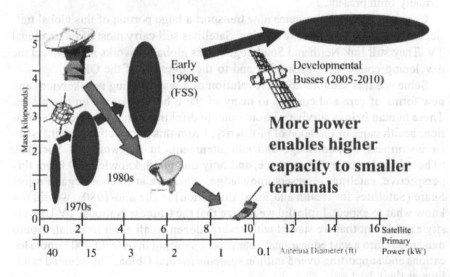

FIG. 1.4. Changing nature of Satcoms.

Furthermore, this international trade of the past involved a much greater element of colonial or neocolonial exploitation than it does today. When satellite communications came to the island nation of Samoa well over two decades ago, the prices of exports increased on the order of 30%, and the price of imports dropped by about 30%. This is because prices could be negotiated via satellite rather than set by the last freighter to steam into port.

We now have a court of world public opinion that stems from worldwide and instantaneous electronic news coverage. We now have worldwide banking, stock trading, and electronic funds transfer at a level that exceeds $100 trillion a year. Global satellite communications, fiber optic networks, and terrestrial wireless have over the past decade begun to create not a "global village," but what might be called the earliest manifestations of a "global brain" or an "E-Sphere" (Pelton, 1999, p. 1).

Part of this new worldwide mind has created an omnipresent Hollywood and MTV culture around the world. Other parts have created global, political, and cultural observatories that span our planet. Yet others have enabled research networks that interlink Nobel scientists in a nonstop quest for new knowledge. America, the land of the Internet and nonstop satellite TV and video streaming, is now at once the primary envy as well as the primary source of perceived evil of the world. Like Janus, our nonstop electronically fed society faces both ways. It marches forward and backward at the same time.

TV broadcasters like Ted Turner, John Malone, Rupert Murdoch, and Kerry Packer of Australia would never have risen to prominence without the satellite. The same is true about MTV, HBO, and a host of other media organizations. Clearly not all change is good nor is it evil. What satellites are today, however, are virtually omnipresent.

Certainly fiber optic systems now transport a huge portion of this global traffic, but satellites started the revolution. Satellites still carry most of international TV. They still link North and South together as global networks interconnect the developing countries to each other and to the countries of the OECD.

Some see this satellite-driven revolution as a way to bring new services and new forms of care and comfort to many of the 6 billion people on our planet. These human beings survive without potable drinking water, electricity, education, health care, or the hope of prosperity. From this perspective, satellites are an instrument of knowledge and enlightenment. In the words of Socrates: "There is only one evil, ignorance, and only one good, knowledge." From this perspective, satellites represent knowledge. When we at Intelsat began Project Share (Satellites for Health and Rural Education) in the mid-1980s, we did not know what to expect. Little did we realize that the Chinese National TV University experiment that we started with several dozen small earth terminals would mushroom into a vast educational enterprise operating in over 90,000 remote locations and supporting over 5 million students in rural China. This seemed satellites at their best.

Others see things differently. These critics look at global entertainment and media "live via satellite" as decadent Western imperialism or at least as "cultural and technological imperialism." They see only a "dark" technology. International satellite transmissions certainly bring to unreceptive audiences such things as scantily clad bodies, "loose morals," and sophomoric "Baywatch" programming. Such "Western" TV and movies disturb societies that follow traditional religions and values. Many perceive this blatant and undiscriminating spread of Western technology as undermining the family and religious authority. Certainly this over-powering satellite and media power were clearly seen by the former Taliban rulers of Afghanistan and the Al Qaida as the "devil" that must be attacked.

In short, there are many—even beyond Osama bin Ladin and his band of ter-rorists—that see satellites, electronic media, computers, TV, and the Internet as Western forces of evil. Yet they embrace the same technology to fight those who violate their belief systems. They see the danger of open and uncontrolled com-munications.

SATELLITES AT THE INTERSECTION OF EASTERN AND WESTERN CULTURES

Arthur Clarke likes to tell the story of the elite and "hip" society attending soirees in New Delhi who say that Hollywood programming is okay for them—the edu-cated elite—to see, but such fare is inappropriate for the rural population of re-mote India. They insist the rural poor should not "see things that they do not un-derstand and would misinterpret." *Noblesse Oblige* is a natural trait for humans the world over.

Religious and other critics see "Western technology" and "Western cultural values" based on scientific rather than religious principles as being increasingly at odds with traditional systems of belief. This is not new. Yet now in the wake of the national U.S. calamity of September 11, 2001, the magnitude of the rift has been exposed. The concerns about the stresses of modernity have been present for some time, but no one in the West has willingly conceded that advanced technol-ogy and traditional societal and religious beliefs are indeed in "fundamental con-flict" in both senses of the word.

In the 19th century, Ralph Waldo Emerson said: "Technology is in the saddle and rides mankind." However, he was warning about the economic and social stresses of lost jobs and regimentation in the face of automation and the deperson-alization of society—not about a cultural invasion of a hostile philosophic, reli-gious, and cultural thought system. Lewis Mumford and Jacques Ellul (noted ear-lier) wrote their own warnings in the 1950s and 1960s, albeit in more contemporary language and concepts. These warnings and concerns, however, were in terms that could be sorted out in Western thought and even in liberal ver-sus conservative political terms.

Today, however, we face something entirely new. We have received some interesting academic insights in recent years about the scope and nature of the problem. We have been able to read about the perspectives provided by Professor Benjamin R. Barber in *McWorld vs. Jihad*, by Samuel P. Huntington in the *The Clash of Civilizations*, and, most recently, by Bernard Lewis in *What Went Wrong: Western Impact and Middle East Response*. These scholars help us understand the wide gap between instant electronic linkage and empathetic understanding across religious and cultural gaps.

We are constantly being shown that instant electronic contact does not eliminate cultural barriers or distrust across religious, social, and economic barriers. However, we are only beginning to recognize that the super-speed change in our technology is leading us not only to new heights of intellectual and scientific understanding, but also to disconnects between cultures that have different values and cultural objectives.

Technology provides a powerful duality. Its positive impacts could be called *telepower*, on the one hand, and the negative impacts might be called *teleshock*, on the other hand. The divisions and emotional rifts that separate our worlds are no longer just economic, but also cultural, religious, educational, and even scientific and technological. Technology can no longer be considered a tool with neutral value until applied. From the perspective of religious fundamentalists, it can be a threat or even a demon.

There are ironies and discontinuities at work here. The Al Qaida and other fundamental extremists actively use the tools of modernity and communications to organize and fight "Western Influence." Nevertheless one should not assume that their anger is directed just at the wealth or religious belief of infidels. They want separation from outside influence at all levels. A Western technology and capitalist marketing system as a "system of influence" is certainly considered a hostile force.

The choice of the World Trade Center and the Pentagon as the focus of the Al Qaida attack was not accidental. These targets were chosen as symbolic "homes" of the enemies. These sites represented the very source of capitalist technology and the U.S. military-industrial complex. Had the capability been there, they also might have tried to flood Silicon Valley and send a missile into the Hollywood Oscar awards.

The events that led to the living horror of passenger aircraft being turned into missiles of destruction on September 11, 2001, undoubtedly have their roots in electronic technology's relentless assault on change. This is a clear and violent backlash against the Western change wrought by communications satellites and outside technology.

There are indisputable clashes afoot in the 21st century. We can no longer dispute the "frightened" and "frightening" reactions by fundamentalists. They have inspired a cadre of followers who resist the miasma of change that modern technology enables and global satellite networks render global and universal throughout our planet. There is fear as well as hostility expressed by those who resist the inevitable changes that modern science and technology make possible and constantly accelerate. This book must address such issues because it is focused on the change, innovation, and progress that satellites have brought to the world. Here is but one example.

Algeria was the first developing country to establish a domestic satellite system. For the Algerian desert people, change came quickly. Algerian government officials leased space segment capacity from the Intelsat global satellite system and then bought earth stations from GTE to link its major regional cities together in 1974 and 1975. The thought was to bring telephone and some evening TV news to remote locations for the first time. For centuries, the bazaars in these desert towns had opened at sunrise and closed at sundown, but in only a few days the markets closed at 5:00 p.m. because that was when the satellite TV shows began. Satellites (as well as VCRs) brought "forbidden fruit" to many traditional cultures, and the religious leaders were the first to feel the attack on their society, culture, and religion. It is no accident that the most rapid growth of VCR ownership in the 1970s and 1980s came in the Middle East. It was such devices that allowed forbidden programming to be seen at last.

Satellite technology has been a "hand maiden" of change and innovation to global society and culture now for four decades. It has enabled not only new technology to evolve, but for it to be broadly shared. It has also been a source of cultural invasion and change more powerful than Alexander the Great and Genghis Khan could have ever dreamed of having at their command. The satellite revolution is thus only a part of a global torrent of change that robotics, biotechnology, artificial technology, fiber optics, computers, the Internet, TV, VCRs, DVDs, and Spandex have brought to an unsuspecting world. During the age of satellites, the world population has expanded by many billions. Fundamentalists from the most traditional of societies are fearful of this change—and for good reason.

THE NEED TO COPE WITH "TELEPOWER" AND "TELESHOCK"

The shape and form of this E-Sphere will change to reflect the stark realities of a global culture in conflict. For years satellites and fiber have created super-speed information networks. These powerful tools have created more and more centralization in the nodes of global civilization located in New York, Los Angeles, Washington, DC, Paris, London, Berlin, Tokyo, Moscow, Beijing, Shanghai, Singapore, and other centers of power.

New concerns about terrorism, urban pollution, energy consumption, traffic congestion, and soaring real estate values could alter the shape of change and modernization. One of the key lessons that should have been learned from 9/11—namely, avoid overcentralization and employ telecommuting more— were largely lost.

Unbridled technological change, instantaneous free-market globalism, and e-commerce on the Internet may very well contain doses of "teleshock" that cancel out the positive gains of "telepower." This book tries to explain not only the scope and type of satellite-driven change, but also where restraint and prudent limits to electronic technology may well be needed. The future of satellites and their opportunity for aiding future generations are thus likely to be conditional. The challenge is to devise "smarter and wiser" systems rather than simply "faster and broader-reaching" networks.

Recent demonstrations at the meetings of the World Economic Forum, the World Bank, the World Trade Organization, and other forums where technology, free trade, and capitalism are promoted have shown that the circle of concern goes beyond fundamentalist religious zealots. There is a wide range of critics who would like to see satellites and information networks in the 21st century being marshaled to create new and more enlightened forms of modern society, but not promote unbridled technological growth as envisioned by pro-technology gurus (Fuller, 1971, chap. 1). Labor unions, environmentalists, social reformers, and others fear "teleshock" tremors still to come.

We can indeed still use satellites, fiber, and innovative wireless technology to create a new type of future that transcends the goals of 19th-century capitalist-led democratic liberalism as defined by Adam Smith and John Locke. As bright and intelligent as those political economists were in their day, their ideas are now out of date.

We could indeed use some innovative thoughts about the interrelationship between democracy and technology. We may well need to invent some new institutions and political philosophies to cope with the 21st century. We probably need to sort out ways to use our most modern of information technologies in a better way. We may need to use our technology to fuel societal and environmental preservation rather than faster growth. We should move toward "electronic decentralization" so that we can move ideas rather than people and avoid the folly of putting all of our people and resources in huge and environmentally unsound megastructures like the World Trade Center.

CHANGING EDUCATION, HEALTH, DEMOGRAPHICS, AND PATTERNS OF WORK

We may need to use our technology to re-instill intellectual freedom in education. Likewise we may wish to use technology to create "smart market" systems that reconnect supply and demand to meet longer term social and economic goals rather than short-term profits. There are many lessons that the Enron collapse, 9/11, and the widening ozone hole in our atmosphere can teach us if we would only listen. Better use of our communications systems is just one of these lessons.

We could use 21st-century satellite and fiber networks to restructure and improve the nature of our cities, reinvent our institutions, and reinvent the fabric of our work and daily lives. The war on terrorism and the problems associated with efficient education and health care systems could help us devise more humane patterns of global growth and sustainable development. We could learn that the most advanced of our satellite and fiber technologies are tools that extend beyond the narrow realm of business. We can use the satellite systems of the future to create interesting and rewarding telecities. These same technologies can help liberate workers via telecommuting systems and make more effective use of electronic immigrants.

We may also invent new forms of electronic and cultural educational and health care systems that replace conventional physical integration with intellectual freedom. These new models of 21st-century development based on electronic and optical technologies could give new meaning to our understanding of words like *neighborhood*, *city*, or *country*. The power of satellites to redefine our world has actually only begun to be tapped in their first 40 years of existence. Out of adversity and questions about the goal of information and media technology in society, new answers and new solutions may well be found.

In the 21st century, we will find workers going to their offices electronically, whether across towns or villages or across the globe. The same electronic restlessness and seeking of security via decentralization, plus the desire to reduce costs, will reinforce patterns of distance education and health care. We will find ways to use satellite and other information technologies to reduce air and water pollution, reduce energy consumption, and even transfer property values. It is here that satellites may create new win–win strategies for workers, capitalists, and environmentalists. New development and economic growth in the 21st century may thus be accomplished by means of satellite and fiber-based electronic propinquity rather than by physical adjacencies.

Constantly evolving technology versus the striving by many for stability and traditional values will be a key part of the story of the 21st century. This book is thus a series of stories and historical facts that are intertwined with some future speculations about the rise of new global systems. Satellites and other new media are slowly but surely revealing a new worldwide environment.

We particularly explore what globalism means in this new age—in this amazing, startling, and frightening new millennium. It seems that both technology and global consciousness are a one-way gate. Once you go through, you cannot easily return to the past. In this regard, satellites have proved—for better or worse—to be one of the most powerful gateways to the future.

In the next few chapters, we relate an interesting, rich, and often untold history, and we also seek some insights into the future. It is the story of scientists and engineers who developed new technologies, sometimes against the odds and other times in the face of fierce competition. It is also the story of international conflict and institutional change and innovation.

Part of the story relates to the cold war and U.S. reactions to postwar conditions and the needs of an increasingly global market place, which puts much more stress on trade in services than in goods and products. It is also the story of Europe and its desire to become a space power and to use satellites to meet its social and economic needs. It is a story of how Japan and Australia have often been the voice of reason, compromise, and plain old good common sense when issues of global satellite power politics have emerged.

It is a story of how the USSR used satellites as a key element of cold war power and influence, sometimes with great effect and sometimes with less than the desired results. All the economically developed countries of the OECD—from Japan, to Canada, to Australia and New Zealand, to Europe—have played roles in satellite system development. Also the developing countries have effectively used satellites to reach isolated populations and escape from the neocolonialist impact that traditional lines of communications imposed throughout the first half of the 20th century and beyond. There are few technologies that have had such a pervasive impact around the world. When CNN produced the satellite-delivered program called the "Day of Five Billion" a little over a decade ago, there was virtually no political entity across the planet that did not play a role, and over 150 countries aired the program. Nearly 80 countries produced some segment of this remarkable program that recognized the significance of the continuing rapid growth of humanity globally. It also demonstrated that more than 200 countries and territories around the world have a vested interest in satellite networks.

CONCLUSIONS

This book is an interrelated and integrated group of chapters that address satellite systems and their impact on the world. These chapters address satellite technology, satellite business and economics, satellites and global politics and regulation, satellites and social services, and even satellites and the future. The attempt is thus to tell the story of satellites from different perspectives and via different disciplines. Too often the world of satellite technology is severed from its social, economic, cultural, and political impacts.

We hope that the reader will first see the satellite story in terms of technology development—past, present, and future. Then it moves to satellite history and regulation. Next, the book morphs into other stories about global business, global TV news, entertainment, and sports—about the many ways satellites have altered our world.

The 12,000 satellite video channels around the world today bring us news, entertainment, the Olympics, and even wars "live via satellite." The Iraq "regime change" of 2003 was the most televised war in history—it was truly "live via satellite."

One of the many stories in this book is that of satellites, the Internet, and the electronic systems we call *cyberspace*. In truth, without satellites, the Internet would not yet have become the global phenomena that it is today. Finally, this book addresses major problems of global concerns about what might be considered negative applications of satellite systems. It provides perspective on a new hope for tomorrow. Satellite systems, by shaping world opinion and allowing teleeducation and telehealth to become available on a global scale, offer new hope for world peace and knowledge. Satellites represent one of the keys to a successful global economy. Satellites offer dual use of networks for commercial and military purposes, for education as well as propaganda and misinformation, for targeted bombs or targeted information.

The forces that shape the world of telecommunications and satellites are more than just technology, business, entertainment, social applications such as health, or education. It is the gestalt of these forces and interests all at once. An attempt to depict the interdisciplinary model that defines the scope and shape of the analysis of this book is shown in Fig. 1.5.

Some believe that, because communication satellites are largely high-tech tools (almost literally supercomputers with specialized software that are deployed in the skies), the stories of these space systems might be dry, straightforward, and devoid of emotion. The truth is that the stories surrounding the development and implementation of satellite communications are filled with elements of intrigue and joy. Just like all forms of human enterprise, the story of satellites sometimes involves conflict, criminal activity, and deception and at other times it offers new hope for the human condition by providing new forms of education, health care,

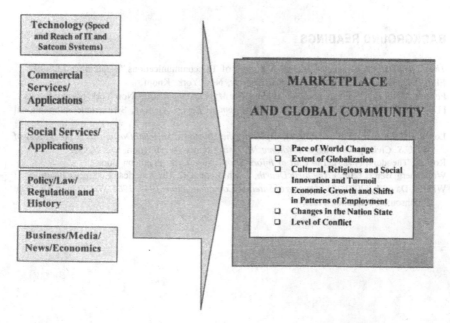

FIG. 1.5. Model for global satellite communications systems and interaction with the global marketplace and political history.

or mitigation of disasters. Change often comes at a considerable price. In today's world of technological innovation, progress can often blend with a sense of loss. The story of communications satellites, in this sense, is certainly no different.

REFERENCES

Barber, Benjamin R., (1996), *McWorld vs. Jihad*, New York: Ballantine.

Braudel, Fernand, (1979a), *The Perspective of the World: Civilization and Capitalism—15th–18th Century (Volume 3)*, New York: Harper & Row.

Braudel, Fernand, (1979b), *The Structures of Everyday Life: Civilization and Capitalism—15th–18th Century (Volume 1)*, New York: Harper & Row.

Braudel, Fernand, (1979c), *The Wheels of Commerce: Civilization and Capitalism—15th–18th Century (Volume 2)*, New York: Harper & Row.

Fuller, R. Buckminster, (1971), *Utopia or Oblivion: The Prospects for Humanity*, New York: Bantam Books.

Huntington, Samuel P., (1996), *The Clash of Civilizations and the Reckoning of World Orders*, New York: Simon & Schuster.

Lewis, Bernard, (2002), *What Went Wrong: Western Impact and Middle East Response*, London: Oxford Books.

McLuhan, Marshall, (1966), *Understanding Media: The Extensions of Man*, New York: Signet Books.

Naisbitt, James, (1980), *Megatrends*, New York: Alfred Knopf.

Pelton, Joseph N., (1999), *E-Sphere: The Rise of the World Wide Mind*, Bridgeport, CT: Quorum Books.

Reich, Robert B., (1991), *The Work of Nations*, New York: Alfred A. Knopf.

BACKGROUND READINGS

The Economist, "A Connected World: A Survey of Telecommunications," September 13, 1997.

Ellul, Jacques, (1969), *The Technological Society*, New York: Knopf.

Ferkiss, Victor C., (1969), *Technological Man: The Myth and the Reality*, New York: Mentor Books.

ITU, (1997), *World Telecommunications Development Report*, Geneva, Switzerland: International Telecommunication Union.

Logsdon, John *et al.*, (Editors), (1998), *Exploring the Unknown: Selected Documents in the History of the U.S. Civil Space Program, Volume III, Using Space*, Washington, DC: NASA.

Roszak, Theodore, (1985), *The Cult of Information*, New York: Pantheon Books.

Wattenberg, Ben J., (2002), *The Birth Dearth*, Washington, DC: The American Enterprise Institute.

Whalen, David J., (2002), *The Origins of Satellite Communications 1945–1965*, Washington, DC: The Smithsonian Institution Press.

PART TWO

TECHNOLOGY

Satellite Technology: The Evolution of Satellite Systems and Fixed Satellite Services

Louis Ippolito
ITT Industries

Joseph N. Pelton
The George Washington University

> *The synchronous satellite is an absurdly simple idea, and involves no real paradox. It is not really motionless, for it is actually moving along its orbit at almost seven thousand miles an hour. At this precise speed it just keeps up with a point on the earth's spinning equator, far below and it just overcomes the pull of gravity as well.*
>
> —Arthur C. Clarke (1967, p. 140)

One can begin the story of satellite technology in many different points in history. For example, in the third century BC, Arcetus of Tarentum developed a working concept of jet propulsion on which today's rockets are still based. This remarkable achievement used steam and his projectiles were wooden pigeons, but the basic physics are still much the same at the start of the 21st century. However, these basic concepts on rocket propulsion were to be lost in the centuries that followed the remarkable discoveries of Arcetus.

Some 20 centuries later, none other than Sir Isaac Newton published the first theoretical account of how an artificial satellite could be launched into earth's orbit. In his discussion about physics and gravitational effects, Newton actually drew a diagram showing how an artificial satellite could be launched into earth's orbit. He maintained that this feat could be accomplished with the "moral equivalent" of a large cannon that had sufficient launch velocity (i.e., muzzle velocity) and the right elevation angle.

Many other visions of satellite networks followed the writings of Newton, both in works of fact and fiction. H. G. Wells, Everett Edward Hale, Jules Verne,

Konstantin Tsiolokowsky, Hermann Oberth, Willy Lev, Hermann Noordung, and Robert Goddard, among other writers, engineers, and scientists, envisioned the possibility not only of rocket launchers, but of communications and navigational satellites as well. In the late 1930s, Hermann Noordung even wrote about a space colony in geosynchronous orbit.

Some of these writings were more flights of fancy than presentations of scientific logic. A few were even both. Some of the more insightful but whimsical thoughts were about the world of satellite communications systems yet to come. The first of these can be found in the writings of Everett Edward Hale in his book about a brick moon. Here Hale envisioned an "artificial moon" built of bricks but inhabited by tropical plants as well as by men (and women) astronauts. These astronauts, without aid of oxygen, would actually provide navigational assistance to ships at sea from a polar orbit. At the time, solving the problem of accurate oceanic navigation was in fact the biggest scientific challenge of the day. The ultimate solution was not navigational satellites, but one that was provided instead by amazingly precise clocks developed by an English craftsman named John Harrison—a story that is well told in a book by Dava Sobel called *Longitude*.

The flight crew in Hale's fanciful book would not only provide navigational aids such as longitude to ships at sea, but would also use Morse code to communicate to the earth below—simply by stamping their feet in unison. The polar orbit he described, however odd his vision seemed a couple of centuries ago, in fact resembles the exact polar orbiting MEO orbits that are used by modern remote sensing spacecraft and GPS networks.

SIR ARTHUR CLARKE: THE FATHER OF THE COMMUNICATIONS SATELLITE

The story of communications satellite technology, however, began seriously in 1945. The overall vision of global satellite communications started in earnest with the writings of a young author and scientist released from service in the British Radar Establishment as the end of World War II drew near. In a privately circulated paper shared with fellow scientists in early 1945 and then in a formal article published in the fall of the same year, this young former air force officer outlined the future of satellite communications. In 1945, his article about "extraterrestrial radio relays" appeared in a now obscure British publication called *Wireless World*. Sir Arthur Clarke, at the time, was neither a knight of the realm, a world personality, nor a science fiction guru. Rather, he was a young Englishman with a strong yen for ocean diving and adventure. He went on to indulge these appetites with vigor in both Australia off the Great Barrier Reef and in Ceylon—now Sri Lanka—where he now lives. In the years that followed, he wrote an increasing array of science fiction novels and scientific papers, but none was ultimately as important as his early writings on satellites.

Arthur Clarke has a truly magical gift for writing solid works of science and engineering while also spinning yarns of science fiction, including *2001: A Space Odyssey*. In 1945 he was an unknown; consequently, his brief but seminal article was largely ignored. Yet the entire vision was there. In succinct but clear terms, he spelled out a remarkably practical vision of what the future of space applications might bring. He described how we might someday establish a global communications system with just three satellites by calculating the speed necessary to launch a "space station" into geostationary orbit and establish radio stations beyond the ionosphere. The satellite application he used as a way of illustration was the relay of TV signals, although TV was then in its infancy. Today, TV relay still remains the most dramatic application of communications satellites.

Clarke described a special orbit, which was the only one whereby a satellite precisely encircles the earth once a day without losing altitude or escaping the world's gravitational field and thus flying off into space. This magical orbit where the "g" force is exactly 0.22 meters/second^2 allows a satellite to remain stationary with respect to a constant point on the earth's equator. Thus, a satellite, once positioned in orbit some 22,230 miles or 35,870 kilometers above the equator (with only modest orbital adjustments from microjets to correct for small north–south or east–west excursions) will stay in place for long periods of time. A geosynchronous satellite, located in what is now called a GEO orbit, thus becomes much like a very, very tall radio relay tower in the sky. The satellite receives the faint up-link signal, filters out noise and interference, translates it to the down-link frequency (to prevent interference), amplifies it, and then retransmits it back to earth (see Fig. 2.1).

With such a high tower, a radio transmission can reach across the oceans and, in fact, "see" 40% of the way around the earth at the equator. Clarke's landbreaking (or rather space-breaking) article explained how three such satellite radio stations equally spaced above the equator could connect the world together and serve the entire planet except at the polar caps.

Due to the earth's curvature, a microwave tower over 500 miles (or over 800 kilometers) high would be required to "see" across the Atlantic Ocean and a much taller one still to "see" across the Pacific Ocean. Of course such towers were not economically or technically feasible. Yet, Arthur Clarke discussed how it was possible to build one that was over 22,230 miles (or 35,870 kilometers) high. Such a relay tower in space could do what a physical tower built up from the earth's surface could not do in a practical manner.

Brilliance and vision is not always immediately recognized. Galileo, Columbus, Kopernicus, Goddard, and many others found their "new ideas" ridiculed or ignored at first. Such was the case with Clarke, whose article was only recognized for its greatness in retrospect. This was, in part, due to Clarke's own beliefs that these radio towers in space would need human crews to constantly replace burnt-out radio tubes, and that space travel and reliable rocket systems were simply too far in the future to be seriously considered in the short term.

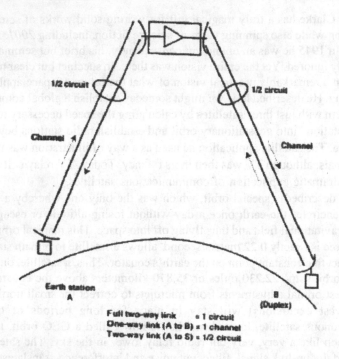

FIG. 2.1.   Schematic representation of a satellite communication link.

Thus, Clarke did not even attempt to patent the idea of geostationary (or more generally geosynchronous) communications satellites for these reasons—something he admitted that he later regretted.

A geosynchronous satellite orbits the earth every 23 hours and 56 minutes to maintain perfect sidereal relationship to the earth and the sun on an annualized basis. A satellite in geosynchronous position, however, can stray above or below the equator in the north and south direction. A geostationary satellite not only revolves with the earth, but also maintains its position over the same exact spot above the equator as well and by means of control instructions from an operating center on earth. A perfectly geostationary satellite is not allowed to rise above or fall below its equatorial position. In practice, all GEO satellites make these excursions or "inclinations" above or below this imaginary plane, but in most cases the deviations are so small that the satellite earth terminals do not have to be repointed to accommodate these minor shifts. The amount of fuel needed to keep the satellite from moving north or south off the equatorial plane is 10 times the amount needed to keep the satellite in position from east to west, and thus tight station-keeping in the equatorial orbit is a highly specialized and difficult task that is hard to execute with perfection.

It was a separate but related research breakthrough—the invention of the transistor less than a decade later—that made possible not only a new generation of reliable communications technology, but also allowed the development of the practical electronic computers. These were needed for applications including calculation of the orbital mechanics for launching a satellite and ultimately the incorporation of solid-state components in the satellites.

THE SOVIETS LAUNCH "SPUTNIK" AND THE SPACE RACE BEGINS

A little over a decade after Clarke's remarkable article was printed, the space age began—not in literature, but in reality. In October 1957, the world was surprised to wake up to the "beep beep beep" of Sputnik. Although the USSR had promised it would launch a satellite as part of the International Geophysical Year global scientific enterprise, most of the world had been skeptical that this feat would be accomplished.

The successful launch of Sputnik was a remarkable scientific and engineering accomplishment. The Soviet Union's launch of an artificial satellite was perceived by the world—and especially the United States—as a cold war challenge of military significance rather than of scientific or business importance. At that time, control of space and the strategic high ground above the earth's atmosphere was seen as the basis of military superiority in the age of atomic weapons and ballistic missiles. In the United States, the race for the presidency in 1960 was dominated by charges that America suffered from a "missile gap" by comparison with the USSR. President Kennedy's razor-thin victory over Vice President Nixon is said by many to have hinged on this one issue, although others attributed it to more mundane matters such as Nixon's sweating and "five o'clock shadow" during the crucial TV debates.

In the wake of Sputnik in the fall of 1957, a scramble of U.S.-sponsored space activities began with the aim of launching an American satellite of any shape or size. The first attempts were by the military. In particular, the U.S. Army and the Signal Corps undertook efforts to launch a satellite into space. In the months that followed, the National Aeronautics and Space Administration (NASA) was created by an act of Congress in 1958. The new agency was hastily formed and headquartered in Washington, DC. In the process, the new NASA displaced and replaced the National Civilian Aeronautics Agency (NACA), then located in Cleveland, Ohio. Priorities were quickly shifted from developing better jets to developing space systems.

These early efforts to create an American space program began with an embarrassing string of failures. Several attempted launches ended in failure, but then the U.S. Army finally launched the Explorer satellite that discovered the Van Allen Belt. On December 18, 1958, the U.S. Signal Corps launched a tiny communications satellite—"SCORE"—into low orbit with a taped message from President Dwight D. Eisenhower that broadcast the message "Peace on Earth, Goodwill to

Men" for about 2 weeks during the Christmas holiday season. This was followed by the simple teletype experimental satellite known as Courier 1B. Although the USSR would move on to launch larger satellites, the United States was initially forced to concentrate on smaller and unmanned spacecraft.

By the 1960s, more serious and practical communications satellite projects were under way. These included the successful, but ultimately rather irrelevant, Echo 1 and Echo 2 experiments that bounced signals off giant metallic-covered balloons, as well as the lesser known military-backed West Ford project. In these "passive satellite" experimental projects, radio signals were simply reflected off satellites. This simple reflection of radio signals from satellite surfaces proved to be a difficult and highly uneconomic way to complete a telephone link. In short, the ECHO experiment, designed by Dr. John Pierce of Bell Laboratories, and the West Ford projects both "worked," just as experiments in the 1940s of bouncing signals off the moon had worked earlier. These trials, however, confirmed that passive signals that bounced off a reflective balloon were simply not an effective way to proceed.

These experiments demonstrated that, to be practical and commercially viable, satellite communications would require radio signals to be received, reamplified, and returned to earth with a significant power boost. Otherwise the signals would be too weak and low in capacity to have any practical value.

The Relay satellite (built by RCA) and the Telstar satellite (built by AT&T Bell Labs), both launched by NASA in 1962, truly set the stage for commercial satellite communications. These spacecraft proved that telecommunications via satellite, including the relay of TV signals, was not only possible, but was achievable on an ongoing basis and likely at affordable prices.

The submarine coaxial cables of the early 1960s lacked the capacity to handle live TV transmissions, and thus these new experimental satellites truly fired people's imagination about achieving broadband telecommunications networks in the sky. AT&T, RCA, and the other telecommunications carriers at this point wanted to seize the initiative and lead the way forward to develop and deploy this technology. However, this particular vision of an industry-led program to create global satellite communications systems was not to be. A full historical analysis of this period in the annals of space is contained in a book by the space historian Dr. David Whalen (2002), *The Origins of Satellite Communications 1945–1965*. This book about the early days of satellite communications explains how the tug of war between the telecommunications industry and the U.S. government (and especially NASA) went on for some time.

The early development of communications satellite technology did not all occur in the laboratory. When President Eisenhower was about to leave office, he attempted on December 30, 1960, to define a national policy on satellite communications development. He outlined how he thought the technology and satellite services should evolve and specifically under whose leadership, ownership, and control. In his policy statement, President Eisenhower essentially outlined an industry-led initiative.

Eisenhower stated,

> The world's requirements for communications facilities will increase several-fold during the next decade and communications satellites promise the most economical and effective means of satisfying these requirements. . . . This nation has traditionally followed a policy of conducting international telephone, telegraph and other communications services through private enterprise subject to Governmental licensing and regulation . . . accordingly the Government should aggressively encourage private enterprise in the establishment and operation of satellite relays for revenue-producing purposes. (Logsdon et al., 1998, p. 42)

Shortly after his election, John F. Kennedy made his famous space speech, in which he set the goal of sending a man to the moon and returning him safely back to Earth within the next decade. This speech led to the Mercury, Gemini, and Apollo Programs and the actual moon-landing mission by Neil Armstrong, Buzz Aldrin, and Michael Collins in July 1969 and is well remembered. Yet at the same time Kennedy likewise set a goal of creating a global organization to establish a communications satellite system "that would benefit all countries, promote world peace and allow non-discriminating access for countries of the world" (Logsdon et al., 1998, p. 42).

These concepts about a global satellite organization were reiterated by JFK in September 1961 in a speech to the United Nations that resulted in a UN General Assembly Resolution known as Res. 1721 P. These Kennedy speeches and the UN Resolution served to move the objectives of establishing a communications satellite system from a mere commercial enterprise to a higher purpose of political and economic equity and world peace. This was not to be a technology developed for commercial profit, but a revolutionary force for global development. In the years that followed, both humanitarian and practical objectives were actively pursued as Comsat and Intelsat were formed.

Whereas Eisenhower and the large communications companies such as AT&T, ITT, RCA, Western Union, and Western Union International were thinking in terms of satellite systems for commercial operations and the generation of new business revenues derived from telephone, telex, telegraph, and even TV relay, Kennedy took a different approach. His initiative implied a role for government and tied communications satellite systems to economic development and political objectives.

Kennedy declared his political goals for communications satellites via a global network as the furtherance of world peace, and he had these concepts endorsed by the General Assembly of the UN. This changed the perspective of how the technology would be developed in the United States and indeed set up a source of conflict between NASA and the telecommunications industry. This likewise evolved into a major dispute within the U.S. Congress about who should develop and use the technology when the Communications Satellite Act of 1962 was debated and finally adopted after months of heated debate.

As the U.S. institutional framework for satellite services was winding its way through political processes in Congress toward final resolution in the form of the Comsat Act of 1962, overseas efforts were just beginning. International contacts were made to begin the discussions that led to Intelsat, the global satellite entity. Negotiations were undertaken by the United States with Europe, Canada, Australia, and Japan, and discussions were even attempted with the USSR. These began in 1962 and ended after more than 2 years of hard bargaining with the signing in August 1964 of a new set of international agreements.

These international agreements, which had the force of treaties in some countries but not in others (including the United States), were signed in Washington, DC, at the State Department. As a result of the many concerns expressed about U.S. dominance, especially in Europe, these documents were made interim arrangements for a 5-year trial period. Thus, as political discussions were inching forward, the technology that would make global communications satellites technically and economically possible was making great speed. (More about the political and institutional meaning of the activity related to the creation, growth, and operation of Comsat and Intelsat can be found in later chapters and books cited in the Reference section.)

HUGHES AIRCRAFT "JUMP STARTS" GEOSTATIONARY SATELLITES

The success of Relay and Telstar inevitably led to the next steps forward. The U.S. Congress was wrangling about whether NASA should be in charge of satellite communications or whether private enterprise would lead the way forward, but outside new technology was being developed apace at both NASA and industry.

The scientists at Hughes Aircraft, particularly a brilliant young engineer named Dr. Harold Rosen, convinced NASA that they could design and build a GEO communications satellite that could realize the 1945 vision of Arthur Clarke. This project was known as Syncom, and the NASA contract called for three of these satellites to be built. The first attempt to launch this new satellite design resulted in a launch failure, but in late 1963 the Syncom 2 was successfully launched, followed by Syncom 3 in 1965. The Syncom satellites represented an immediate triumph for Hughes Aircraft Company, NASA, and the U.S. communications satellite program. The Syncom launch proved that GEO satellites could work, and it also demonstrated that global communications satellite systems were indeed technically feasible. Further, the cost of the satellites, at a few million dollars each, also strongly suggested they were economically viable as well.

The launch of these experimental spacecraft was accomplished in two stages. The first stage positioned the satellite into a highly elliptical cigar-shaped orbit with the apogee (or high point of the orbit) some 22,230 miles (35,870 kilometers) into space and the perigee (or lowest point of the orbit) only a few hundred miles above the earth's surface. Second, after this highly elliptical transfer orbit was established, an apogee kick motor was fired near the apogee of the elliptical orbit at exactly the right moment, and this served to transfer the satellite into its circular GEO orbit. This GEO orbit represents a perfect circle 22,230 miles distant from earth—almost a tenth of the way to the moon. This is some 40 times farther away from the earth than so-called low earth orbit satellites (LEOs), which are positioned below the Van Allen radiation belts and some four times farther away from the earth than so-called medium earth orbit (MEOs), which are positioned above the Van Allen belts. The relative positioning of GEO, MEO, and LEO satellites are represented (but not to full scale) in Figs. 2.2 and 2.3.

The Syncom 2 and Syncom 3 satellites were the first practical demonstrations of how to launch and operate GEO satellites. They also demonstrated that a rather

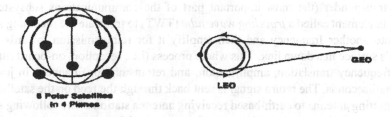

FIG. 2.2. Graphics showing the relative positioning of LEO and GEO satellite orbits above the earth.

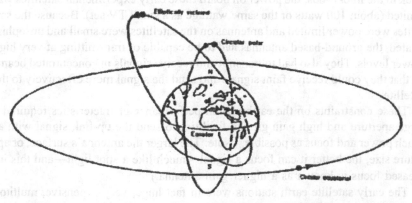

FIG. 2.3. Relative orbital positions of LEO, MEO, and GEO satellites (not to scale).

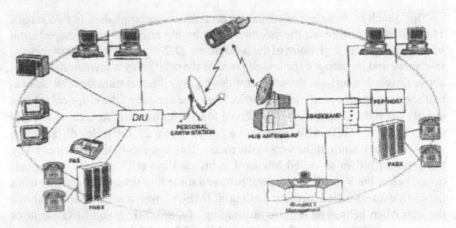

FIG. 2.4.   A satellite network with various types of ground earth stations.

modest, squinted omni-beam satellite antenna could receive a radio signal at this great altitude and then pass the signal through a filter to an on-board transponder. This transponder (the most important part of the communications subsystem) used an element called a *traveling wave tube* (TWT) to translate the incoming signal into another frequency and then amplify it for retransmission towards the earth's surface in a down link. This whole process (i.e., reception, on-board filtering, frequency translation, amplification, and retransmission) happens in just a few milliseconds. The return signal is sent back through the feed on the satellite's transmitting antenna to earth-based receiving antenna stations. The following satellite network diagram shows how signals are relayed from user terminals on the earth's surface to the satellite and back to another earth station in a modern network. This diverse network is shown in Fig. 2.4.

The power on today's satellites can vary from 2 to nearly 15 kilowatts. Yet back in the mid-1960s, the power on board these early experimental satellites was limited (about 100 watts or the same wattage as a small TV set). Because the satellites were power limited and antennas on the satellites were small and unsophisticated, the ground-based antennas had to be capable of transmitting at very high power levels. They also had to transmit and receive signals in concentrated beams so that they could receive faint signals and send the signal more effectively to the satellite.

These constraints on the earth station performance characteristics required a large aperture and high gain ground facilities to send the up-link signal with as much power and focus as possible. (Note: The larger the antenna's surface, or aperture size, the better it can focus a signal—much like a spot light—and this increased focus is known as a higher gain antenna.)

The early satellite earth stations were in fact huge, very expensive, multiton parabolic dishes some 30 meters in diameter. The one exception was a large horn antenna built by AT&T in Maine to support their experimental work with Echo and Telstar, but it was soon verified that the parabolic dish was more efficient and cost-effective.

These costly and sensitive ground facilities, equipped to both transmit and receive signals, had to be high performance to collect the faint signals (i.e., a few millionths of a watt) that the earth station had to detect at the earth's surface due to the spreading of the satellite signal over its 22,230-mile journey. (This phenomenon is known as *path loss*.) Thus, the early satellite earth stations were almost like radio telescopes, and thus required special cryogenic cooling systems for the high-performance amplifiers. This in turn meant the earth stations had to be staffed by around-the-clock crews of 40 to 50 people each. Ultimately, the cost of these earth stations ended up costing as much or more than building and launching the satellite.

It is remarkable that satellite technology has evolved so rapidly over the last 40 years from this rather crude beginning. Today, dishes the size of cereal bowls can receive TV signals from the highest power direct broadcast satellites, and mobile satellites can work with hand-held units.

The success of Syncom 2 set several actions in process. The officials of the newly formed U.S. organization Comsat signed a contract with Hughes Aircraft Company, on behalf of Intelsat, to build an improved version of Syncom. This satellite was described in the contract as the HS 303 vehicle. This satellite ultimately became officially known as Intelsat-I. When it was launched just over 1 year later in April 1965, this first commercial satellite also became affectionately known as "Early Bird" in headlines across the world.

THE RUSH TO BUILD OUT THE EARTH SEGMENT

Another important activity as noted earlier was the construction of the massive and expensive earth stations that would work with Early Bird. There was already the huge horn-shaped antenna built by AT&T in Andover, Maine, to work with the passive Echo and the active Telstar satellites that could be employed to work with Early Bird. Yet there was a need for similar facilities on the other side of the Atlantic because the first satellite was to be positioned over the Atlantic Ocean. Consequently, the British Post Office (which later became British Telecommunications) constructed a facility on Goonhilly Downs in England, the German Post and Telecommunications Ministries constructed facilities in Raisting, Germany, and so on for Pleumeur Boudou, France, Fucino, Italy, and elsewhere. From this modest beginning with just a handful of large earth stations, the industry has come a long way in just short of four decades. There are now over 100 million receive-only satellite terminals in use for satellite TV reception. Soon there will also be over 1 million very small aperture terminals (VSATs) for interactive business and small office/home office (SoHo) use and millions more number for personal mobile satellite services and GPS radio navigation services.

A mini space race to build satellite earth stations was thus created in 1965. As noted earlier, the most exotic of these early designs was an earth station built by AT&T in Andover, Maine, in the shape of a huge horn antenna. It was more striking in its design that in its performance, however, and the United States and overseas satellite antennas that followed assumed the more traditional parabolic dish shape that (in a variety of sizes from huge to small) has become commonplace today.

The world's first operational communications satellite was what is today called *power limited*. That is, its power was less that that used by a small TV or even a good-sized light bulb. The received signal strength from "Early Bird" satellite was equivalent to a few millionths of a watt when it reached the earth's surface. Therefore, to operate successfully, the first earth stations had to be extremely sensitive and possess what is called *very high gain*. Also the satellite signals had to be separated from other radio transmissions sent on adjacent frequencies. It is for this reason that the locations of these first earth stations were built well away from cities or other population centers. The early locations such as Andover, Maine, and Goonhilly Downs, United Kingdom, were about as isolated (and as electronically silent) as one could find in the respective countries.

This may seem strange in today's world of powerful communications satellites, which can operate with small dishes or VSATs positioned on the top of buildings or mounted on houses (and these dishes are often less than a meter in size). Even more remarkably, there are now hand-held user terminals for mobile satellite communications. This is possible because in nearly 40 years satellites have increased their net transmit power in the range of 10,000 to 100,000 times. Instead of satellites being able to generate something like 100 watts of power, there are now spacecraft in orbit with solar arrays that can generate many kilowatts.

Yet increased power is only part of the story. Even more important, today's satellites also have extremely large, high gain antennas operating out in space that allow radio frequency (rf) power to be focused into powerful beams. The net result is that they can generate rf signals that are, in net effect, a million times more powerful than the Syncom or Early Bird satellites, at least when measured in terms of the signal's flux density when the satellite signal reaches a user terminal on the earth's surface.

SATELLITE TECHNOLOGY TRENDS

The development of satellite technology is fascinating to telecommunications engineers, but not always to the general public. However, the changes that have taken place in technology for both satellites and user terminals are important to understanding how new networks are able to be deployed across the world and new broadband services to expand while the cost of providing these services plummets.

The general pattern of development in the satellite industry is actually straightforward. The "big picture" does not require a degree in rocket propulsion or electrical engineering to understand. The big trends that reflect how satellite communications technology has evolved and expanded over the last 40 years can be summarized as follows:

- *Satellites have expanded the performance of their power systems; instead of producing less than 100 watts, these spacecraft now can produce 2 to 15 kilowatts depending on the type of service provided.* This power increase has been accomplished in several ways: (a) by making the solar cell arrays larger; (b) by using higher performance solar cells that convert the sun's energy more efficiently; (c) by creating new ways of pointing the solar arrays so that they are always facing the sun; and (d) by developing and using larger, high-performance batteries with a longer operating life to sustain satellite operations during solar eclipses.

- *Satellites now have more rf spectrum to use. There are more radio frequencies that have been allocated for operational use. Newer satellites also have ways to "reuse" the frequencies many times over without causing interference. This primarily involves the use of a number of narrow and highly focused "spot beams."* Frequency reuse is largely the result of new high gain, multibeam space antennas. Altogether the usable spectrum for satellite communications has, by a variety of means, increased nearly a hundred times. This expansion of the available frequencies has been done in several ways. New frequency bands, typically in ever-higher frequency bands, have been allocated for satellite use. These spectrum allocations for commercial services start with frequencies as low as 137 MHz and 400 MHz. In the early days of satellite service, the most important spectrum was in the 4,000 MHz to 6,000 MHz bands (known as C-Band). This allocation has limited bandwidth (i.e., 500 MHz), and throughput capability is not too large, but there is a limited problem with rain fade. Next comes the 12,000 MHz to 14,000 MHz (or Ku-band), which has more spectrum (and thus greater throughput capability), but increasing problems with rain fade. Today we are moving into the highest of the commercially available spectrum known as the Ka-band. These frequencies in the 18,000 MHz to 30,000 MHz bands are transitioning from experimental systems into the first commercial services via SES Astra (in Europe) and Eutelsat with Spaceway, Wild Blue, and others to follow soon. These new Extremely High Frequency (EHF) bands are difficult to use because radio waves at these extremely demanding frequencies and small bandwidth are "bent" by heavy rainfall. This creates what is called *precipitation attenuation*, and this creates problems when transmitting and receiving signals during rainstorms.

In addition to this expansion of the frequency bands by new allocations made by the International Telecommunications Union (ITU) and their assignment under national licenses by governments around the world, there are a variety of techniques to expand available spectrum for commercial satellite systems. The most important of these procedures follows a process similar to that used in terrestrial cellular and PCS mobile telephone systems. This is to create separate spot beams that can form a pattern of cells that allow the satellite system to reuse the same frequencies again and again. Elimination of the interference is accomplished by creating highly focused spot beams. Each of these beams is separated from each other and aimed at different locations to prevent radio frequency interference by spatial isolation. Another way to eliminate interference is to use polarization techniques. These isolation techniques minimize radio frequency interference between various beams. The circular or orthogonal polarization isolation process is much like using Polaroid sunglasses to screen out the sun's rays in one direction, but allowing natural daylight into your eyes. The net effect of more frequency allocations, frequency reuse via spot beam cells and spatial separation, as well as frequency reuse via cross-polarization is that there is now more than 100 times more spectrum to use for satellite communications than in the early years. Further, new digital modulation and efficient multiplexing systems also increased spectrum efficiency as well.

• *Bigger and more reliable rockets have been developed that can launch much bigger satellites; there are also better thruster-jet systems onboard the spacecraft to keep the satellites pointed precisely where they need to be.* These new rockets can either launch several smaller satellites at a time or use a single launch for placing the largest satellites into orbit. The rockets have become more reliable, but their costs have not greatly decreased. Thus, any innovation that allows a satellite's antennas or the power system to be less massive helps reduce the cost by lowering launch costs.

• *The use of carbon-epoxy elements that are stronger and lighter than metal have created cost savings.* The design of satellite systems has become more efficient by using lower mass but stronger materials. We have also seen lower mass antenna and communications systems with longer life. This has created less demand for rocket launches. (This means that sometimes design gains made on one technical or economic front can result in an adverse impact on the other side of the business. This is why systems analysis, which includes the user terminal, the TTC&M network, the satellite design, and the launch cost, must be undertaken to show the most cost-effective economic solution.)

• *Reliable and higher performance microelectronics and solid-state equipment on board allows satellites to be less massive, more reliable, and to more easily convert to new and more efficient digital communications systems.* The new electronic subsystems in satellites have continued to shrink in size and resemble electronic computers with specialized communications-oriented software. A modern satellite is thus very much like a mainframe computer in the sky with software that enables it to relay TV, data, or telephone calls through its central processing unit (CPU).

• *New ground antenna systems called very small aperture terminals (VSATs) also use microelectronics to work with the high-powered satellites.* These VSATs

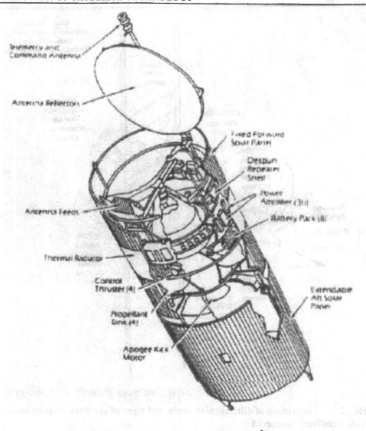

Telemetry and
Command Antenna

Antenna Reflector

Antenna Feeds

Thermal Radiator

Control
Thruster (4)

Propellant
Tank (4)

Apogee Kick
Motor

Fixed Forward
Solar Panel

Despun
Repeater
Shelf

Power
Amplifier (30)

Battery Pack (8)

Extendable
Aft Solar
Panel

FIG. 2.5. Comparison of different size, scale, and types of satellites—"spinning"
spacecraft.

can be manufactured in high volume and at an increasingly lower cost. Specifically, receive-only terminals such as those used with Direct Broadcast Satellite (DBS) services have dropped in cost to only a few hundred dollars when produced in millions of units. There are a number of excellent books on satellite communications, as listed in the reference section, that describe these many technical innovations as well as other more detailed improvements.

• *New ways have been developed to maintain satellites precisely in orbit. These satellites are thus always pointed exactly to the same position on the earth.* This new three-axis stabilization technique typically uses momentum or inertia wheels (similar to the spinning motion used in a gyroscopic top) and has eliminated the need to "spin" the entire spacecraft. The differences in the size of satellites and the differences between three-axis stabilized "birds" with solar wings and the can-shaped "spinners" are illustrated in Figs. 2.5 and 2.6.

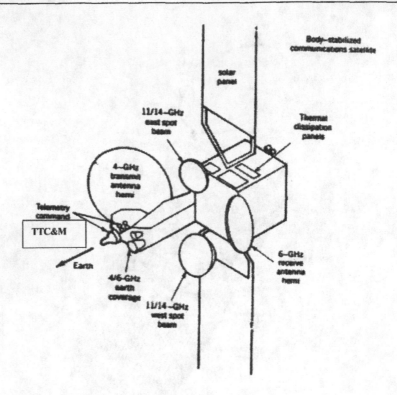

FIG. 2.6.   Comparison of different size, scale, and types of satellites—"three-axis body stabilized" spacecraft.

THE ROLE OF THE ITU IN THE REGULATION OF SATELLITE SYSTEMS

When the first communications satellites began operations, there was only one type defined by the International Telecommunication Union (ITU), and this was known as fixed satellite services (FSS). The Geneva-based ITU is the specialized international agency within the UN structure that provides regulatory oversight, recommended performance standards, and frequency allocations for worldwide telecommunications, including satellite systems. The definition of FSS included telephone, telegraph, data, radio, and TV services—when provided between the large fixed satellite ground stations. Today, the number of satellite services allocated frequency by the ITU has expanded to a sizable number and now number almost 20.

International frequency allocations are now specified by the "radio wave section" of the International Telecommunication Union (the ITU-R), and these allocations are based on the type of service provided by the telecommunications system. The allocations include those needed for terrestrial, mobile, satellite, and intersatellite delivered services. The fixed satellite service refers to point-to-point communications links, which means for transmissions to and from ground terminals that remain fixed in place.

The other types of satellite services are precisely defined by the ITU and are discussed in depth in chapter 3. The most important of these include mobile satellite services (i.e., point-to-point communications links where one or both ends of the link are moving during transmissions) and broadcast satellite services (i.e., point-to-multipoint where a program is radiated over a "service area" to many receive-only terminals). In recent years, however, these distinctions have started to become less clearly defined. For instance, some satellite operators have used FSS frequencies to provide broadcasting and mobile satellite services. Even more recently, it has become possible to provide thin route return channels associated with broadcast services, which is known as Digital Video Broadcast with Return Channel Service (DVB-RCS).

Historically speaking, FSS started with the Early Bird satellite, which was the first type of space communications service to be implemented for domestic and international applications. As noted earlier, satellite ground antenna systems had to be large. These were also difficult to point and steer, and they were somewhat cumbersome to operate to achieve adequate power levels to communicate with the GEO satellites. In the early days of operations, satellite ground antennae required reflectors that were literally hundreds to thousands times the size of most user terminals today in terms of surface area.

Frequency bands allocated by the ITU for the FSS services have continued to expand from the days when the first serious allocations were made at the Extraordinary Administrative Radio Conference held in Geneva, Switzerland, in 1959. Today, the most important allocations are those in the so-called *super high frequencies* (SHF). This band ranges from 3 GHz to 30 GHz, but there are now also opportunities to use the so-called *extremely high frequency* bands from 30 GHz to 300 GHz with allocations in the Ka, Q/V, and W bands.

The FSS allocations are defined as follows: C-band (6 GHz uplink/4 GHz downlink), Ku-band (14 GHz/12 GHz), Ka-band (30 GHz/20 GHz), Q/V-band (48 GHz/38 GHz), and W-band (92 GHz/82 GHz). A GigaHertz (GHz) is a billion cycles per second. This means that all the frequencies used for satellites are operating at small wave lengths of a meter to a centimeter in size. As the demand for frequency increases, it seems possible that satellites will actually start to use frequencies above 30 GHz and thus start to use wavelengths that are so small that they are in the millimeter range.

First-generation FSS satellites operated in C-band (6 and 4 GHz) beginning in 1965. New and replacement systems began to utilize Ku-band (14 and 12 GHz) in the 1980s. Today, there are over 200 communications satellites offering FSS in either the C-band, Ku-band, or hybrid systems operating in both. Only a limited number of Ka-band systems (30 and 20 GHz) are so far deployed with the Astra system beginning service in Europe at the start of 2002 and with

Eutelsat and others following in 2003. Several other Ka-band systems are in various stages of development and will be deployed over the next few years. A major FSS move to Ka-band, which was anticipated with much excitement in the mid- to late 1990s, has been deferred or slowed because of technical and cost challenges as well as limited market demand. These problems are still to be fully overcome. Briefly, the challenges currently being faced with regard to the use of the Ka-band frequencies and higher include the following:

• *Rain Attenuation or Rain Fade*: The higher the satellite operating frequency, the greater the problem with so-called *rain attenuation*. In simple terms, the closer the wavelength used for satellite communications approximates the size of rain drops, the greater is a tendency for the rain to act as a lens that distorts or bends the radio signal and weakens the signal at the receive site. During heavy rainfall, this can even cause the receive antenna to lose the signal. To combat this problem, high power, improved signal processing, and other digital processing techniques must be used to make Ka-band service reliable. The Astra Ka-band system in Europe has the advantage of the best climatic conditions in the world for this service.

• *Satellite and Terminal Cost and New Technology Development*: The development of these new technologies involves a good deal of research and development (R&D) expense to develop both the satellites and new ground antenna systems. The ability to design components that can reliably operate at such high frequencies and extremely small wavelengths is a major technical challenge. Thus, the earliest systems, both satellites and ground systems, tended to cost more money. Some electronic components that are designed to operate at these extremely demanding frequencies are difficult to design, build, and operate reliably. For the most part, radio tube technology is still required to deliver high-power performance at these extremely high frequencies. Q/V-band systems (i.e., 48 GHz and 38 GHz) have been filed with the ITU and with the FCC in the United States. However, none of these satellites is yet in the construction phase, and deployment will most likely be deferred until well after the Ka-band has been utilized and tested. In short, operational satellites in the Q/V bands are at least a decade or more away. The newly adopted FCC licensing provisions (mid-2003) that require the posting of bonds and milestones that must be met to have multimillion dollar bonds returned to filing organizations will likely serve to slow the use of these new satellite bands even further.

THE DEVELOPMENT OF FSS SYSTEMS

The FSS category consists of four general classes of systems based on the service area provided: domestic, international, regional, and global.

• *Domestic FSS*. This type of satellite system provides fixed services within the borders of a single country and generally have contoured antenna beams (footprints) designed to radiate over or cover only the geographic area of interest. More than one beam may be provided for large domestic coverage areas. Domestic FSS servicing the United States, for example, usually have at least three or four antenna beams, plus one for Alaska, one for Hawaii, and one or two extra. Some of the highest capacity systems operating in the high-frequency bands can have many dozens of beams just for the lower 48 states. Domestic FSS satellites are typically designed to work with VSAT networks. These are with small, low-cost antennas of 1 to 3 meters in dish size, and they can provide high-capacity video distribution and Digital Video Broadcast (DVB) service users. These satellites are typically located at orbital locations on the equatorial plane, but with longitudes that are directly over the country to allow reasonable elevation angles for ground terminals (usually no lower than 5–10 degrees).

• *International FSS*. These satellites provide fixed service links between countries and are allocated specific frequencies for international FSS. International FSS often place their satellites over ocean locations to allow for intercontinental links with reasonable ground station elevation angles. International FSS satellites are also designed for larger gateway antennas, as well as VSAT networks and high-capacity video distribution. These satellite systems often include a significant amount of backbone links that can serve voice, data, and Internet-related circuits as well. These systems can also provide domestic FSS.

As noted earlier, Intelsat became the first provider of worldwide international FSS in 1965, with full global coverage across the Atlantic, Pacific, and Indian Oceans dating from July 1969. Intelsat and now others such as Panamsat, Loral Cyberstar, SES Astra, and soon Eutelsat have satellites that provide transoceanic links. Intelsat was the sole provider of worldwide international communications for over two decades until 1992. At that time, an open skies policy was implemented, allowing competition for international services. This has led to the new environment in which there are over 200 geosynchronous satellites in operation that are owned and operated by many dozens of international, regional, and domestic entities.

• *Regional FSS*. These systems were introduced in the late 1980s, starting with Eutelsat and then followed by Arabsat, Asia Sat, APStar, Europestar, Brazilsat, the Indonesia Palapa system, Koreasat, the Australian Optus system, and others. These systems are designed to meet the domestic and regional fixed-service needs of smaller countries in a common geographic area. Regional satellites provide both domestic and international communications to the countries served. Space segment costs are shared to reduce costs per country. Regional systems often provide a cost-effective alternative to international systems for small compact regions and for countries with common economic and cultural structures.

• *Global FSS.* These systems are differentiated from international systems by the additional claim of full global coverage. GSO systems, as traditionally implemented, cannot cover the polar regions because of beam coverage limits from the GEO orbit. GEO coverage for satellites with low-inclination angles is typically limited to the earth's surface between 70° North and 70° South latitudes. Therefore, a truly global system that is satellite-based must utilize nongeosynchronous orbit (NGSO) satellite constellations to achieve coverage in the high-latitude regions.

The cost of deploying a full constellation of satellites to provide broadband FSS has proved to be much more expensive than first thought. The French-backed LEO system known as SkyBridge formally abandoned its original multiple-satellite constellation plan as of April 2002. A reduced SkyBridge service will now lease capacity from Alcatel Espace and offer its broadband services to VSATs using existing Ku-band frequencies and GEO systems.

The other major proposal was to offer broadband FSS from an NGSO. This was a system called Teledesic. As of 2003, Teledesic abandoned its filing and returned all allocated frequencies to the FCC. The concept of a satellite system that would be completely global (over land, sea, and air) and could provide at low cost broadband "Internet in the Sky" services to quite small microterminals, ranging down to as small as 0.6 meters, remains an interesting prospect for the future, but for the time being the idea has been abandoned as too costly.

The story of FSS continues to unfold, and exciting new technologies are still to come. A major liability of FSS, especially with regard to voice services, but also with Internet Protocol (IP) services, has been that of transmission latency or delay. This is the time it takes (about 250 microseconds) for the signal to travel to and from a GEO satellite that is located a tenth of the way to the moon. It was thought that LEO and MEO satellites in low or medium Earth orbit might be the best way to overcome this delay problem and could also provide a better direct look angle straight to receivers, also avoiding interference from buildings or trees.

The economics of these LEO and MEO systems (i.e. their high cost and a lack of business success with this approach) have tended to steer the satellite industry back to the conventional GEO orbit approach. However, the latency problem still remains. In later chapters addressing future systems, other ideas of how the delay and look angle problem might be solved are covered. These include totally new solutions, such as the High Altitude Platform System (HAPS) that flies over earth in the stratosphere and remains stable in its location. For the time being, however, the GEO satellite retains its supremacy.

The FSS story is, however, just part of the satellite story. Some of the next steps in the field of space communications involve mobile (MSS), broadcast (BSS), and broadband satellite services. These exciting new satellite service areas have now joined the FSS satellite world in a serious way and are also addressed in chapter 3. Mobile, broadcast, and new broadband satellite services represent new billion dollar satellite markets that will redefine 21st-century satellite communications, but fixed satellite services will remain a key part of that future as well.

REFERENCES

Clarke, Arthur C., (Editor), (1967), *The Coming of the Space Age*, New York: Meredith Press.
Logsdon, John *et al.*, (Editors), (1998), *Exploring the Unknown: Selected Documents in the History of the U.S. Civil Space Program, Volume III, Using Space*, Washington, DC: NASA.

BACKGROUND READINGS

Clarke, Arthur C., (1945a, October), "Extra-Terrestrial Relays: Can Rocket Stations Give World-Wide Radio Coverage?" *Wireless World*, 305–308.
Clarke, Arthur C., (1945b), "The Space Station: Its Radio Applications," British Interplanetary Society, Private Paper.
Gordon, Gary D., and Morgan, Walter L., (1993), *Principles of Communications Satellites*, New York, John Wiley and Sons.
Iida, Takashi, (Editor), (1997), *Satellite Communications: System and Its Design Technology*, Burke, VA: IOS Press.
Pelton, Joseph N., (1978), *Global Communications Satellite Policy*, Aalphen aan der Rhyn, Netherlands: Sijthoff and Noordhoff.
Pelton, Joseph N., (1995), *Wireless and Satellite Telecommunications*, Upper Saddle River, NJ: Prentice-Hall.
Pelton, Joseph N., (2002), *The New Satellite Industry: Revenue Opportunities and Strategies for Success*, Chicago, IL: IEC.
Pelton, Joseph N., and Alper, Joel, (Editors), (1984), *The Intelsat Global Satellite System*, New York: AIAA.
Whalen, David, (2002), *The Origins of Satellite Communications 1945–65*, Washington, DC: The Smithsonian Press.

CHAPTER THREE

The "New" Satellite Services: Broadcast, Mobile, and Broadband Internet Satellite Systems

D. K. Sachdev
President SpaceTel Consultancy

Dan Swearingen
Consultant; formerly COMSAT Mobile Communications

> *Satellite communications have revolutionized global communications. For an investment much less than the cost of a medium-sized jet airplane, even the smallest countries with modest telecommunications networks could as of the 1970s link to every other country in the world, effectively and at low cost.*
> —Santiago Astrain, First Secretary General of Intelsat, Speech at UN, 1983

In the early days of satellite communications, it was Intelsat and a few regional and domestic satellite systems that defined the world of commercial satellite communications. This early phase of the satellite world and the fixed satellite technologies that allowed satellite systems to get started was explained in some detail in the chapter 2. The last 20 years have brought enormous change to satellite technology and services. The world of satellites has expanded as new technology defined new markets and as new markets created the need for new space technologies. The most important of these new satellite technologies and services in terms of revenues and global reach have been broadcast satellite systems (both for TV and now direct broadcast radio) as well as mobile satellite systems for maritime, aeronautical, and land mobile services.

Most recently, new satellite technology has evolved to support broadband Internet services for business enterprises, Internet Service Providers, and service to the desktop for small office/home office (SoHo) applications or even to the sophisticated home consumer. With the latest digital technology, some high-powered and sophisticated satellite systems are moving toward multipurpose type satellites that can provide a complete blend of integrated service offerings. The story of these new satellite services is best told, however, by tracing the technology and services as they have evolved over time. Billions of dollars have been made—and billions lost—as this remarkable satellite-based journey into new consumer-oriented services has unfolded.

BROADCAST SATELLITE SERVICES

In many ways, direct broadcasting was perhaps the most natural type of service for a truly global medium that is as ubiquitous and far reaching as satellites. However, several important technology advances had to take place in the designs of spacecraft, launch vehicles, and receiving equipment before this objective could be realized. In short, the first generations of satellites of the 1960s and early 1970s were simply too weak in power, too limited in modulation and multiplexing schemes, and too limited in radio frequency spectrum to provide true direct broadcast services to consumers. There were also problems that went beyond technology. Some key policy and regulatory factors also somewhat delayed the ability of the new Direct to Home (DTH) or Direct Broadcast Satellite (DBS) services or—in the official terminology of the International Telecommunication Union—Broadcast Satellite Services (BSS) to get started.

The first problem or obstacle to broadcast satellite services was the simple fact that the initial satellite systems evolved as part of international consortia. The primary aim of the then "single global satellite system," namely the Intelsat consortium, was to provide heavy international route telecommunication trunking (or backbone) for telephone and telex, plus some TV and radio program distribution, but only between official or national earth stations operated either by governments or government-sanctioned monopoly organizations. BSSs were simply not part of what the governmental parties that owned Intelsat wanted this consortium to do. There was no entity charged with such a mission as providing direct BSS. The regional Eutelsat consortium that followed Intelsat's formation had the same types of constraints.

The second delaying mechanism was what might be called the "the Big Brother" factor. Many governments, organizations, and public interest groups were concerned that powerful broadcasts could directly reach homes across political and geographical boundaries. Many believed this would cause cultural pollution, political damage, or even revolution in the case of totalitarian states. The countries of the Soviet Bloc, in particular, were strongly allied against the creation and use of such technology in the 1960s, 1970s, and 1980s.

Despite these hurdles, serious efforts to provide direct broadcast satellite TV started in the 1970s on both sides of the Atlantic. The proven satellite broadcast technique at the time was analog and required at least 27 MHz of radio frequency bandwidth for each TV channel plus extremely high satellite transmission power (i.e., about 65 dBW of Effective Radiated Power [EIRP]) to reach small-receive antennas. This transmit capability was at least 100 times more powerful than that which was required for ordinary fixed satellite services that organizations like Intelsat required for their service. The net result was that if one designed and built a direct broadcast satellite that used all of the spacecraft power available, the result was not attractive as a business proposition. Such a broadcast satellite service would only produce a maximum of about five channels, and this would cover only a medium-size market. Using this type of early analog technology would, for instance, require that the U.S. or European market be divided up into three or four zones even if the largest available and most expensive spacecraft bus were to be used.

The first operational Direct Broadcast or DTH system in Europe was a joint German–French program. This was called TV-Sat in Germany and TeleDiffusion Francais (TDF) in France. This program consisted of three spacecraft, of which two were launched, covering mainly France and Germany and a few neighboring areas. This system required significant efforts for the development of high-power radio tubes called Traveling Wave Tube Amplifiers (TWTAs), with some 230 watts of power to operate at the officially assigned BSS frequency bands of 11 GHz in downlink and 18 GHz in uplink.

The TDF/TV-Sat direct broadcast satellite program had several objectives. In addition to the development of new technology, the programs were expected to provide new employment opportunities and add to the overall prestige in the space technology field. Certainly the fact that the European Space Agency had earlier selected a contractor from the UK for its experimental DBS satellite program spurred the decision to proceed with a commercial DBS system in Germany and France. This experimental satellite program was variously called H-Sat, L-Sat, and ultimately Olympus. Certainly, when Germany and France learned their industry had lost the contracts for what ultimately became Olympus, they decided to launch their own broadcast satellite program with home-grown contractors selected to engineer the satellites for their own national systems. The fact that the ESA's Olympus satellite's autonomous control system froze up and disabled the spacecraft for some 2 years was not widely grieved in France and Germany.

On the U.S. side of the Atlantic, the Comsat Corporation attempted a similar venture even earlier. It created a subsidiary, called the Satellite Television Company (STC), through its Comsat General Corporation and attempted to deploy a Direct Broadcast or DTH system for the U.S. market in the late 1970s. This system was designed to offer a handful of value-added channels of programming, but was limited in both the diversity of the programming and the number of channels it could offer subscribers. The satellites were ordered from the then RCA Corporation (now Lockheed Martin), but these satellites were eventually abandoned when Comsat was not able to conclude the necessary business arrangements to make the venture profitable.

POLITICAL CONCERNS AND INNOVATIVE SOLUTIONS

Although these limited capacity DTH satellite TV broadcast systems were advancing on the U.S. and European sides of the Atlantic, the direct broadcast satellite planners also needed to resolve associated thorny political issues, primarily arising from apprehensions around the world, with respect to "Big Brother" and "propaganda" broadcasts that could reach directly across national boundaries into many millions of homes. Furthermore, there was concern that developing countries would be left with no spectrum when they got around to investing in their own national broadcast systems.

The solution that was finally adopted by the International Telecommunications Union (ITU) in 1983 led to a "planned orbit planning system," where each nation—at least in the ITU Region 1 (Europe and Africa) and ITU Region 3 (Asia-Pacific)—was allocated a certain number of channels in the appropriate portion of the suitable arc in a band exclusively set aside for satellite broadcasting for that country. Eventually, the ITU Region 2 (the Americas) also came up with a plan that was somewhat more flexible and allowed for longer term expansion.

Concurrent with this detailed spectrum planning, however, better receiver technology and more innovative link budgets began to substantially lower the technical barriers to BSS or DTH satellite services. In the early 1980s, this new technology led some satellite executives in Europe to consider whether they could use the spectrum allocated to fixed satellite services to provide TV to the home and avoid the higher cost and higher technical difficulty of using the so-called BSS frequencies.

This innovative idea was developed by Clay "Tom" Whitehead, an American expert from Hughes, who found a supportive administration in the Duchy of Luxembourg. His plan was to use the so-called Ku-band frequencies and a medium power traditional "fixed service" satellite to broadcast TV directly to small and low receive-only dishes and to do so over a reasonably sized coverage area in Europe.

This concept was pioneered initially as Coronet and then later acquired in 1984 by SES Astra. The early Astra satellites were still analog, but a larger number of TV channels could be provided with medium power satellites than with the earlier "true" DBS satellites, although the receive-only dish had to be somewhat larger to compensate for the lower transmit power from the satellite. This was so successful that additional satellites operating in different segments of the nominal 11 GHz downlink transmit band were collocated as needed over the following years to meet market demands. This approach continues even today, with the SES Astra now providing digital services to over 75 million homes in

Europe either directly, via cable TV systems, or via so-called Satellite Master TV terminals on the top of wired apartments.

The same type of approach was adopted in a somewhat unplanned fashion in the United States. In the early 1970s, HBO, Cinemax, and other movie services were distributed on fixed satellite networks in the United States. to cable TV stations. As always seems the case in entrepreneurial America, a number of electronics outlet stores saw an opportunity to sell "backyard dishes" so that people in isolated areas could receive the satellite signals and obtain premium movie services and superstation programming as well. (These simple TV receive-only [TVROs] terminals were initially available just in the lower C-band frequencies, but later as premium channel services moved on to Ku-band transponders, these dishes and descramblers were available in the Ku-band frequencies as well.)

The legal problem that emerged was that many of these "backyard" dish owners bought illegal descramblers and essentially began to obtain premium cable movie programming service for free. Soon there were millions of these so-called "illicit" backyard dishes (some 3–5 meters in size) violating copyright law. There was a great debate as to whether this service could be legitimately offered to these consumers and at what price and under what formal type of collection process. Ultimately the advent of better and more powerful digital satellite technology nearly a decade later provided a definite answer to this ongoing debate as many rural users have migrated to the superior technology. Many rural subscribers are still angry about paying for what seemed to them to be an entitlement to get free programming once they bought a dish and a descrambler.

THE ADVANCE FROM ANALOG TO DIGITAL

Although "tens" of analog satellite channels were adequate for several markets, and especially so in Europe where programming options were much more limited, they still could not compete against the much larger number of channels provided by cable-TV systems. But then two almost simultaneous advances in the late 1980s and early 1990s enabled realization of much higher capacities for satellite systems and at the same time provided the service at affordable cost to the consumer.

The first of these advances was digital compression. This technique now enabled several TV programs to be squeezed into a single transponder without too much impact on subjective reception quality. The second was the development and manufacture at low cost of Application Specific Integrated Circuits (ASIC) chipsets for the home receivers. This new technology brought the cost down and increased the performance. These developments finally made DTH-TV with many more channels a reality. Most important was the introduction of new digital compression standards, particularly the so-called Motion Picture Expert Group (MPEG)-2 standard that allowed the transmission of a high-quality digital TV channel at only 6 megabits/second. With this innovation, DTH-TV satellites can actually now provide hundreds of channels. Today, DTH-TV is a growth engine for the satellite industry in several parts of the world. Over 50% of these transponders were in the Ku-band. Overall it is estimated that there are some 10,000 satellite TV channels in operation over some 2,000 video-related transponders (with up to 14 digital channels being transmitted on a single satellite transponder). This large quantity of channels represents a significant number for distribution to cable TV stations, for network news and sports organizations, and a larger number for direct broadcast access for home consumers.

Nevertheless, despite these gains in the number of DTH-TV channel available worldwide, these video channels are still largely in the developed world, although services are now in operation in South America and Asia, and systems are planned for Africa. On both sides of the Atlantic, there are competing DTH operators—Astra and Eutelsat in Europe, AsiaSat and APStar in Asia, NHK and Perfect TV in Japan, and competitive systems pending in Korea. In the United States, the two main providers are EchoStar and DirecTV. Recently, SES Americom announced its intention to enter this market through its Americom2 Home service, and Cablevision has now launched the first of its Rainbow satellites to initiate DTH service.

In summary, DTH-TV has finally arrived as a successful consumer business via satellites. Starting with systems like Astra, all successful DTH-TV systems have indeed given high importance to customers' needs. Significant investments were made in Application Specific Integrated Circuits (ASIC) chipset development to make the ground equipment affordable in the target markets. Until the service had grown enough, traditional transparent transponders were used wherever possible, thus minimizing risk of having specifically dedicated assets for a new service. Even after the market had grown, modular addition of co-located satellites has spread the investments to match revenue growths. The financial markets have recognized these sound business management practices through continued market capitalization even when other parts of the satellite industry were experiencing financial difficulties.

Now that these markets have been firmly established, the DTH and BSS operators are now moving toward new high-powered and highly concentrated satellite beam patterns that resemble the complex antenna patterns found in terrestrial cellular beam systems. These spot-beam architectures can be used to cater to local content that can thus be exclusively provided in each highly focused beam area.

THE FUTURE OF DTH AND BROADCAST SATELLITE SERVICES?

Despite its impressive growth, DTH-TV has still not captured significant markets in developing countries. This may be due in part to concerns about "Big Brother" or "neo-Colonialist" factors. A much more important factor is the cost of the consumer equipment and the cost of the service in developing markets. Although start-up DTH service is often given away in developing markets against the prospect for long-term service contracts, the cost in emerging markets is still too high (especially with high tariffs applying to the receiving dishes). Thus, high cost remains a barrier to entry in several developing markets, particularly when compared with cable-TV. These costs must be brought down substantially through low-cost production and a fresh look at the overall system design, starting with consumer needs in each and every specific market.

DTH-TV continues to be in the middle of the ongoing market competition in the satellite and allied industries to become the "information hub" in every home. Apart from DTH-TV operators, the other players in this broadband information race include cable-TV operators, computer manufacturers, and online services operators like AOL, MSN, and others. The so-called imminent arrival of set-top boxes has been around the corner for much too long and often tends to divert the attention from the core business fundamentals for individual services. Every year or so, we are told that the ultimate universal box has finally arrived. Although the motivation to expand an established market is well justified, we should not lose sight of the fact that two-way services make significantly different demands on the space segment than pure broadcast services. This ongoing fascination with the illusive set-top boxes has the potential to make the core DTH-TV services more expensive than they need to be. The advent of the new Digital Video Recorders may actually represent a breakthrough for the future that gives DTH technology an edge over cable-TV systems. Only time will tell.

SATELLITE DIGITAL RADIO SERVICES

It was radio that started the broadcasting and telecommunications revolution over a century ago. Radio technology is at the core of the electronics, wireless, telecommunications, and entertainment industries. Yet in the DTH field, satellite radio was a latecomer compared with TV. There were two reasons for this time lag. In the developed world, radio was perceived as a way to get information and entertainment while doing other things at home, at the office, or in the car. The thousands of local radio stations serving these sizable markets developed a vested interest in maintaining this status quo and did not relish a big area-wide system eroding their individually captive markets.

In the developing world, in addition to the focus on local markets, there has been a long-standing, but small, market for short-wave radio for bringing news and other features, mostly state-sponsored, from distant and developed areas. The technical quality of these broadcasts has always been marginal and was identified early enough as an ideal candidate for replacement by much higher quality and stable satellite-based radio systems. However, this transition lacked the business impetus because a majority of these radio services were government owned with proverbially limited budgets. So what finally unlocked this stalemate to make satellite radio a reality? Three factors stand out over the others.

The first development was a new spectrum allocation in 1992. The issues were somewhat different from those faced by DTH-TV. After considerable debate, three segments in L- and S-bands were designated for digital radio. These were 1452 to 1492 MHz at L-band and 40 MHz and 130 MHz segments at 2.3 and 2.6 GHz, respectively. The allocations in the L-band were on a shared basis with existing services, thus putting an upper limit on the radiated power. All the allocations were to be further shared between terrestrial and satellite parts of the digital radio systems.

The second was a major technological push, largely from several European research and industrial organizations, to make digital radio work satisfactorily in multipath environments. The underlying technologies were developed under the aegis of the now well-known Eureka 147 program. These efforts were successful and led to several terrestrial single-frequency networks in different parts of Europe. It was well recognized that a wide-area operation would require a hybrid solution with satellites covering the sparsely populated areas and terrestrial networks catering to urban areas. However, the satellite part encountered some concerns. These included considerably higher power for Eureka 147 waveform, leading to uneconomic satellite payload design and genuine concerns about relatively low elevation angles in Europe from the geostationary orbit.

The third development was the start of a satellite-based digital radio system for the developing countries by a new company, WorldSpace. This project was initially targeted at Africa with a view to bridging an awesome information gap in remote and less developed areas. The system adopted the well-known Time Division Modulation/Quadra-Phase Shift Keyed (TDM/QPSK) downlink format instead of the Eureka 147 waveform to realize an economical satellite design. Due to significantly lower multipath concerns in sparsely populated areas and high-elevation angles, this change was acceptable from a performance point of view. The WorldSpace system concept was soon extended to both Asia and Latin America with a common spacecraft design.

At the end of 2002, the WorldSpace system was operational in Africa and Asia with several uplink stations in the two continents. Several versions of the portable satellite receivers are being marketed. Although the production cost for these receivers is typically in the $50 to $100 range, in many cases they are subject to high import tariffs that drive their price sharply upward. Although these costs are reasonable in developed countries, they are still considered high in the developing world. In the United States, several proposals were put forward in the early 1990s concurrently with the ITU spectrum allocations. However, it was only in 1997 that FCC finally awarded two slices of the 2.3 GHz S-band spectrum to two of the contenders, now known as XM Radio and Sirius Radio, respectively. Both systems are operational now, and their subscriber base is growing rapidly.

Unlike the shared allocation in L-band, the U.S. allocations in S-band are exclusive to digital radio. This has enabled the system design to take full advantage of the concurrent advances in spacecraft bus power capabilities. Even more notable is the fact that, taken together, the two systems have made good and effective use of practically all the preceding efforts in this field.

The XM Radio system uses two operational geostationary satellites in a simultaneous space and time diversity modes. Use of the established time division modulated (TDM) together with QPSK technology in the downlink, for areas with minimal multipath, allow full use of previously established hardware. For urban areas, XM Radio satellites use a variation of well-established Coded Orthogonal Frequency Division Modulation (COFDM) technology. On similar lines, the Sirius system has taken full advantage of the prior work on nongeostationary orbits to improve elevation angles for northern locations. It has three satellites in the so-called Tundra 24-hour orbit, and it provides high-elevation angles throughout the U.S. landmass. Due to this feature, it envisions a lesser number of terrestrial repeaters than for the XM system. The Sirius repeaters also use a version of COFDM to counter multipath effects.

The initial results from the U.S. systems are encouraging both in terms of technical performance and market acceptance. XM had over 1 million subscribers at the end of 2003. XM equipment is being sold by General Motors dealers as part of new car packages, and Sirius, which lags XM in subscribers, hopes to have a parallel arrangement with the Ford Motor Company as of the 2004 model year. Both companies have similar arrangements with other car manufacturers as well. Japan has just ordered a system of its own. The Global Radio project for direct broadcast audio services, however, has ended its bid to start such a service in Europe from its base in Luxembourg.

By and large, the operating digital radio systems vindicate the direct consumer service principles. The system designs have given high importance to the consumer equipment right from the start. In terms of meeting consumer price barriers, the WorldSpace system has faced a far more difficult challenge. In developing countries, a stand-alone portable radio faces a significantly higher price resistance than the minor add-on cost of a satellite radio to a $20,000+ car in the U.S. consumer market. However, the U.S. systems have to compete with fairly attractive media alternatives, such as FM radio and CD players, as well as the digital terrestrial systems recently introduced by Ibiquity. Overall, it is the quality and uniqueness of the content that may dictate the market shares.

So what are the key factors that will influence the future growth of the digital radio technology?

• The ITU allocations in L- and S-bands are viable for digital radio for both terrestrial and satellite broadcasts. The dedicated S-band allocations in the United States are attractive; however, this band needs more power margin.

- Unlike TV, digital radio did not start with an installed base. This lack of a firm revenue base has created a challenge during the initial phases in terms of early advertising revenues until a certain minimum threshold is reached in terms of number of listeners.
- Subscription services are the main revenue option in the beginning. However, for an audience used to free radio, initial customer resistance can be expected and could be countered with unique and ubiquitous content.
- Will there be real growth in new markets? Japan is already on the way, but the European system has now folded. The developing countries in Asia and Africa already have the Worldspace option available.
- Through the Original Equipment Manufacturer (OEM) channel, we will soon see an effort to create synergy between other electronics, entertainment, and communications systems in the car, somewhat akin to the set-top box fascination for DTH-TV.

Will this change the next generation of digital radio systems? Will two-way capabilities in automobiles lead to completely fresh thinking in system design? It is too early to guess. However, the nearly $4 billion upfront investments already made in the three systems need undivided attention to make current systems an economic success. Taking the sum total of the experiences to date, satellite digital radio can indeed be a growth engine for coming decades, provided customer-based business models prevail and the appeal of WorldSpace type services in emerging markets takes hold.

BROADBAND AND INTERNET-BASED SERVICES

In the foreseeable future, the success of practically any consumer-oriented telecommunication or broadcasting system is likely to be judged by its direct or indirect relevance to Internet. No wonder the satellite systems and associated industries are all aggressively trying to be an integral part of this new business opportunity for space communications.

To the extent that the Internet backbone network is resident on global telecommunications networks, international satellites are already providing Internet-related services. In fact satellite-based Internet backbone service is the single most rapidly growing source of revenues for the FSS industry.

The second role or niche that several new satellite-based enterprises have developed has come to be known as "broadcasting-to-the edge" or "multicasting" service. National and international switched networks often encounter serious congestion particularly at times of popular media casts over Internet. Today's satellites, through their multidestination links, not only relieve such network congestions, but in nominal conditions have also proved to be a more efficient means to transmit large identical files for rich media content to many different ISPs. The largest sector of revenue growth for the past 3 years for Intelsat, Eutelsat, Panamsat, and other FSS systems has been to support backbone Internet distribution services, but these systems have limited ability to provide broadband services directly to consumers.

The real competitive battleground in the Internet world, however, is what is more commonly referred to as broadband, or direct provision of high-speed, two-way links to homes (DTH-I) and small and large businesses. The competitive media here are DSL, high-speed modems over cable-TV systems, satellites, and wireless.

On the face of it, DSL and cable modems should take away most of the market because they utilize the ubiquitous copper telephone wires and coaxial cables, respectively. However, current state-of-the-art permits DSL services only within a certain distance from the telephone exchanges. Cable modems are rapidly becoming attractive options in the United States. However, the penetration of cable-TV systems varies considerably in different parts of the world. In this respect, the U.S. market, with nearly 70% of homes on cable-TV routes, is the hardest to penetrate for other media.

Wireless technology is building some attractive niches in high-speed Internet markets as well. These applications range from wireless LANs within individual homes or business premises to broadband easy access at airports and other public places via Wi-Fi (or the 802.11g standard). Clearly, the rapid development of standards is key to keeping pace with maturing capabilities in both satellites and broadband terrestrial wireless systems. However, until some entrepreneur stitches these nodes into area-wide networks, wireless remains only a good niche service in specific bounded areas.

A medium like satellites could still become an important alternative by accessing all users and allowing the flexible connection of Wi-Fi cells. Here are the key factors to consider in terms of future growth potential.

The first group of DTH-I operators are now in operation and are piggy-backing their service on DTH-TV. In principle, all they had to do was allocate part of the huge throughput that is available as the TV downlink to become an Internet link for downloading data to the users. In practice, for capacity and other reasons, they often end up using another nearby satellite for the downlink. Initially, the return link was only through a dial-up phone line accessing an Internet Service Provider (ISP). In addition to the significant costs for the new equipment and service, the users still had to pay for the dial-up ISP and phone line as before. Nevertheless, this is a good start and a placeholder for satellites in the broadband sweepstakes.

The prior operators have now upgraded their DTH-I offerings to genuine two-way through the use of satellites. Each user now has a small Ku-band uplink accessing a separate dedicated return-link capacity or full transponder. The extra running cost of the return transponder is currently controlling the operating economics of the entrepreneurs. In addition, the user equipment is not only more expensive, but now has to meet uplink clearance requirements at millions of locations.

Long before these Ku-band DTH-I solutions became a reality, the satellite broadband campaigns had begun. In fact they started way back in the early 1990s when *Internet* was not even a familiar word to consumers.

The first system announced with considerable enthusiasm was the Craig McCaw and Bill Gates' backed Teledesic Ka-band (30/20 GHz) satellite system. Its initial proposal was to launch many hundreds of LEO satellites and provide broadband integrated services to rich and poor alike around the world. This system, however, has now been abandoned and the frequencies returned to the FCC.

Highly publicized deliberations on 30/20 GHz band allocations raised high expectations of ubiquitous broadband around the corner and around the world. Today, however, most of the systems have been abandoned. Astrolink, Celestri, and Millennium, along with Teledesic, are now canceled and the Loral Skynet Ka-band satellite is on hold. All that remain at this time are Hughes Spaceway, Wild Blue, Telesat's Ka-band satellite, and some hybrid Ku-band and Ka-band systems of Eutelsat and SES Astra.

Only a small percentage of the global filings for Ka-band services were ever likely to lead to real systems. Out of the nearly 15 systems filed from the United States, only 2 or 3 systems are likely to be completed. The majority of the rest stand to lose their licenses, and several have already returned their frequencies to the FCC.

Why has there been such slow progress despite the phenomenal potential for DTH-I worldwide? It would be easy to blame it all on the general dotcom/telecommunications meltdown, but there are other issues as well. These issues include cost, technical performance, rain attenuation, access to low-cost user terminals, and marketing constraints.

In principle, Ka-band equipment costs should eventually be lower, but it is still too early to predict what the final DTH-I customers will be asked to pay. Although some of the new systems have gone for reasonably modest capacities in the space segment via hybrid Ku-Ka-band systems, others have placed greater emphasis on maximizing the space segment throughput even at the cost of making the technical and financial risks high for the first few years of implementation. Such large-capacity systems, such as Wild Blue and Spaceway, can become serious financial risks if adequate market size either does not materialize or cannot be captured with the customer pricing based on genuine cost recoveries. In other words, the "build-it-and-they-will-come" approach may once again be fraught with danger.

DTH-I is too important a segment to be dropped from the future of satellite systems. However, can we expect DTH-I to become another growth engine? Only the next few years and the success of the early Ka-band systems will tell for sure.

Satellite Internet business, DTH-I, is often considered a natural extension of DTH-TV. However, the system risk factors are quite different. Whereas the DTH-

TV satellite cost is independent of audience size, the total user count controls the whole satellite size and system complexity for Internet applications.

Satellites with full-mesh connectivity on-board processing appear to be necessary only for large multisite enterprises. Hybrid Ku-Ka- and Ka-Ka-band bent-pipe architectures can provide the needed capacity and flexibility at acceptable risk for both smaller enterprises and consumers. The economics for two-way services improve as system capacity increases. However, there should be matching credible market size to justify the large investments. The North American high-speed Internet market is expected to rise to 23 million households in 2007. Satellite-based two-way systems have the potential to capture at least a reasonable share of that market, although it may well be at the lower end of the projected 5% of the total market that some market forecasters have foreseen.

The cost of Customer Premise Equipment (CPE) is today a major factor in market penetration for two-way Internet services via satellites. This cost is sensitive to volume, and adoption of uniform standards can bring down the costs to levels comparable to that for competitive media. Return channel spectral efficiency is key to the overall economics of such services. Newer techniques are providing increased efficiencies that will offset CPE expense by reducing recurring space segment costs.

Above all, DTH-I is another crucial test of the satellite industry's ability to make a business success in the consumer sector. Will it build the systems from customers' needs upward or from satellite technology downward? Will it match system capabilities to the needs of individual markets in a timely manner? The answers to these questions and the judicious adoption of direct consumer service principles could determine the ability of the industry to effectively compete with alternative media. The fact that high-technology satellite networks have tried to use "technology push" in the past, rather than relying on solid "market pull" to spearhead their ventures into new markets, has not always been reassuring to investment bankers and market analysts. Thus, many in the investment community continue to be skeptical and see DTH-1 broadband services via satellite as "bleeding edge" technology.

MOBILE SATELLITE SYSTEMS

The fact that a single satellite can serve as a radio relay to span great distances has supported great improvements in telecommunications—not only for conventional telephone, TV, and data services, but also for ships, aircraft, and mobile travelers all around the world. Developing a mobile satellite service (MSS) system to serve mobile users and provide key safety and rescue services, while simple in concept, also (like broadcast satellite services) confronted great technical and operational challenges, as well as a few policy and regulatory issues.

An earth station for a ship, aircraft, or land vehicle has to be small (and reasonably low cost) to fit the constraints of the particular vehicle as well as meet the basic requirements of affordability. From a radio engineer's point of view, this means that the effective radiated power coming down from the mobile satellite in space has to be much stronger than that used to support an FSS channel. If intended for use while in motion, such as on board a ship or aircraft, the mobile earth station antenna has to be either nondirectional or omnidirectional (with approximately equal sensitivity in all directions) or else be designed to automatically correct its pointing as the vehicle moves.

Despite the technical difficulties, there is a real need on the high seas and in the air for better radio communications and improved safety services by both civilian and government users. Starting in the 1960s, various groups around the world, including Comsat, conducted studies to establish the viability of civilian aeronautical and maritime communications satellites for oceanic communications.

The maritime community began to study the use of satellites to improve radio communications for ships under the auspices of the Inter-Governmental Maritime Consultative Organization (IMCO), now the IMO because the word *Consultative* was dropped. These studies led to consideration of a possible international maritime satellite system to be created through an intergovernmental agreement and with an organization similar to that of the Intelsat organization. The primary purpose of Intelsat did not include the provision of mobile services, so the key questions from the outset were not only the technical feasibility, but also which organization would actually provide the service.

The ITU processes first allocated frequency bands for maritime and aeronautical satellite communications in the L-Band (near the 1.6 GHz frequency range) at the 1971 World Radio Conference. A pair of bands, each 7.5 MHz wide, was then specified for the Maritime Mobile Satellite Service (one band for ship-to-shore transmissions and the other for shore-to-ship transmissions), and another pair was allocated (each 15 MHz wide) for the Aeronautical Mobile Satellite Service. A 1 MHz-wide band for each direction was designated for generic Mobile Satellite Service and restricted to distress and safety communications usage.

The first commercial mobile satellite system was introduced as a component of the Marisat system launched in 1976. With this system, the economic challenge that every new mobile satellite system faces (i.e., the challenge of building a large base of customers to provide revenues sufficient to pay for a capital-intensive multimillion dollar satellite all at once) was first overcome by having two pools of customers and two ways to grow usage of the satellite network. This evolved from two totally different sets of needs and spread costs in a helpful way.

In 1972, the U.S. Navy was looking for some sort of gapfiller capacity that could be leased in the interim before they could launch their own new satellite system and to do so at a relatively low price. The solution turned out to be a hybrid satellite concept (developed by Comsat and Hughes). This maritime satellite system (Marisat) was designed to support both a small L-Band commercial maritime payload and a UHF payload to satisfy some, if not all, of the Navy's stated capacity requirements from a geosynchronous orbit.

A fortunate combination of similar coverage requirements and complimentary capacity requirement schedules enabled Marisat to offer cost advantages to both the anchor client or user (i.e., the U.S. Navy) and other civilian maritime users. Initial coverage was needed by both to link the continental United States with mobile users in the Atlantic and Pacific.

The Marisat satellite, built by Hughes, initially provided a small amount of L-band capacity operating in parallel with the full UHF capacity required by the Navy. Later, as the Navy's capacity requirement was reduced or disappeared, more power could be switched to the L-band payload to satisfy growing commercial traffic demands. The available L-band transponder RF power thus depended on the number of required UHF channels.

MSS FREQUENCIES AND ANTENNA SIZES

The commercial MARISAT system links from the shore stations to the satellite made use of frequencies in the FSS bands at 6/4 GHz. The 6 GHz uplink signals from each coast station were translated by the satellite to approximately 1.5 GHz (L-Band), amplified, and transmitted to the ships through a global coverage antenna beam. The 1.6 GHz signals received from the ships were translated by the satellite to approximately 4 GHz, amplified, and transmitted via a global coverage antenna beam to the coast station. Because initially only a single-coast earth station was needed per satellite to service many ships, relatively large diameter antennas (i.e., 13 meters or 40 feet) could be justified for the coast earth station so that relatively little satellite power was needed for the ship to shore links.

Practical considerations limited the ship terminal antenna size (on the order of 1.2-meter diameter), so much higher power from the satellite was required for the shore to ship links than for conventional FSSs. The size considerations included wind loads, difficulty of shipping and installation, and the few suitable shipboard locations available for installation. To provide a consistently clear view of the sat-

FIG. 3.1. Tanker with MARISAT ship terminal (note white radome on mast).

ellite for various headings, the antenna needed to be installed high up on the ship's superstructure (or one of its masts) where space was limited.

Despite the need for small size for ship board operations, the gain of the ship antenna was dependent on the size of the dish to get the needed performance. Thus, the antenna needed to be as large as practicable to minimize the satellite power required to support a telephone service. A compromise sensitivity requirement (antenna gain to noise temperature ratio) was chosen (G/T = –4 dB/K), which allowed a 1-meter diameter antenna to support a single-voice circuit. Automatic pointing was provided to keep the antenna looking toward the satellite as the ship rolled, pitched, and yawed, as well as when it simply adjusted its course.

The telephone channel design choice for MARISAT was based on a need to minimize the required receive carrier power to maximize the number of simultaneous phone calls that could be supported with the limited total amount of satellite power. Another requirement was for relative transparency of the channel to data modem signals. Subjective listener comparison tests of the candidate schemes led to the selection of companded frequency modulation (FM), which coincidentally was also adopted at about the same time for the original cellular systems.

By the early 1970s, telex had become a nearly ubiquitous mode of telegraphy and was well standardized. Consequently, each ship terminal was to include a standard 50-baud (or bits per second) telex machine. This system was designed to support automatic interworking in both the shore-to-ship and ship-to-shore directions.

The MARISAT system was designed to enable calls to be set up quickly in either direction and in a manner as close as possible to the quality of terrestrial telephone and telex calls. For prompt ship response to calls from shore-side parties, the ship terminal always listens to a call announcement channel for possible call

FIG. 3.2.   MARISAT ship Earth station antenna subsystem (without radome).

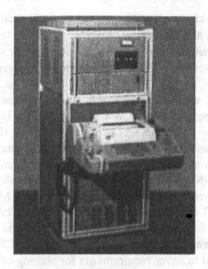

FIG. 3.3. MARISAT below-decks user terminal.

alert with its address when not already engaged in a call. In addition to a ship terminal identification code, each assignment would contain codes for the type of call (e.g., telephone or telex) and terminal instructions (e.g., frequency/time slot tuning for the call).

Rapid ship-to-shore calling was ensured by the use of a separate ship-to-shore frequency on which the ship could send a short request message when it wished to place a new call. This message would be within a single, short, digitally modulated burst. The use of the request channel frequency would be on a random basis by which all ships accessed the satellite. If the first burst failed, a second burst followed until access was achieved. The concept of using a random access technique for MARISAT call requests was an adaptation of the "Aloha" packet protocol developed only a few years earlier at the University of Hawaii.

The ship-to-shore request burst contained only minimal information needed by the coast earth station. This information included the ship terminal identity code, the type of channel requested, and priority. The coast earth station would validate the ship terminal's eligibility for service and assign a channel frequency (and time slot if telex) via the shore-to-ship Timed Division Multiplexed (TDM) assignment channel.

The system design provided for ship-to-shore calls to be routed automatically by the coast earth station signaling and switching subsystems. Call routing in the shore-to-ship direction required the cooperation of the terrestrial telecommunications carriers to recognize the new codes established for the three satellite ocean region coverage areas. New country codes were designated for each ocean region. For example, a telex call to a ship terminal in the Atlantic would be dialed by starting with the country code 581 followed by the seven-digit ship identification number.

Special care was taken to ensure that maritime distress calls had absolute priority over routine traffic and were set up promptly. Such call requests were also specifically alarmed so that a telephone (and/or telex) operator was alerted to monitor the call and assist in routing it to the appropriate safety agency (usually the U.S. Coast Guard).

While MARISAT was being deployed, an international maritime satellite organization was being planned that would be called the International Maritime Satellite Organization (INMARSAT). The details of the establishment of INMARSAT are discussed later when the institutions of satellite communications are examined. The preparatory group charged with planning INMARSAT decided that the initial INMARSAT system specifications should be based on the MARISAT to minimize transition problems for existing MARISAT users.

When INMARSAT came into existence on July 16, 1979, there was considerable work yet to be done before an operational transition from MARISAT to INMARSAT could be accomplished. Coast earth station (CES) technical requirements were developed to cover requirements for sharing of satellite capacity by multiple CESs. These included requirements for a Network Coordination Station (NCS) to arbitrate the use of frequencies and process channel assignments to and from the other CESs in each ocean region. Other preparations included conversion of the MARISAT Ship Terminal Technical Requirements, Commissioning Procedures, and Type Approval Procedures into the INMARSAT documents that were to define the INMARSAT Standard A System.

INMARSAT began operations in early 1982 by initially leasing the L-band capacity of the three MARISAT satellites. Yet this was only part of the new system. The master plan was to use the newer MARECS satellites that had been developed by the European Space Agency (ESA) as well as employ the maritime packages on INTELSAT V satellites (the F5–F8 models) that were available later that year. As the newer capacity was launched and made available, the MARISAT satellites were relegated to backup roles.

ESA leased two MARECS satellites to INMARSAT, one over the Atlantic and one over the Pacific, each with a projected capacity of 35 telephone channels. INTELSAT leased the L-band add-on payloads on four of its INTELSAT-V satellites, with one over the Atlantic, one over the Pacific, and two over the Indian Ocean. Each payload had a projected capacity of 30 telephone channels.

The locations of the first-generation INMARSAT satellites provided coverage from three nominal locations with two payloads near each, one in an operational role and one or more payloads in a standby backup role as shown in the Fig. 3.4.

EVOLUTION OF THE EARLY INMARSAT SYSTEM

Having started operations by building on the initial MARISAT startup, INMARSAT continued to add to capacity and capabilities to its system.

Role\Region	Atlantic	Pacific	Indian
Primary	MARECS @26°W	MARECS @177.5°E	INTELSAT-V @63°E
Hot Standby	INTELSAT-V @18.5°W	INTELSAT-V @180°E	INTELSAT-V @66°E
Residual Spare	MARISAT @15°W	MARISAT @176.5°E	MARISAT @72.5°E

FIG. 3.4.   INMARSAT First-Generation Satellite Locations (1980s).

The first-generation satellites had relatively low L-band power capabilities, so in the mid-1980s, INMARSAT ordered a new satellite series with significantly higher power and wider bandwidth to replace the first-generation satellites. The specifications for the new INMARSAT-2 series included not only higher power, but wider bandwidth that would support maritime services and aeronautical services. The INMARSAT organization at this time, in fact, changed officially changed its name to the International Mobile Satellite Organization, although it renamed the INMARSAT as its shorter name.

In the mid-1980s, INMARSAT also defined requirements for a new all-digital ship earth station class known as INMARSAT Standard B. Standard B was specified to be spot-beam compatible and capable of being used with future satellites having different spot-beam configurations and frequency plans. Flexibility was built into the design via a system bulletin board that would broadcast essential information regarding the current spot-beam configuration and signaling frequencies. The scheme was similar to that used with personal computers of that time, which used the Disk Operating System (DOS) track on a floppy disk to inform the PC of the DOS to be used with the disk.

The innovative system bulletin board scheme developed by INMARSAT has been adapted to numerous other systems since then and enabled different generations of satellites to be introduced without obsolescing existing mobile earth stations.

In addition to the full capability Standard B ship earth station, there was also a need for a compact ship earth station standard that would be affordable and fit on a wider range of seagoing vessels, down to small yacht size. The answer was Standard C, which was specified to operate with a nondirectional antenna and support two-way messaging that could be used anywhere in the world via the satellite global beams. Because of its small size and low cost, it was expected that Standard C could be expected to play a fairly ubiquitous role in supporting not only basic store and forward messaging services, but other services like distress alerting, data reporting, and shore-to-ship group calling. For high seas areas, Standard C was adopted by the United Nations specialized organization, known as the

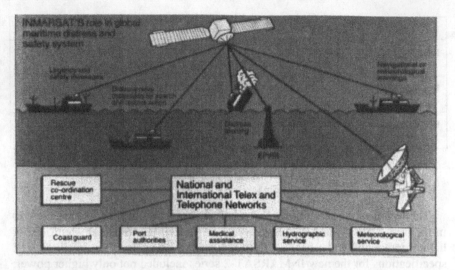

FIG. 3.5.   INMARSAT role in global maritime distress and safety system.

International Maritime Organization, as a minimum communications capability component of the new Global Maritime Distress and Safety System (GMDSS) standards. The GMDSS relies heavily on the INMARSAT system for its satellite component.

The third new mobile earth station set of standards developed in the first 10 years was the aeronautical communications system. The aeronautical standards were produced in collaboration with the aeronautical industry standards setting bodies, especially the Airline Electronics Engineering Committee (AEEC). Two types of standards were developed: the "Low Gain AES" to support core low-speed data capabilities and the "High Gain AES" to support both telephony and data communications. A driving factor for the introduction of aeronautical mobile satellite service was the desire of airlines and traffic controllers to have more reliable communications to and from the cockpit and to introduce automatic position reporting for high seas flights. Consequently, the earliest airliner installations were for the "Low Gain AES," which could be most readily implemented and still support the needed core data communications.

Substantial development was required for the multichannel "High Gain AES" terminals. However, once available, the more capable terminals were procured by the airlines as part of newly constructed aircraft so that telephony and data services could be available to passengers as well as crew. Many airlines offered the satellite telephone capability as a supplement to terrestrial airphone capability, which usually lost its coverage when more than 200 miles offshore on transoceanic flights.

By the late 1980s, INMARSAT's traffic was growing rapidly, and it was becoming clear that the Atlantic satellite was likely to saturate before the other

TABLE 3.1
INMARSAT Second-Generation Satellite Locations

Role\Region	Atlantic-East	Atlantic-West	Pacific	Indian
Primary	INMARSAT-2@15.5°W	INMARSAT-2@54°W	INMARSAT-2@178°E	INMARSAT-2@64.5°E
Spare	MARECS @15.1°W	INTELSAT-V @31.5°W	INTELSAT-V @180°E	INTELSAT-V @64.9°E
Residual Spare		MARISAT @15°W	MARISAT @177.7°E	MARISAT @72.5°E

ocean regions. In addition, the coverage from the three operational satellite locations (26° W, 63° E, 177.5° E) left a small coverage gap between the Atlantic and Pacific coverage areas. The four INMARSAT-2 satellites being procured offered the possibility of four region coverage, so a new western Atlantic operating location was selected near 54° W to complement the eastern location near 15° W. The new four-region constellation covered a gap in the coverage off the western shore of South America. This constellation deployment became as shown in Table 3.1.

INMARSAT EVOLUTION IN THE 1990S

To provide additional satellite capacity (higher rf power and bandwidth), the INMARSAT-3 series of satellites was deployed in the mid-1990s with L-band transmit and receive spot-beam coverage for up to five zones in each region.

The INMARSAT-3 matrix power amplifier design allows the L-band radio frequency transmitter power to be continuously variable between beams depending on the number of communications carriers transmitted toward each beam. Each of the five INMARSAT-3 satellites also includes an add-on navigation transponder that serves to relay a GPS overlay signal and to support the GPS Wide Area Augmentation Service for aeronautical applications.

In the early 1990s, INMARSAT introduced a new standard, INMARSAT M, that could operate with a smaller antenna and lower power than Standard B. A more efficient speech-encoding design was selected known as Improved Multiband Excitation that operated at only 6.4 kbps. With such a low bit rate, the maximum required power was significantly reduced, and the typical portable Standard M design could operate by battery and still be only the size of a standard hardback briefcase.

Another major shift in the mobile satellite industry was now exploding on the scene. The Motorola Corporation suddenly surfaced a revolutionary plan in 1990 to create a total new type of satellite network. The concept, known as the Iridium system, would create a global constellation of personal mobile communications satellites that could operate not to user dishes, but actually provide service to hand-held units. Suddenly the whole paradigm of mobile communications had shifted. Under this new type of satellite system, users could talk via something like a cell phone and connect to a satellite anytime and anywhere. This was truly revolutionary, but would it work? Would it be cost-effective? Would others try to deploy parallel systems?

When the FCC asked for competing proposals, Space Systems/Loral and Qualcomm proposed another low earth orbit system known as Globalstar. This system had less satellites, did not cover the highest and lowest latitudes near the poles, and did not employ cross-links like the Iridium satellite system, but it was still planned to offer a new range of personal mobile satellite services. The details of these new global personal communications satellites systems are addressed in greater depth in the following sections.

INMARSAT, in light of this new and to some extent unexpected competitive threat, undertook to perform its own studies and planning for the best technical means of providing hand-held capabilities. As a result of these studies, INMARSAT created a separate new company with investments from a mix of interested INMARSAT owners and outside sources. The new company, ICO Global Communications Ltd., included in its name the acronym that described its preferred orbit—ICO (intermediate circular orbit).

The INMARSAT organization continued its work to develop an even more portable version of Standard M, called Mini-M, by taking advantage of the greater sensitivity of the INMARSAT-3 satellites. A significantly improved speech-encoding technique called Advance Multiband Excitation was selected for Mini-M, which operated at a low bit rate for voice service of only 4.8 kbps. The lower bit rate and the greater sensitivity of the INMARSAT-3 satellites enabled the Mini-M terminal to be only the size of a small laptop computer (see Fig. 3.6) and operate for much longer between battery recharges.

FIG. 3.6.   INMARSAT Mini-M portable earth station (antenna is raised lid).

The Mini-M terminal development turned out to be well received and nearly as widely used as the compact Standard C. Various forms have been optimized for not only portable land mobile applications, but also for maritime and aeronautical applications.

NATIONAL MOBILE SATELLITE SYSTEMS

As cellular services grew in popularity in the 1980s, certain companies and agencies in the United States and elsewhere undertook studies of satellite systems to serve land mobile users. Communications satellites were viewed as the best way to supplement the limited geographical coverage of cellular systems and provide nationwide and even region-wide coverage for such applications as truck fleet management.

Ideally, such applications would operate in the mid- to lower UHF frequency range, where nondirectional antennas are most effective and frequency allocations were sought in that range. However, the 1987 World Radio Conference for the mobile services decided to reallocate and augment the existing mobile satellite L-band allocations to add the land mobile satellite service (LMSS) to the maritime and aeronautical allocations.

New L-band satellite systems were subsequently planned in several countries including Australia, Mexico, Canada, and the United States. Although the original systems were conceived as being primarily for land mobile use, they were planned to be supporting of serving maritime and aeronautical services as well and were therefore classified as operating in the more general Mobile Satellite Service (MSS).

The satellites for these national systems were all designed to fly in GEO orbit and employed relatively narrow satellite antenna beams focused on the land areas of their respective countries with some overlap across adjacent national borders. As a consequence of their narrow focus, the new satellites were capable of operating with mobile earth stations having little directivity.

The first of these to be put into service was Australia's Optus System in 1992, followed by Mexico's Solidaridad in 1993. American Mobile Satellite Corporation (AMSC) and Telesat Mobile Incorporated (TMI) coordinated their satellite specifications so that their satellites, launched in 1995 and 1996, respectively, would be capable of mutual backup in case of failure in one or the other.

Interestingly enough, one of the more successful land mobile satellite systems developed for the United States was introduced without the launch of any new satellite, but instead made use of an existing Ku Band domestic satellite designed for the FSS. That system, known as OmniTracs, was a two-way data-messaging system designed specifically for truck fleet management. Because the 12/14 GHz frequencies were not compatible with the use of omni-directional antennas, the truck antennas were made directive in azimuth and provided with auto-pointing capability. Because the message lengths were short and intermittent, a large number of vehicles could be tracked with relatively low bit rate transmissions. Special spectrum spreading coding and efficient modulation techniques were used to meet regulatory constraints on radio power spectral density.

GLOBAL MOBILE PERSONAL COMMUNICATIONS SATELLITE SYSTEMS

Motorola, Globalstar, and then ICO Ltd., as well as several other would-be systems (i.e., Ellipso, Odyssey, Aries Constellation, etc.) in the space of a few years, began to define a whole new type of satellite service based on personal hand-held voice units and pagers.

To provide such a hand-held capability to mobile users, Motorola decided to use a constellation of 77 and then after redesign 66 low-orbit satellites with moderately narrow antenna beams instead of using just a few geostationary satellites with many more smaller antenna beams. To minimize the number of fixed gateway stations, the Iridium satellites were designed with onboard remodulation and radio cross-links to other satellites in the constellation so that communications packets could be relayed from satellite to satellite if needed to reach the satellite above destination location.

Shortly after the Iridium project was announced, and in response to the FCC's call for alternative filings, several other satellite projects to provide global mobile personal communications services (GMPCS) were announced, including Globalstar, Odyssey, and ICO.

Like Iridium, the Globalstar satellites were planned to fly in low earth orbits, but at somewhat higher altitude (1413 km instead of 780 km) and with less inclination (52 deg instead of 86 deg), which enables global coverage with fewer (48 instead of 66) satellites. Unlike Iridium, the Globalstar system was also a much less technically advanced (or, put another way, a less complex) system. Thus Globalstar was not designed to employ onboard remodulation or satellite to satellite cross-links. Instead each Globalstar satellite was designed with a pair of frequency translating repeaters that serve to merely extend the coverage of existing terrestrial cellular/PCS systems via fixed gateway earth stations.

Unlike Iridium or Globalstar, the two other systems—the TRW-backed Odyssey and INMARSAT spin-off ICO—planned to use fewer (10 or 12) satellites flying at a much higher intermediate circular orbit that provided a longer period of visibility and higher average elevation angles toward the satellites. The Odyssey project was in time merged with the ICO project, and most recently ICO merged with the Globalstar network.

All of the Leo and Meo systems—Iridium, Globalstar, and ICO—have experienced major financial difficulties, and all have been reorganized after bankruptcy. Iridium and Globalstar, which were fully deployed, suffered from much lower than expected mobile terminal subscriptions and slow take-up rates. Iridium and ICO filed for bankruptcy in August 2000, and Globalstar filed in 2002. The $4.5

billion Iridium system was bought out of bankruptcy for $25 million and now operates as New Iridium. Its services have grown steadily, and revenues were boosted by the Iraq invasion in 2003.

ICO was bought for about $1 billion by Craig McCaw to become New ICO. Then in April and May, New ICO acquired Globalstar out of bankruptcy as well. There are still questions as to when these systems can become true market successes, but services on New Iridium and New ICO/Globalstar are both now growing well and seem headed toward market viability. It now seems that the medium earth orbit network, designed to operate in the 2GHz band, will become the follow-on system to the original Globalstar LEO network.

REGIONAL MOBILE PERSONAL COMMUNICATIONS SATELLITE SYSTEMS

The virtual bent-pipe digital processor technology used in the ICO design has also been used in geosynchronous regional mobile satellites to route calls among over 200 spot beams. With approximately one-degree diameter spot beams, the satellites are capable of supporting large numbers of users with mobile terminals no bigger than ordinary terrestrial cell phones. Two such satellite systems are the Asian Cellular Satellite System (ACeS), based in Indonesia, and Thuraya, based in the United Arab Emirates, each currently with a single operational satellite. Another proposed system, known as Agrani, which was to have been based in Bombay, India, and serve the Indian Ocean market, was not able to complete its capitalization, and thus manufacture of their satellites by Hughes was halted several years ago.

ACeS, which recently encountered market viability issues, uses L-band and operates at 123 deg East. From that location, ACeS has the technical capability of serving an area that includes all the countries of Southeast Asia, India, China, and Japan. Thuraya also uses L-band and operates at 44 deg East. From that location, Thuraya (see Fig. 3.7) has the technical capability of serving an area that includes all the countries of the Middle East, North Africa, and most of Europe. Thuraya will soon deploy its second satellite and now has several hundreds of thousands of subscribers. Both ACeS and Thuraya use communications designs based on the widely used terrestrial GSM standard. Consequently, users with dual-mode cell phones can easily switch to satellite service whenever they travel outside the range of their terrestrial service.

LONG-TERM PROSPECTS FOR MOBILE SATELLITE SERVICES

Mobile satellite services are here to stay because the need for ubiquitous communications will always remain. Despite the coverage expansion of terrestrial mobile radio systems, there will always remain areas of the world that will be more economically covered by satellite.

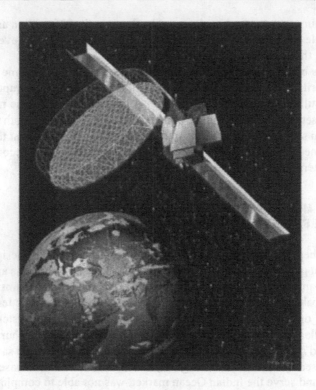

FIG. 3.7.   Thuraya regional mobile satellite with 12.2-meter L-band antenna.

Reducing the cost of mobile satellite services to users will continue to be as big a challenge as the technology. For mobile telephony and Internet communications, mobile satellite systems that have wide coverage with favorable economics are likely to be the source of most future capacity. A key technology enabler will continue to be the recently introduced virtual bent-pipe designs, which allow power and bandwidth to be quickly shifted from region to region as required to support the changing traffic patterns of mobile users.

It is unlikely that any new constellations using LEO will be deployed to provide telephony and high-speed data communications for mobile users in the near term. Those already in service—namely, Iridium and Globalstar—are likely to remain in service for the lifetimes of the existing satellites, which have been reconfigured to extend their usable lifetime from 5 to 7 years for much longer. Still unanswered is whether the revenues will increase sufficiently to support the costs of replacement satellites at the end of the first-generation space segment's lifetime.

Satellites known as "little leos" provide messaging service to support such applications such as car and bus navigation systems, ship and truck fleet control, pipe-line monitoring, and Supervisory Control and Data Access (SCADA) operations at electric power stations, remote labs, environmental monitoring stations, and buoys. In short, these small and low-power satellites support applications that require the infrequent transmission of small amounts of data from remote or mobile locations. Because of their infrequent transmission of small data units, these systems can support a large number of inexpensive terminals. The energy required to put such satellites into orbit is relatively small, therefore the economics can be favorable with an appropriate choice of launcher and orbit injection plan. The prime example is the Orbcomm packet message system with 32 satellites, each weighing only 40 kg and flying at about 800-km altitude.

Because mobile satellite terminals are, by their very nature, small and designed for mass markets, they are attractive not only for mobile applications, but also for applications where an inexpensive earth station is needed at a fixed location. Strictly speaking, the satellite service provided in such a case is an FSS. A question to be addressed in the future is whether the present distinction between fixed and mobile satellite services is helpful to spectrum planners. Clearly, there has been and continues to be a willingness to look the other way when mobile systems are used for fixed communications applications such as remote monitoring or village phone booths in developing countries.

For the aviation and maritime communities, however, there will continue to be a need for distinctive mobile satellite communications capabilities. Their requirements include high seas coverage, reliable real-time voice and data communications, provision for handling distress and safety communications, and accessibility by ships and aircraft from all nations. Up to the present, provision of satellite capacity to provide distress and safety services has been undertaken by INMARSAT working hand in hand with the bodies responsible for safety communications (i.e., IMO for maritime and ICAO for aviation). It remains to be seen how—or if—other mobile satellite systems begin to support those requirements in any real tangible way and to what degree international standardization of satellite safety communications can be achieved in the future.

BACKGROUND READINGS

Abramson, Norman, (2000, July), "Internet Access through VSATs," *IEEE Communications Magazine.*

Bever, M. et al., (2001), "TRW Broadband payloads for Energing Ka-band market," International Astronautical Federation (IAF).

Butrica, Andrew J., (Editor), (1997), *Beyond the Ionosphere: Fifty Years of Satellite Communication* (NASA SP-4217), Washington, DC: Chap. 21 by E. J. Martin.

Evans, John V., (1998, July), "Satellite Systems for Personal Communications," *Proceedings of the IEEE.*

Keane, L. M., and Martin, E. J., (1974, November/December), "MARISAT, Gapfiller for Navy Satellite Communications," *SIGNAL*, pp. 6–12.

Linder, H. et al., (2000, June), "Satellite Internet Services Using DVB/MPEG-2 and Multicast Web Caching," *IEEE Communications Magazine.*

Lipke, D. W., and Swearingen, D. W., (1994), "Communications System Planning for MARISAT," International Communications Conference, IEEE Cat. # CHO 859-9-CSCB.

Lodge, John H., (1991, November), "Mobile Satellite Communications Systems: Toward Global Personal Communications," *IEEE Communications Magazine*.

Pelton, Joseph N., (2002), *The New Satellite Industry*, Chicago: International Engineering Consortium.

Richaria, M., (2001), *Mobile Satellite Communications, Principles and Trends*, New York: Addison-Wesley.

Sachdev, D. K., (2002, May), "Three Growth Engines for Satellite Communications," AIAA 20th International Communications Satellite Systems Conference, Montreal.

CHAPTER FOUR

Launch Vehicles and Their Role

Eric J. Novotny
American University

> *Let us create vessels and sails adjusted to the heavenly ether, and there will*
> *be plenty of people unafraid of the empty wastes.*
> —Johannes Kepler, in a letter to Galileo (1610)

Each and every one of the hundreds of communications satellites that provide their services around the globe today were launched into orbit atop a rocket or, as it is known in the industry, a launch vehicle. Although there are highly imaginative concepts emerging about how we might lift satellites into orbit by different means—such as tethers, transmitted radio frequency power, or even more exotic techniques—the reliance on expendable launch vehicles to lift communications satellites into orbit will remain at the core of the industry for at least the next few decades. Thus, the story of launch vehicles and how they evolved is a critical part of the communications satellite story as well. Launch systems have always been a key enabling technology of satellite communications advancements, and they will remain so as long as there are requirements to deliver payloads into space.

Serious speculation about reaching space—sending a vehicle outside the earth's atmosphere and into orbit or to another celestial body—began with several imaginative mid-19th-century writers and inventors. Those who led the way included Jules Verne, Konstantin Tsiolkovsky, and, in the 20th century, Hermann Oberth, Robert Goddard, and Wernher von Braun. Experiments with rocketry in the mid-20th century culminated in the first space vehicle to leave the earth's atmosphere—the A-4 single-stage rocket known during World War II as the V-2. This was accomplished on October 3, 1942, from Peenemünde, Germany.

This V-2 design was the forerunner of what are known today as expendable launch vehicles (ELVs), which perform their launch mission once and are not recovered and reused. Today, the Ariane 5, Atlas, Delta, Proton, and Zenit launch vehicles are well-known examples of ELVs used in commercial and government satellite applications by the operators of satellite systems around the world. Launch systems such as the U.S. space shuttle are known as reusable launch systems (RLVs), in which all or part of the launch vehicle returns to earth in a condition allowing for refurbishment and relaunch. Although today the U.S. space shuttle is the only RLV in routine operation, it has not launched commercial satellites since 1986. Concepts for advanced RLVs (including systems that could launch and return to earth with or without a human crew), new propulsion systems, and other advanced technologies may supplement expendable launch vehicles in the future. NASA is actively studying successor designs to the space shuttle. Some of these, such as the orbital space plane design, may even employ an expendable launch vehicle during its main boost phase. For the present, reliable, affordable, and versatile expendable launch vehicles provide the surest access to space for communications satellite operators in both government and private sectors.

LAUNCHING A SATELLITE: BASIC PRINCIPLES

Placing a satellite into a specified orbit around the earth requires the satellite, or *payload* as it is called in the launch business, to be attached temporarily to the launch vehicle from liftoff to its appointed destination in space. The flight path of each launch system placing its payload into orbit can differ depending on the mission requirements and the payload's orbit-raising capabilities.

The launch vehicle in turn consists of multiple operating stages, a payload adapter, and a fairing to protect the payload during its ascent into the atmosphere. The major components of a launch vehicle system are: its aerodynamic outer shell, propulsion systems (including engines, pumps, and propellant tanks), and an electronic guidance and control system (see Fig. 4.1).

The actual launch sequence consists of one or more periods of powered flight, in which the payload and launch vehicle are lifted together up and out of the earth's atmosphere. The payload fairing is jettisoned after most of the atmospheric drag is gone. Solid rocket motors used to boost performance during the initial phase of the mission are also jettisoned by this time (see Fig. 4.2). The launch vehicle will typically have two or more stages, such as in the case of the Atlas V, a two-stage vehicle. The Atlas main stage begins the flight, and the Centaur upper stage propels the payload to a geostationary transfer orbit.

As the launch vehicle stage continues its ascent, velocity and altitude increase to the point where the vehicle and its attached satellite payload attain a preliminary orbit around the earth. The work of the launch vehicle concludes when the desired orbital position, or parking orbit, is attained—a process that can include

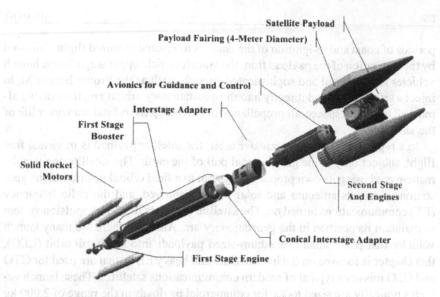

FIG. 4.1. Launch vehicle components. Launch systems to carry communications satellites to geostationary transfer orbits or directly to the geostationary arc are typically made up of multiple stages, whose constituent parts include propulsion systems, guidance and control systems, and systems for deploying the satellite payload.

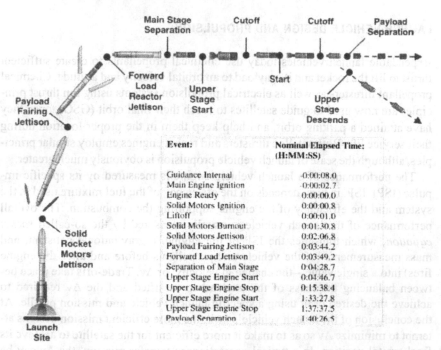

Event:	Nominal Elapsed Time: (H:MM:SS)
Guidance Internal	-0:00:08.0
Main Engine Ignition	-0:00:02.7
Engine Ready	0:00:00.0
Solid Motors Ignition	0:00:00.8
Liftoff	0:00:01.0
Solid Motors Burnout	0:01:30.8
Solid Motors Jettison	0:02:06.8
Payload Fairing Jettison	0:03:44.2
Forward Load Jettison	0:03:49.2
Separation of Main Stage	0:04:28.7
Upper Stage Engine Start	0:04:46.7
Upper Stage Engine Stop	0:15:38.4
Upper Stage Engine Start	1:33:27.8
Upper Stage Engine Stop	1:37:37.5
Payload Separation	1:40:26.5

FIG. 4.2. Nominal mission profile for an expendable launch vehicle. Drawing adapted by the author from a Lockheed Martin Atlas V Mission Profile.

85

periods of coast and re-ignition of the engines to resume powered flight, followed by the separation of the payload from the launch vehicle upper stage. Some launch vehicles are powerful and sophisticated enough, such as the Proton Breeze M, to inject a satellite payload directly into the geostationary orbital arc, thus saving almost all on-board spacecraft propellants for extending the orbital maneuver life of the satellite.

In a typical geostationary transfer orbit, the satellite payload is in virtual free flight, subject only to the gravitational pull of the earth. The satellite can then be maneuvered using its own propulsion system to a final orbital position in the geostationary arc, its antennas and solar panels deployed, and the radio frequency (RF) communications turned on. The satellite then uses its own propulsion system to maintain its position in the geostationary arc. Although there are many launch vehicles that place small to medium-sized payloads into low earth orbit (LEO), this chapter is focused on the intermediate and heavy ELVs that are used for GTO and GSO missions typical of modern communications satellites. These launch vehicles typically are sized today for commercial payloads in the range of 2,000 kg to 6,000 kg of separated mass, the customary way to measure the size of the payload. Separated mass includes the mass of the satellite hardware and its on-board fuel.

LAUNCH VEHICLE DESIGN AND PROPULSION

Expendable launch vehicles today use chemical propellants to create sufficient thrust to lift the rocket and its payload to an orbital velocity and altitude. Chemical propellant thrusters as well as electrical propulsion systems using ion thrust principles are now used to guide satellites to reach their final orbit (GSO), once they have attained a parking orbit, and help keep them in the proper location during their service life. Both satellite thrusters and rocket engines employ similar principles, although the scale of launch vehicle propulsion is obviously much greater.

The performance of a launch vehicle engine is measured by its specific impulse (ISP). ISP, in turn, depends on the efficiency of the fuel mixture used in the system and the efficiency of the engine supporting the combustion. The overall performance of the launch vehicle is usually measured by the so-called *rocket equation*, which translates the ISP of the engine, a gravitational constant, and mass measurements (of the vehicle and propellants before and after the engine fires) into a single value—the change in velocity, or Δv. Trade-offs take place between balancing the mass of the payload to be lifted and the Δv required to achieve the desired orbit using a specific launch vehicle and mission profile. At the conclusion of the launch vehicle's task, the more efficient mission designs attempt to minimize Δv so as to make it more efficient for the satellite to achieve its final orbital position. In a typical geostationary transfer mission, the Δv can be anywhere between 1,500 and 1,800 meters per second.

Launch vehicle propellant systems can be categorized as either solid or liquid, and some launch vehicles use a combination of both types to produce the required thrust in various stages of a launch mission—usually the boost phase originating at lift off. The most common propulsion systems are liquid, and these in turn can be categorized as storable, cryogenic, or hypergolic.

Liquid propulsion systems combine a chemical fuel with an oxidizer. The resulting mixture burns in an engine or combustion chamber to produce thrust, the ISP, which measures how much thrust or lifting power (e.g., in kilograms) is produced by burning one kilogram of propellant in one second. The fuel and oxidizer are carried separately in tanks mounted inside the walls of the launch vehicle, pumped to the combustion chamber or engine, mixed, and then burned so that the resulting exhaust produces the required thrust. Because the lifting power of the burning propellant is greater than its own mass and the mass of the launch vehicle, the vehicle with its payload will gradually overcome the earth's gravitational pull and ascend into space.

Storable propellants are those that can be kept for reasonably long periods in an uncompressed liquid state, such as alcohol or kerosene. Today, most storable propellants are hydrocarbon-based fuels typically refined from petroleum, such as a certain grade of kerosene commonly known as rocket propellant one, or RP-1. Storable propellants can also be chemically enhanced or synthetically produced. These types of fuels are typically combined with liquid oxygen as the oxidizer. Storable systems have a long heritage in launch vehicles, including providing the propulsion of the first stage of the Lockheed Martin Atlas family of launch vehicles and the first stages of the Saturn launch vehicles used in the Apollo moon program.

To gain further efficiencies and greater thrust, and thereby raise the ISP of the propulsion system, launch vehicle designers turn to cryogenic propellants. Although more difficult to handle and keep stable for long periods, their efficiency can be greater than storable propellants because the density (or compactness) of the chemical fuel can be greater within the same volume. A number of launch systems today employ cryogenic propulsion for their upper stage engines, such as the Centaur upper stage of the Atlas. Liquified hydrogen gas provides the fuel in these cases, which is kept in such a state at very low temperatures and high pressures. Such a cryogenic system also powers the main stage engines of the U.S. space shuttle and the Delta IV, both using hydrogen and oxygen.

Hypergolic propulsion systems employ fuel and oxidizer elements that ignite spontaneously on contact with each other. Their advantage is that they require no other source of ignition, and are therefore reliably able to virtually guarantee ignition in the course of a specific mission to orbit. As one might surmise, hypergolic substances are more dangerous to handle than storable propellants and must be treated with additional care. The most commonly used hypergolic fuels are hydrazine and certain hydrazine derivatives, and the most common oxidizers are nitric acid and nitrogen tetroxide. Hypergolic systems are used today in a number of Russian, Chinese, and Indian launch vehicles.

Solid propellants are also employed in modern launch vehicles. Here the fuel and oxidizer are solid substances that are mixed and suspended during production in a matrix, usually of a rubber-like substance, and poured into a cylindrical casing where the mixture solidifies. When ignited, the hot gases produced from the controlled burning are expelled out an exhaust nozzle to produce thrust. Solid propellants have the advantage of being reliable, but they cannot be shut down and restarted. When combined with liquid systems on the same stage, typically the liquid system is ignited first; when the required level of thrust is achieved, the solid propellants are then ignited and liftoff occurs.

Today, most solid rocket boosters are used as strap-on supplemental engines, which are added to the main stage liquid booster to provide added thrust to lift heavier payloads. A number of solid rocket motors can typically be added to graduate the amount of additional performance required. Several contemporary launch vehicles, such as the Ariane 5, require external, solid rocket boosters in their basic configuration. Others, such as the Atlas V, employ modular strap-on boosters to optimize performance depending on the size of the payload and the mission requirements. One launch vehicle, the Indian GSLV, employs a solid propellant main stage surrounded by smaller liquid strap-on boosters (see Fig. 4.3).

FIG. 4.3.   Liftoff of an Indian Geostationary Launch Vehicle (GSLV) on its inaugural flight from the Satisk Dhawan Space Centre, Srihavikota (near Chennai). Photograph courtesy Indian Space Research Organisation (ISRO).

With the 1940s' demonstration that spaceflight was technologically feasible, in the 1950s both the United States and USSR began to invest substantially in the research and development of long-range rocket technologies, stimulated by the potential delivery of nuclear weapons by intercontinental ballistic missiles. Given the demonstrated progress made in Germany during World War II with medium range missiles, speculation about achieving orbital and even interplanetary space-flight in the immediate future followed quickly. Space became the new high ground in studies of future military applications, including the idea of establishing populated installations on the moon or space stations orbiting the earth. In Europe, policymakers were quick to identify access to space as a necessary condition for military and scientific independence. Jacques Servan-Schreiber, author of the widely read book *The American Challenge*, emphasized the importance of space to the future of technically advanced societies. The critical connection to be made was between space as a strategically vital zone of national competition and access to space as a source of national power and security.

The feasibility of launching payloads into orbit around the earth has therefore spawned a remarkable variety of space-based applications, including geostation-ary communications satellites. Space launch capabilities played a fundamental role in the feasibility of communications satellites. Launch vehicles operated both as an enabling technology and a constraint on the size and design of satellites that could be launched and the orbits they could attain. The size and mass of a satellite has a direct bearing on the communications capabilities it can deliver.

First, communications satellites need to be placed in a specific orbit that corre-sponds to the services to be delivered. Launch vehicles are selected to optimize that flight path, placing the satellite in its intended orbit. Selecting a launch vehi-cle is therefore governed by its capability to lift and carry a specific maximum payload mass to a specific destination at a specific velocity. Generally speaking, the more a satellite gains in total mass, the larger the launch vehicle needs to be to achieve the same end result. As launch vehicles grow in total performance, satel-lites can be made larger and therefore more powerful. Satellites can be built to contain larger and more powerful batteries, solar arrays, antennas, and communi-cations electronics to meet more demanding requirements. Likewise, the demand for ever more complicated, powerful satellites has stimulated the development of more capable vehicles that can meet such requirements.

ORBITS FOR COMMUNICATIONS

Launch vehicle systems are capable of placing communications satellites into var-ious types of orbits to accomplish a variety of communications services, as ex-plained in chapters 2 and 3. As noted in these earlier chapters, practical orbits can be grouped into four major types: geostationary (GEO), high earth orbit (HEO), medium earth orbit (MEO), and low earth orbit (LEO). The most common orbit associated with modern communications satellites is the geostationary orbit, and different launch vehicles can be used to achieve cost-effective results for each of these generic orbital types. Some smaller launch vehicles are not designed with the capability to travel to a geostationary transfer orbit and so are confined to missions that place satellites in lower earth orbits.

A typical launch mission may keep the launch vehicle attached to the satellite for as little as 30 minutes or as much as 6 hours. During this period, the upper stage of the vehicle alternately fires its engine and then coasts as the mission progresses, adjusting the orbital parameters with each successive burn. After separation at the end of the vehicle's mission, the satellite then uses its on-board propulsion to remove inclination to zero degrees and then circularizes its orbital parameters to the proper distance from the earth required for geostationary communications operations.

Second, a satellite must fit inside the payload fairing of the launch vehicle. Most communications satellites have their antennas and solar power arrays folded to meet the spatial clearances inside a launch vehicle fairing. These devices are then mechanically deployed or extended after the satellite separates from the launch vehicle. The satellite will operate on battery power until its solar power arrays are deployed and directed toward the sun. Today, payload fairings exceed 4 or even 5 meters in diameter, allowing satellite operators to launch ever larger, more complicated, and powerful spacecraft. The satellite must also be designed to tolerate a variety of environmental conditions, including certain temperatures and vibrations when inside the launch vehicle's fairing and at liftoff. Each satellite is also built to be compatible with a specific adapter and separation system, which is used to mate the satellite to the launch vehicle and release the satellite into orbit at the end of the flight.

Launch vehicles and their satellite payloads are therefore paired to gain the optimum performance from each of them as the path to the final orbital destination progresses. As launch vehicles are able to lift larger and heavier payloads, so satellite manufacturers are able to build larger and more powerful satellites to fulfill a wider diversity of communications applications. Over time the technologies advancing both launch vehicle capability and satellite sophistication reinforce one another, as each requires an additional increment of performance with respect to the other.

When either launch vehicle or satellite technology is limited in some critical way, technical solutions need to be found to overcome these temporary limits. In the early years of geostationary communications satellites, when miniaturization of electronics, efficiency of solar power, reliability of batteries, and so on were not perfected, satellite performance could only be increased by adding size and mass, which in turn was constrained by the performance of available launch vehicles. For example, when the Soviet Union built and launched the first communications satellites, the Molniya series, the orbits selected were not geostationary. Highly elliptical orbits allowed these satellites to have better coverage at higher latitudes in the Northern Hemisphere, but also allowed a larger, more powerful spacecraft to be lifted by the performance constraints of the launch vehicles then available to orbits short of geostationary.

LAUNCH VEHICLE TECHNOLOGY AS A STRATEGIC GOOD

Ballistic missiles provide a most effective means to deliver nuclear weapons. They have the advantage of long, fast, and increasing accurate range targeting. A country defending against an attack by ballistic missiles would not find it possible today to stop them; an effective wide-area antiballistic missile defense system has yet to be developed and deployed by any nation. Attempts are being made by the United States to develop such capabilities, and recently European governments, including the Russian Federation, have been invited to participate in research and development toward ballistic missile defense. Today, however, the only effective military means of defense against missiles is a countervailing capability to retaliate in kind against the attacker. Thus, ballistic missile development, coupled with nuclear weapons, presents a priority capability for any nation wanting to join the ranks of the major powers and to protect itself against nuclear extortion. Such missile development is similar to technologies used for launch vehicles to place satellites into orbit.

SURVEILLANCE AND INTELLIGENCE FUNCTIONS

Space-based satellite systems serve an important corollary function in an environment of nuclear weapons. Satellites can be used effectively for early warning of missile launches and to gather routine photographic and electronic intelligence. A major nation lacking the technical capability to build and launch surveillance satellites is at a distinct intelligence disadvantage prior to a major military engagement as has been seen in such confrontations as the two Persian Gulf wars, the U.S. anti-terrorist campaign in Afghanistan, and so on.

COMMAND AND CONTROL OF MILITARY FORCES

Space-based communication also serves a vital function for the direction of military forces. Satellites are inherently difficult to destroy, even for space-faring nations with the technical capability to intercept and disable orbital satellites. At present, only the United States and the Russian Federation have engaged in practical research to deploy anti-satellite weapons, although other space-faring countries have reportedly begun work on similarly related technologies. Thus, having satellites available for military communications provides a reasonably robust link in the overall control of forces. Especially if nuclear conflict is anticipated, satellite systems can function reasonably well with mobile ground forces after fixed installations are destroyed.

The United States, the United Kingdom, France, the Russian Federation, and India have deployed communications satellites to assist in the control of their strategic nuclear forces. A number of other countries maintain access to space by developing and deploying launch vehicles to preserve options for the use of satellites in military applications. Today, modern military forces are highly dependent on space-based communications, and therefore launch vehicle technology is a major component of assured access to space.

STRATEGIC MISSILE DEFENSE

A new element in military space is the application of space-based weapons to intercept and destroy ballistic missiles. At present, no active defenses have been demonstrated to be effective against an incoming warhead delivered by an intercontinental ballistic missile, although the United States is leading efforts, with its allies, to deploy such defenses in the near future. As more countries around the world acquire the ability to launch weapons of mass destruction aboard intercontinental missiles, the need to consider such active defense will grow. Launch vehicles will play a crucial role in maintaining space access critical to the deployment of active space-based warning systems and defenses.

NATIONAL PRESTIGE

Beyond the purely military dimension of national power, access to space represents a political form of national prestige and scientific prowess for nations. The development and perfection of space launches reflect on a nation's inherent scientific and technical capabilities, and they signify that major technological power status is achieved. It is also an expression of national wealth to be able to afford a national space program that pays dividends in scientific knowledge, space-based imaging, and builds a national infrastructure for satellite-based telecommunications. Arguments in favor of national prestige by governments are important in obtaining public support for governmental space expenditures. Especially when the space budgets are civil in character (i.e., based on primarily scientific research missions), national scientific prestige is an important factor in maintaining support among the population. Nevertheless, maintaining a capability to access space in the form of reliable and plentiful launch vehicles still remains a priority for the space-faring nations of the world, and it represents an important dimension of national power. Thus, there remains an important relationship between military space programs and civilian space activities. In the space launch arena, the interplay between governmental support for launch vehicle technology and the use of launch vehicles for commercial applications will always be closely linked. This relationship is addressed in some detail in the chapter 7, which addresses so-called *dual use* civilian and military communications systems.

INDIGENOUS HIGH-TECHNOLOGY CAPABILITIES

Another factor in space launch research and development is the potential for attaining and maintaining a national proficiency in leading edge, high technology. Space industrial development depends on the ability of related industries to achieve new levels of performance and capability. For example, composite materials are crucial for high-strength, lightweight aerodynamic components for launch vehicles and satellite structures. Better performance through the development of microprocessors and other semiconductor technologies gives satellites and launch vehicle guidance systems the ability to perform better information-processing tasks using lower power and reduced mass. Advanced fuels and improved combustion techniques may have applications in a variety of areas of space launch and satellite operations.

These clusters of technological development applied to launchers may in turn have collateral applications and benefits in other industries. In all maintaining a national space capability is a way to nurture important national intellectual assets. Because many aerospace companies have been compelled to look outside government for growth opportunities, the proved capabilities of satellite systems have spawned new commercial applications. Likewise the appeal of augmenting governmental capabilities of launch systems with commercial applications leads several of the world's launch suppliers to seek markets for commercial launches.

THE GLOBAL LAUNCH VEHICLE INDUSTRY

In the United States, the development of the manned Space Transportation System, the space shuttle, began with flights in 1981. It was intended to replace expendable launch systems and help make the shuttle program financially independent of the federal treasury. The shuttle was intended to be self-financing through launches for government agencies other than NASA, such as the U.S. Department of Defense, and through commercial contracts for launches with private satellite operators and international organizations such as Intelsat. Indeed it has been Intelsat that has led the way for larger, more complicated satellites and therefore more capable launch systems. The demand for larger, more powerful satellites, and the consequent supply of more capable launch vehicles, are shown in Fig. 4.4, which depicts the growth of satellite payload masses through time.

With the combined trend of larger satellites and the need to justify the expenditures of a national manned launch system, commercial satellite operators began to

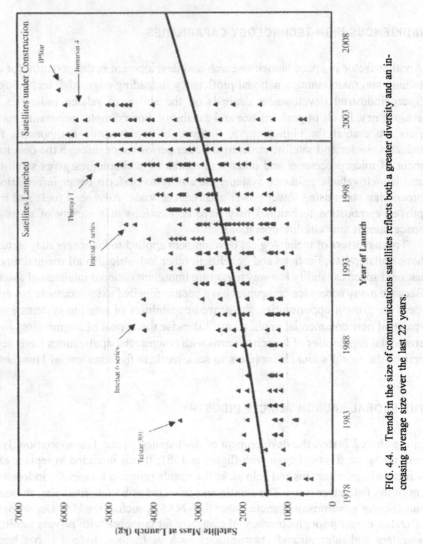

FIG. 4.4. Trends in the size of communications satellites reflects both a greater diversity and an increasing average size over the last 22 years.

schedule numerous launches on the U.S. Space Shuttle, which was supposed to be the dominant U.S. launch system for both commercial and governmental satellite missions. Given that the shuttle was able to take satellite payloads only to LEO, each communications satellite destined for the geostationary arc had to include an upper stage, most often a solid rocket motor, to kick the satellite to a higher transfer orbit. This combination is shown in Fig. 4.5, where the combined satellite and upper stage of one of NASA's Tracking Data and Relay Satellites (TDRS) is deployed from the shuttle cargo bay.

The last flight of the *Challenger*, in January 1986, brought this concept to an abrupt and tragic end, with the failure of one of its solid rocket boosters shortly after launch and the subsequent loss of the orbiter and all seven astronauts aboard. Companies in the satellite business had previously contracted with NASA for multiple launches of commercial payloads. These contracts were terminated after NASA exited from the commercial launch business. Major segments of the commercial satellite business were then faced with finding expendable launchers for their requirements, as satellite operators would not enter into new contracts for new spacecraft without the means available to launch them. Russian (at that time the USSR), Chinese, and Japanese launch vehicles were either not adaptable

FIG. 4.5. A view from inside the cargo bay of the U.S. Space Shuttle during the deployment of a tracking data and relay satellite.

quickly or unavailable to commercial users for policy reasons. Therefore, commercial users and manufacturers turned to the Ariane 4 launch vehicles (Fig. 4.6) that the European Space Agency (ESA) developed and then to other U.S. systems, such as the commercial Atlas and Delta II vehicles.

THE ARIANESPACE SOLUTION

Today, Arianespace is the commercial marketing and operating organization that is owned by a number of aerospace, engineering, banking, insurance, and other commercial organizations in Europe. Yet it depends on the ESA to develop the technology and provide substantial funding for the Ariane launch programs.

FIG. 4.6.   An Ariane 4 launch vehicle begins its ascent from Kourou in French Guiana.

Arianespace is currently owned by 53 shareholders, including European aerospace and engineering companies that supply and support the launch vehicle production. A majority (over 53%) are combined French governmental and industrial interests. These ownership arrangements of Arianespace are currently in flux as the European Union and the members of the ESA reallocate responsibilities for Europe's launch systems among Arianespace, the national space agencies of each country, ESA, and the major contractors, such as European Aerospace and Defense Systems (EADS). A new arrangement of owners was implemented in 2003, and Arianespace assumed mostly marketing functions.

The launch vehicles themselves are built by the prime contractor EADS and launched from ESA's main launch center in Kourou, French Guiana, operated by the French space agency *Centre National d'Etudes Spatiales* (*CNES*). Arianespace received its major commercial opportunity after the *Challenger* accident in 1986, when NASA withdrew from the commercial launch business. Commercial satellite manufacturers and satellite operators were without capable launch vehicles at the time NASA withdrew from launching commercial payloads. Referring back to Fig. 4.4, note that there were only two launches of communications satellites in 1986 largely due to the fact that the U.S. shuttle fleet was at once grounded and had no plans to carry anymore commercial satellites.

Arianespace had been flying the Ariane 2 and 3 models to geosynchronous orbit since 1979, but the heavier satellites required by commercial customers and the demand for alternative launch vehicles led directly to the development of the Ariane 4 series. The Ariane 4 fleet was retired in early 2003 after many commercial flights, and the company is now flying exclusively the Ariane 5 series mostly in its dual satellite configuration. Originally designed to lift heavy payloads for a European RLV called Hermes, the Ariane 5 has been modified to carry commercial satellite payloads. The development of the Ariane 5 series has had to contend with the development of new major vehicle components, with increasingly heavier lift vehicles superseding the earlier configurations in the series. The ultimate objective will be to develop a vehicle capable of placing about 10 metric tons into geostationary transfer orbit (GTO).

A typical mission will wait until two satellites are delivered to the launch site and then be launched together. Various configurations will have to be tested and flown reliably before this level of performance is achieved. The Ariane 5 is slated to launch about five times per year, and it is designed to carry two average-size satellite payloads at once as it has demonstrated with the six-ton capability known as the Ariane 5G. Heavier satellites can be flown in a single-mission configuration, but these will be the exception rather than the rule. The future success of the Ariane 5 program depends on the development of a reliable vehicle with a heavy lift capability with a sufficient number of satellite pairs to keep up a minimum launch tempo.

The technique of launching two satellites requires the development of an inner and outer encapsulation and separation system, which allows the satellite carried in the upper position to be deployed first and the second satellite in the lower position to be deployed shortly thereafter. Clearly both satellites must be destined for the same orbital plane because the trajectory of the current upper stage of the Ariane 5 is fixed. Increased efficiencies and performance could be obtained if a launch vehicle upper stage could be restarted to allow coasting between deployments and possible maneuvers to alter orbits. ESA is developing such enhanced upper stages for initial test flights on the Ariane 5 by 2005 or 2006.

INTERNATIONAL LAUNCH SERVICES

The U.S.-based General Dynamics Company, later acquired by the Martin Marietta Company, first developed the Atlas family of launch vehicles beginning as a ballistic missile program in the 1960s. Lockheed Martin Corporation is now the builder of the Atlas family of launch vehicles after the merger between Martin Marietta and Lockheed Corporation. Atlas launch contracts were first placed with Intelsat, which began launching the Intelsat 5 series of satellites in 1980. The Atlas family of launch vehicles was upgraded throughout the 1980s largely to serve the commercial market for communications satellites and meet increasingly demanding U.S. government requirements. The U.S. Congress passed legislation authorizing the direct commercial sale of U.S. launch services, such as Atlas, which allowed Lockheed Martin to sell launch services directly to commercial satellite operators. Throughout the 1990s, the vehicles (Atlas II, IIA, and IIAS) became widely accepted for the reliable launch of both commercial and governmental communications satellites.

In the early 1990s, the U.S. government was concerned about the next generation of launch vehicles to serve governmental requirements and to compete effectively with the Europeans, Russians, Chinese, and Japanese for commercial business as well. Martin Marietta made a brief debut in the commercial launch market with the Titan II launch vehicle and retained its dominance of the heavy lift missions for the U.S. government. After a series of studies, the U.S. Air Force began a program to develop a new generation of launch vehicles called the Evolved Expendable Launch Vehicle System (EELV).

Two industrial teams were formed to develop these vehicles—one from Lockheed Martin and the other from Boeing. Governmental investment was combined with matching investments from each company, the end result giving the U.S. assured access to space by maintaining two independent launch vehicle suppliers. The two vehicle families that launch government satellites under this program, which began in 2002, are known as the Atlas V and Delta IV.

The Atlas team from Lockheed Martin took an evolutionary approach by bringing forward the proved technologies from the Atlas II and IIAS programs and developing an intermediate vehicle known as the Atlas III. This latter vehicle and the Atlas V employ a main stage propulsion system developed and built in the Russian

Federation by NPO Energomash, known as the RD-180 engine. It is a testament to the end of cold war competition that an American launch system with substantial defense and civilian space usage would be the product of cooperation between these two countries. The Atlas V had its successful debut in August 2002 carrying a satellite for Eutelsat.

Today, Lockheed Martin operates its launch services through International Launch Services (ILS), which was formed in 1995. ILS is responsible for all launch services for the Atlas family of launch vehicles, including the Atlas V, and the Russian-built Proton, which came into ILS by virtue of an earlier Russian partnership with Lockheed, known as Lockheed-Khrunichev-Energiya, formed in 1993 to market the Proton commercially. Today, ILS is the leading supplier of launch services worldwide, and it uses its two independent launch systems—Atlas and Proton—as a schedule assurance system in which a commercial satellite customer can be launched on either of these two, overlapping, compatible launch systems.

To move Soviet space industry from military to commercial and civil applications, the Gorbachev government began serious efforts to develop linkages with Western space industry beginning in 1989. One of the first ventures to be formed was between the state industrial organizations Khrunichev and Energia, and the design bureau Salyut. All of these organizations had a history going back to the founding years of the Soviet space program. Subsequently, the Salyut design bureau was absorbed into Khrunichev. The three-stage Proton launch vehicle built by Khrunichev, coupled with a fourth stage designed and built by Energia, was a powerful new entrant into the marketplace, beginning with its first commercial launch in April 1995. Due to the requirements for increased performance, the Proton vehicle has been modernized to carry heavier commercial payloads, ultimately in excess of 6 metric tons. An enhanced upper stage, called the Breeze M, has flown successfully and is now standard for Proton's commercial satellite missions (see Fig. 4.7).

To assist with the political and marketing tasks associated with bringing the Russian state organizations into commercial practice, Khrunichev sought Lockheed Corporation as a partner. This successful partnership, established in 1993, evolved into International Launch Services after the Lockheed merger with Martin Marietta. It is certainly one of the most successful Russian–American joint ventures in any area of commerce between the two countries. ILS also acquired the marketing and mission management rights to the Angara family of launch vehicles, now being developed by Khrunichev for flights beginning in 2005.

The political arrangements for the commercial use of Russian and Ukrainian launch vehicles were made pursuant to intergovernmental agreements between the United States and the Russian Federation, the United States and Ukraine, and the European Union and the Russian Federation. These agreements originally placed a limit on the number of launches that U.S. or European operators could purchase and at what prices. Due to the increased demand for commercial launch

FIG. 4.7. A spectacular nighttime launch of a Proton launch vehicle from the Baikonur cosmodrome.

vehicles, these limits or quotas have been raised repeatedly. Finally in the year 2000, such market-restricting limits were eliminated altogether for the Zenit and Proton vehicles.

THE BOEING COMPANY AND ITS LAUNCH SERVICES

The McDonnell-Douglas Company developed the successful Delta II launch vehicles primarily to serve the lower end (i.e., the smaller) satellites to be launched into geostationary orbit. McDonnell-Douglas also began development of the Delta III launch vehicle after receiving interest from Hughes Electronics to serve as an anchor customer for a block purchase of launch vehicles in 1995. The Delta III production has since been halted by Boeing due to the lack of demand for satellites in its class and a succession of launch failures that created concerns by customers as to the vehicle's reliability. When McDonnell-Douglas and Boeing merged, the combined launch vehicle businesses included the Delta II, Delta III, Delta IV, and Sea Launch consortium described later. The Delta IV made its first flight in 2002 and, along with Atlas V, is one of the two vehicles in the Air Force's EELV program. In mid-2003, Boeing announced that the Delta IV would not be marketed to commercial customers until at least 2008.

Together with a Norwegian ship-building company, Kværner, a Russian state company, Energia, and the Ukrainian state company, NPO Yuzhnoye, Boeing has made a substantial investment in using the two-stage Zenit launch vehicle with a modified Energia upper stage to produce a vehicle capable of launching nearly 6 tons to a geostationary transfer orbit when launched from the equator. The Sea Launch consortium, as it has become known, constructed a launching platform on a converted off-shore oil drilling platform off the coast of Christmas Island in the central Pacific at or near the equator. The platform and its associated control ship are based in Long Beach harbor near Los Angeles, California.

China

Beginning with the development of ballistic missiles and a small satellite launch capability, the People's Republic of China began to offer the Long March launch vehicles to commercial customers in 1988, attracting them with comparatively low prices. The U.S. government sought an agreement with the Chinese in 1989 to impose limits on the number of launches and to attempt to raise its prices closer to other Western launch vehicles. American sanctions against China from 1989 to 1991 also prevented a number of Chinese commercial launches by the U.S. government's refusal of export licenses to China by firms such as Hughes Electronics and Loral, which built its satellites in the United States and were therefore subject to U.S. export control laws.

A series of launch failures in China with commercial customers also dampened some of the enthusiasm for the use of these vehicles, and therefore the Long March launch vehicles are not actively offered in the commercial marketplace today. Nearly all commercial satellites have some U.S. components and are thus subject to prohibitions by the U.S. government to launch on Chinese systems. As European satellite manufacturers become independent of using U.S. components, there may be renewed interest by some commercial customers in using the Long March vehicles.

Japan

The Japanese H-2 rocket has given Japan a limited launch commercial capability, although the vehicles developed under this program are now almost exclusively used for governmental launches for the Japanese space program. Its launching facility at Tanegeshima is not usable for extended periods of the year, and therefore the number of launches is quite limited. Japan recently curtailed further development of the H-2A heavy lift vehicle and is now studying alternatives for its national launch capabilities, including a cooperative development with Lockheed Martin for the development of an intermediate launch vehicle.

India

As an outgrowth of its development of a wide variety of missiles and research rockets, the Indian Space Research Organisation (ISRO) developed a series of launch vehicles with increasing capabilities, now being able to place payloads into geostationary transfer orbits up to 2 tons. The first flight of this vehicle, known as the GSLV-D1, had its successful inaugural launch in October 2001. Additional GSLV configurations are being planned and tested.

THE CURRENT LAUNCH VEHICLE MARKET

Clearly the launch vehicle market is driven by the demand for orbital satellites. The burgeoning demand for launch capacity has not, however, been met with corresponding increases in demand. The launch vehicle market today is characterized by oversupply and falling prices. Commercial launches of intermediate and heavy payloads to geostationary orbits number about 12 to 15 a year, beginning in 2002, down from about 18 to 20 in previous years. The desire to maintain market share has further driven down prices and increased competition, especially between ILS and Arianespace—the two companies that now account for most all the commercial satellite launch business.

LAUNCH SITES

The choice of a launch site is determined by a number of factors, including proximity to the equator for launches into the geostationary arc. The closer the launch site is to the equator, the more efficient will be the launches of satellites for geostationary communications. A launch that takes place directly along the equator can have the advantage of both less inclination in its orbit and less absolute distance from the launch site to a suitable geostationary transfer orbit. These advantages can also be attained by launching a higher performing launch vehicle from a higher latitude. When the satellite is separated from the launch vehicle in a geostationary transfer orbit, inclination must be removed to place the satellite into the proper orbit by using the satellite's own propulsion system. Inclination can also be removed by increasing the performance of the launch, employing a more powerful launch vehicle to overcome or at least match the performance of a comparable launch vehicle at or near the equator. Thus, launch vehicles at sites of increasing distance from the equator tend to require more performance to place an equivalent payload in the same orbit; or, conversely, the payload needs to carry more fuel to achieve the same performance. With the wide range of performance available on many launch systems today, proximity of the launch site to the equator has not reaped major advantages for any one supplier.

Launch sites are also chosen with other geographic constraints in mind. The most important of these is the safety aspect of launching at or near a coastline so that the launch vehicle will drop its expended stages and strap-on motors over water. In the event of a launch failure, the vehicle can be destroyed over water as well, minimizing the risks to life and property. Thus, a number of major launch sites around the world are located near oceans, such as Cape Canaveral, Florida

(23° N; see Fig. 4.8), Tanegashima (30.4° N), and Kourou (5° N). The major GSLV launch site of India is located on the eastern coast of India at Sriharakota (13.7° N). For the Sea Launch system, the platform and command vessel are positioned in the central Pacific Ocean at or near the equator.

Other countries are less fortunate by reason of their geography and must design their launch vehicles to perform at higher latitudes. For launches destined for geostationary orbits, the Russian Federation uses a launch complex known as Baikonur. This site, once inside the USSR, is now in northwestern Kazakhstan. This base was originally sited there because of its remoteness and more southerly location within the former USSR. The base is now operated under an agreement between the Russian and Kazakh governments, which allows the use of Baikonur for Russian governmental and commercial launches. Launch vehicles expend their stages along certain established drop zones in the deserts east of the launch site away from populated areas. Efficiencies in the consumption of fuel also minimize the chances of unexpended fuel being deposited on the ground, especially when utilizing the Breeze M upper stage on the Proton.

Launch sites support a number of important activities that lead up to the launch event. Final assembly of each launch vehicle and its integration with the satellite payload takes place at the launch site. Each satellite payload is also processed at the launch site where final electronic checking is performed and the satellite is fueled, mated to the separation system of the launch vehicle, and then encapsulated inside the launch vehicle's fairing.

The launch site also supports the launch operations activities. Here a team of engineers and technicians from the satellite operator, the launch vehicle supplier (and its subcontractors), the satellite manufacturer, and the ground support organizations combine capabilities and specialties at a launch operations center. Thus, the launch site must not only be in an appropriate location in terms of safety and launch speed and direction, it must also be able to support a fairly large number of people in the launch crew and be easily accessible to carry out technical operations during a launch campaign (see Fig. 4.9).

HOW SATELLITE OPERATORS CHOOSE LAUNCH SUPPLIERS

No discussion of the role of launch vehicles in satellite communications would be complete without examining how commercial satellite companies choose their launchers. Some smaller operators who do not have the expertise or desire to manage a launch vehicle contract will choose to order their satellites delivered in orbit from the satellite manufacturer. Some satellite manufacturers keep block purchases of launch services available for themselves to serve deliver-in-orbit (DIO) circumstances. Others contract individually for each launch as the need arises.

Most satellite operators purchase their satellites in a mode known as on-ground delivery (OGD). Under this contracting scheme, the satellite manufacturer deliv-

FIG. 4.8.   Aerial view of Atlas Launch Complex 41 at Cape Canaveral, with an Atlas V Vehicle on the pad. Photograph courtesy of Lockheed Martin Corporation.

FIG. 4.9. A modern, state-of-the-art launch operations center at Cape Canaveral, Florida.

ers the satellite to the selected launch site for launch on a vehicle chosen independently by the owner of the satellite. Insurance for the launch event as well as for on-orbit performance of the satellite is also purchased independently.

A number of factors are at play in the selection of a launch service by a satellite operator. Clearly the satellite and launch vehicle must be technically compatible. Standards for performance, launch environments, and interfaces are established between the launch suppliers and satellite manufacturers in the form of compatibility agreements. Most satellite operators want to maintain some flexibility in choosing a launch supplier, and therefore they specify that their satellites must be compatible with more than one launcher.

Concerning the choice of suppliers, first and most important is the reliability of the launch vehicle. Reliability is measured by the vehicle's launch success record. As an example, the Atlas family of launch vehicles has the highest overall reliability for vehicles presently flying, at 100% over 68 consecutive successful launches as of mid-year 2003. Satellite operators sometimes establish thresholds of reliability for their launch suppliers in the form of absolute percentages of successful launches, which are typically at 90% or higher. Some satellite operators specify that the launch system must have two or more prior successful launches to be qualified to carry its next payload. Launch suppliers with vehicles that are either making a debut or returning to flight after a failure sometimes offer price discounts or other financial incentives to encourage commercial customers. Less fortunate launch systems have resorted to test flights to prove reliability, although test flights are obviously more expensive for the supplier and do not demonstrate the utility of the launch vehicle as well as a *bona fide* satellite payload and a paying customer. Although launch failures can be mitigated somewhat by purchasing insurance, most satellite operators do not want to risk further business interruption by having to procure a replacement satellite and launch—a process that might stretch to 2 years or more. Reliable launch vehicles are clearly the most important factor to the commercial operator. Reliability also affects the insurance rates the operator will pay to cover the costs of the launch and of the satellite in the event of a launch failure.

A second important dimension is schedule assurance. Delays in payload processing, especially for launch systems that require two payloads to be launched at once, can cause operators to face unanticipated opportunity costs in delaying a launch. Having an available vehicle and being launched on time are clearly major concerns for a commercial operator who has to have its assets in orbit to make money.

Satellite operators also pay attention to terms and conditions with a launch supplier's contract. These factors may include delay rights, penalties, termination rights under certain conditions, and so on. The overall value of the launch service, reflected in price, payment terms, credit, insurance, and other financial terms, clearly affect the choice of a launch service as well.

Finally, because satellites and launchers are coupled to their political environments, political concerns can sometimes influence the choice of a launch system.

LAUNCH VEHICLES AND SATELLITE TECHNOLOGIES

The expansion of global telecommunications services is fueled by a powerful combination of technological advances and deregulation. Once under the exclusive control of governments, many telecommunications organizations that were formerly monopoly operators are now subject to competition and privatization. New international agreements are also codifying trade principles into telecommunications transactions among nations. These agreements, under the banner of the World Trade Organization (WTO), are steadily creating competitive markets for a wide variety of wireless services, including space-based systems. The International Telecommunication Union (ITU), which is still responsible for allocating radio frequencies and establishing standards for interconnection, is now reinforcing these trends toward open markets and global competition just as the ITU has been challenged by competitive regional regulatory bodies. These trends are discussed in greater detail in a later chapter on changing the institutional structure of satellite communications.

The launch vehicle industry has exploited these trends toward deregulation and competition by promoting enabling technologies and in some cases offering themselves to invest in the operations of many new competitive telecommunication satellite systems. The rapid increase in digital applications, such as electronic commerce made possible by advanced data networks such as the World Wide Web, has demonstrated a shortfall in aggregate supply of network facilities and services. Although established telecommunication operators are moving to fill this gap, it is expensive and time-consuming to provide compatible facilities to households and small enterprises exclusively by terrestrial telecommunications networks. Hence, new satellite systems are proposed to take advantage of space communications' inherent economies—mobility of its users, insensitivity to distance between service points, and wireless connections to remote locations. These trends were addressed elsewhere, especially in chapter 5.

THE FUTURE OF THE LAUNCH VEHICLE INDUSTRY

As the industry continues to evolve, questions will arise about the needed investments to further the development of new launch systems, including RLVs. There are currently experimental programs underway to develop new commercially oriented RLVs in Europe by the ESA, and several NASA-funded projects that have recently been curtailed due to budgetary constraints as well as several independent projects funded privately by aerospace corporations. There is also the ongoing X-Prize initiative that offers a $10 million prize to the first aerospace organization that can develop and fly a small manned crew to LEO (above 60 miles), land, and then repeat the exercise again within 2 weeks. The X-Prize is as much oriented toward space tourism as it is toward providing low-cost access to orbit, but the results in terms of a cost-effective space plane would presumably be able to perform multiple tasks.

Beyond space plane technology, there are others working on new or improved means to achieve orbit or improved propulsion systems once LEO is achieved. These include electronic or ion propulsion systems, mass-driver systems, rail-gun systems, improved fuels (such as metallic hydrogen), solar sail systems, nuclear-powered (fission or fusion) systems, and a variety of tether-based technologies to achieve more reliable, powerful, or cost-effective ways to achieve access to space. In the following, there are a number of references that provide further information on these systems. Within the next 30 years, dramatic advances in launch vehicle technologies could lower the cost of space access by an order of magnitude, allowing greater applications of satellite-based technologies. Until that time, satellite operators will employ ELVs to meet their space access requirements.

REFERENCES

Baumgardt, Carola, (1951), *Johannes Kepler: Life and Letters*, New York: Philosophical Library.
Servan-Schreiber, Jacques, (1968), *The American Challenge*, New York: New American Library.

FURTHER READINGS ON LAUNCH VEHICLES

Giget, Marc, (2002, May), "Long-Term Trends in Demand for Launch Services," Paris: Euroconsult.

Goldman, Nathan, (1992), *Space Policy: An Introduction*, Ames, IA: Iowa State University Press.

Gouré, Daniel, and Szara, Christopher, (Editors), (1997), *Air and Space Power in the New Millennium*, Washington, DC: Center for Strategic and International Studies.

Hall, Rex, and Shayler, David J., (2001), *The Rocket Men*, Chichester, UK: Praxis.

Handberg, Roger, (1995), *The Future of the Space Industry: Private Enterprise and Public Policy*, Westport, CT: Quorum Books.

Harvey, Brian, (1996), *The New Russian Space Program: From Competition to Collaboration*, New York: Wiley.

Hickman, John, (1999, November), "The Political Economy of Large Space Projects," *Journal of Evolution and Technology, 4*, 1–14.

Isakowitz, Steven J., (1999), *International Reference Guide to Space Launch Systems* (3rd ed.), Washington, DC: American Institute of Aeronautics and Astronautics.

Johnson-Freese, Joan, (1998), *The Chinese Space Program: Mystery within a Maze*, Malabar, FL: Krieger.

Kay, W. D., (1995), *Can Democracies Fly in Space? The Challenge of Revitalizing the U.S. Space Program*, Westport, CT: Praeger.

Kiselev, Anatoly I., Medvedev, Alexander A., and Menshikov, V. A., (2003), *Astronautics*, New York: Springer-Verlag.

Lambakis, Steven, (2001), *On the Edge of Earth: The Future of American Space Power*, Lexington, KY: University of Kentucky Press.

McDougall, Walter A., (1985), *The Heavens and the Earth: A Political History of the Space Age*, New York: Basic Books.

Oberg, James E., (1999), *Space Power Theory*, Washington, DC: U.S. Government Printing Office.

Reeves, Robert, (1994), *The Superpower Space Race: An Explosive Rivalry through the Solar System*, New York: Plenum.

Saxer, Robert K., Knauf, J. M., Drake, L. R., and Portanova, P. L., (2002, March–April), "Evolved Expendable Launch Vehicle System: The Next Step in Affordable Space Transportation," *Program Manager* (Defense Systems Management College), pp. 2–15.

Vaughan, Diane, (1996), *The Challenger Launch Decision*, Chicago: University of Chicago Press.

Von Bencke, Matthew J., (1996), *The Politics of Space: A History of U.S.—Soviet/Russian Competition and Cooperation in Space*, Boulder, CO: Westview Press.

HISTORY AND POLITICS

CHAPTER FIVE

The Geopolitics and Institutions
of Satellite Communications

Robert J. Oslund
The George Washington University

> *The development of international organizations has been, in the main, a re-*
> *sponse to the need arising from international intercourse rather than the*
> *philosophical appeal of the notion of world government.*
> —D. W. Bowett (1963)

Communications satellite technology became an enabler of globalism four decades ago when technology made possible the establishment of the first truly multinational global commercial entity of any sort—Intelsat—by bringing together, around the same table, countries of different political persuasions, economic systems, cultures, and levels of development to own and operate a worldwide satellite system. Other competing global, regional, and domestic communications satellite systems have since joined Intelsat as enablers of the globalization process by opening up information links to virtually every national capital and market in the world (see Nye & Donahue, 2000).

Satellite communications have been—and remain—a powerful force in reshaping and integrating our world in new and unforeseen ways. In turn, these have—and are—requiring a redefinition of the concepts of global trade, global competition, global power, and global governance along the way.

To understand just how powerful of a force the tetherless technology has become, one need only turn back the calendar one century when imperialism, the antithesis of globalism, was the international norm, and the British were nearing their apogee as an imperialist power. The "Union Jack" was prominent in a worldwide

The views expressed herein, unless indicated otherwise, are solely the views of the author.

chain of colonies, and the "Brits" were finalizing the underwater deployment of "Big Red." This submarine telegraph cable, which linked the western coast of Canada with Australia, gave London the capability to communicate with the "down under colony" in a westerly direction via a transatlantic submarine cable, an overland trans-Canadian telegraph wire, and "Big Red," and in an easterly direction through submarine cables under the Mediterranean Sea and the Suez Canal to India and from India through the Straits of Malacca to Australia. The estimated time to go halfway around the world point by point was measured in minutes.

Today, in stark contrast, nations, including former colonies, can instantaneously communicate via telephone, data, videoconferencing, or Internet with one or many of the 220-plus nations and possessions accessing Intelsat, Eutelsat, PanAmSat, New Skies, Loral Cyberstar, Astra, or other satellites hovering above the Atlantic, Pacific, and Indian Oceans and the land masses that separate them. If so inclined, any one nation can send itself a message around the world by "triple hopping" signals through satellites spanning the three ocean regions in under 10 seconds.

This global electronic leap from point-to-point communications in minutes to point-to-multipoint communications in seconds is the result of succeeding innovative stages in transoceanic telecommunications technologies over a century and a half—submarine telegraph cable, high-frequency radio, submarine telephone cable, geosynchronous communications satellites, submarine fiber optic cables, and "LEO" satellites. As is shown, their time and mode of commercial introduction, for the most part on a global system scale, coincidentally, correlate with changes that were taking place in the geopolitical milieu (see Table 5.1).

The evolution of these transoceanic telecommunications technologies is analyzed in terms of "innovation cycles" that were affected by the geopolitical milieu and the interaction of the forces of technology, industry, and government at the national and international levels (see Frieden, 1996, 2001; Galloway, 1972; Hudson, 1990; Kinsley, 1976; Maddox, 1972; Oslund, 1975, 1977, 1984, 1990; Pelton, 1974). Sometimes these forces converged, sometimes they collided, and sometimes they diverged. Their overlap ebbed and flowed as national economic, defense, foreign, national security, trade, industrial, space, and regulatory poli-

TABLE 5.1
Correlation of Geopolitical Milieu and Technologies

Geopolitical	Technology	Operational Scope
Post-Civil War to WWI	Submarine telegraph cable	Global and regional
Post-WWI to WWII	HF radio	Global and regional
Post-World War II to Cold War (WWIII)	Submarine telephone cable	Regional
Cold War (WWIII) (Post-"Sputnik")	Satellites (GEO)	Global and regional
	Fiber optic cable	Global and regional
	Satellites (LEO)	Global
September 11, 2001 (WWIV)		

cies changed, usually incrementally, but on occasion abruptly, as was the case in the United States with communications satellites. These innovative cycles also affirm Kennedy's (see Kennedy, 1987) contention that there is no one single force or determinant in international relations. The reactive and proactive policies and regulations of the United States that paralleled these developments are discussed in detail in chapter 6.

INNOVATION CYCLE OF SUBMARINE TELEGRAPH CABLE

Submarine telegraph cable was commercially introduced between the end of the American Civil War and World War I. Internationally, this was the age of imperialistic expansion by the foreign powers. In contrast, the initial focus of the United States was domestic, although it did change dramatically toward the end of the century with imperialistic forays into the Caribbean and the Pacific basins.

Geopolitical Dimension

This age of imperialism was marked by maneuvering for colonial expansion by Britain, the European powers (namely, France, Portugal, Spain, Belgium, Italy, and Germany), and also Japan in varying degrees around the world. At the outbreak of WWI, 79% of the world's land surface was under colonial control (see Palmowski, 1997). In the years immediately following the Civil War, U.S. focus was inward as the nation sought to recover from the "War Between the States," continued to pursue its "Manifest Destiny," and embarked on coast-to-coast industrialization. Toward the end of the 19th century, however, influenced in part by Captain A. T. Mahon's twin theories of naval and world power, American strategists shifted their focus to the Caribbean, mainly Cuba and Puerto Rico, and to the Pacific, as far away as Guam and the Philippines, as evidenced by the Spanish–American War.

Technology

Undersea telegraph cable evolved from overland telegraph. Its obvious strength was the capability to electronically transmit information between continents. The initial technical drawbacks of the loss of signal strength over long distances and of cumbersome ways to protect the copper wires from a hostile ocean floor environment were eventually overcome. Deployment of submarine cables by cable ships proved to be a continuing complex and costly undertaking. The British completion of the "Big Red" took U.S. strategists by surprise because the length of the cable exceeded what was considered then to be technically feasible. Submarine telegraph systems operated as a primary mode of communications through the 1950s before being abandoned in favor of more efficient submarine telephone cables.

Interaction Among Technology, Industry, and Government

Domestic dichotomies in ownership of telecommunications in the United States, Europe, and Great Britain appeared during the submarine telegraph era. These dichotomies influenced the domestic and international development of telecommunications technologies for decades thereafter.

Ownership Dichotomies. The principle of private ownership of telecommunications in the United States was established in the 1840s. The Post Office turned down Samuel Morse's offer to sell the patent rights to his telegraph invention, after which he turned to the private sector to finance and commercially operate the invention (see Herring & Gross, 1936). Western Union Telegraph Company eventually emerged as the domestic telegraph monopoly and expanded its telegraph business internationally, including leasing circuits in British submarine cable systems. One other domestic telegraph system appeared—the Postal Telegraph Company—and formed an international extension—the Commercial Cable Company. A handful of other American companies owned and operated their own international submarine cable systems. Later when Alexander Graham Bell invented the telephone, government ownership was never seriously considered. AT&T was incorporated in 1884 around this technology and went on to become a national and international monopoly.

In Europe, governments owned and operated all communications—domestic and international—from almost the beginning. Telegraph and later telephone systems were merged with the Post Offices to establish post, telegraph, and telephone (PTT) government entities, in which the foreign governments became one-stop owners, operators, and regulators.

In Britain, ownership was in the hands of industry initially, but by 1911 the domestic telegraph and telephone industry was completely nationalized whereupon the British Post Office (BPO) became a PTT. The worldwide British submarine telegraph cable systems also were industry owned and operated in the beginning. Although Cyrus Field, an American entrepreneur, laid the first successful transatlantic submarine telegraph cable in 1866, the undertaking was funded primarily by British capital. A number of privately owned British cable companies subsequently were formed, with political encouragement from the government, to launch a cable offensive worldwide. As stated earlier, at the turn of the century, they managed to establish the first global transoceanic telecommunications system, with the completion of "Big Red," although its construction required some government subsidization. The European colonial powers built less extensive cable systems to link up with their colonies.

Cable Collisions.   The British won the cable race with the United States in the Pacific basin when "Big Red" was completed. As early as 1873, Congress had provided funds for the Secretary of the Navy to take scientific soundings in the Pacific Ocean in anticipation of an eventual transpacific cable. Greater urgency was attached to the planned project when British plans for a transpacific cable became known. Yet it was not until the same year that the British finished circling the globe with submarine telegraph cables that the Commercial Cable Company was given the go ahead to lay a U.S.-owned trans-Pacific cable, which was completed in 1904 (see Herring & Gross, 1936). In the latter stages of World War I, Western Union, at the private request of the Wilson Administration, sought to establish an independent cable link to Latin America to offset reliance on British cables for transmitting sensitive commercial, diplomatic, and military communications. However, when blocked by a British interport monopoly (i.e., an exclusive multiyear agreement signed with Brazil), Western Union opted to lease circuits from British submarine cable operators (see U.S. Congress, 1921, *Hearings on Cable Landing Licenses*). When the company sought to construct a cable between Bermuda and Florida to link the United States with a British cable running northward from Latin America, President Wilson intervened, turned down the company's request to land the cable in Florida, and dispatched a U.S. Naval vessel with orders to stop the British cable ship that was starting to lay the cable with force if necessary. Fortunately, the only force that proved necessary were a few shots fired by the U.S. Navy across the bow of the British cable ship (see Bamford, 1983). Western Union mounted a legal challenge to this presidential action, which prompted Congress to pass the *Cable Landing License Act of 1921* to protect this executive privilege.

International Institutions

The international regulatory framework for telegraph and submarine telegraph cables was European in origin. European nations sharing common borders formed cross-border telegraph agreements bilaterally in the 1850s, then regionally, and, eventually, intracontinentally. When membership was opened up to other regions in 1865, the first intergovernmental organization (IGO)—the International Telegraph Union (ITU)—came into being and became one of many IGOs that would subsequently provide the international regulatory infrastructure for telecommunications (see Table 5.2). Because government ownership of telegraph systems was a criterion for membership, the British initially were ineligible to join; the United States never did qualify for membership (see Clark, 1931). In the 1880s, France hosted two international meetings to address the maritime regulatory regime for submarine cables that resulted in *A Convention on the Protection of Submarine Cables* in 1884, which the United States signed. The treaty's basic concepts withstood the test of time and were incorporated in the current *Convention on the Law of the Sea.*

TABLE 5.2
The Rise of the International Regulatory Structure

Technology	Year	Organization
	1865	International Telegraph Union
International submarine telegraph cable		
	1874	Universal Postal Union
	1906	International Radiotelegraph Union
	1919	League of Nations
HF radio		
	1934	International Telecommunication Union
	1944	The World Bank
		The International Monetary Fund
	1945	United Nations
	1946	UNESCO
	1947	International Civil Aviation Organization
	1948	International Maritime Organization
International submarine telephone cable		
	1959	UN/COPOUS
Geosynchronous satellites		
	1970	World Intellectual Property Organization
Fiber optic cable		
	1994	World Trade Organization
LEO satellites		

Note. From *Basic Facts About the United Nations* (1995) and IGO's Web sites.

The Universal Postal Union was established in 1874. It is cited only to demonstrate the long-standing importance that the international community has attached to the concepts of *universal service* and the *international flow of information*.

INNOVATION CYCLE OF HIGH-FREQUENCY RADIO

Transoceanic high-frequency (HF) radio was commercially introduced after the Paris Peace Conference that negotiated the end of World War I. It remained the preferred mode of transoceanic telecommunications, especially after the introduction of radio telephone, up to and through World War II and the decade thereafter. During the latter stages of the Peace Conference, the Wilson administration moved decisively to win the "HF race" with the British that was part of the president's postwar strategic vision to expand U.S. presence in the international community.

Geopolitical Dimension

The Peace Conference, convened in Versailles to negotiate a treaty to end World War I, also drew the blueprint for the League of Nations, the first attempt at world government. Unfortunately, the League never achieved its lofty goals. This was due, in large part, to the United States' sudden, unexpected postwar retreat into a period of isolationism and steadfast refusal to become a member of the new international organization that President Wilson had championed. From then until the months immediately preceding the outbreak of World War II, the League devolved into a debate-prone, ineffective entity unable to prevent acts of aggression by the Nazi and Fascist regimes in Europe and the Pacific Basin. When the major Western democracies finally decided to act collectively against the totalitarian governments, it was too late. The outbreak of World War II signaled the end of the first attempt at world government.

Technology

HF radio's strengths were its capabilities to transmit information through the ether to multiple points, thus making HF radio systems easier and cheaper to construct and operate than submarine telegraph cable systems. Its weaknesses were loss of signal strength initially, reliance on limited portions of the radio spectrum, and susceptibility to atmospheric disturbances that caused disruptions in service. The Alexanderson alternator, named after its developer who worked for General Electric, was unveiled prior to World War I with the potential to dramatically increase radio transmission over long distances.

Prior to the turn of the century, Gugliemo Marconi had transformed wireline telegraphy into wireless telegraphy to build what became a powerful international maritime radio monopoly that provided short-range radiotelegraph services to the maritime shipping community. In 1901, he launched the era of transoceanic HF radio when he transmitted the first radiotelegraph message—"S"—across the North Atlantic. International radiotelephone service was commercially introduced in 1926.

Interaction Among Technology, Industry, and Government

The era of submarine telegraph cable ended with President Wilson's overt intervention to prevent a British cable ship from landing a cable in Florida, and the era of HF radio began with his overt intervention to prevent Marconi from acquiring the GE-owned HF radio technology.

Marconi had initiated discussions with GE in 1912 to acquire exclusive rights to the Alexanderson alternator, only to have them interrupted by the outbreak of World War I. He sought to reopen discussions in 1919 while President Wilson was attending the Peace Conference. Upon informally being advised by GE executives of Marconi's renewed efforts, the president immediately dispatched his telecommunications adviser, Admiral William Bullard, the Navy's Director of Communications, to meet with GE senior management in New York City at the corporate headquarters. In his discussions with GE's senior executives, the Admiral referred to Wilson's ambitious international intentions, including both the critical role that international communications would play in post-WWI politics and the need to ensure that the British would not be allowed to dominate international communications in radio as they had done in submarine telegraph cables. It was agreed that the objective of establishing American leadership would be an extremely complex one—"the whole picture puzzle had to be put together as a whole in order to get an effective national instrument." That chosen instrument was established in 1919 as the Radio Corporation of America (RCA), overseen by a Board of Directors that included an observer from the Department of the Navy. RCA then pursued and attained a dominant position in international communications (see U.S. Congress, 1929, *Hearings on Commission on Communications*; Archer, 1938).

In London, an undaunted Marconi plunged ahead. In 1919, he proposed to build a second worldwide British telecommunications system using HF radio technology, but was unable to complete it for 10 years due to vacillation by the British government and opposition by the colonies. The delayed date notwithstanding, the completion of the worldwide radio system marked the second global telecommunications system that would be built, owned, and operated primarily by British industry. The 1929 Imperial Telegraph Act established Cable and Wireless (C&W) as a holding company of common stocks for existing British cables and of a major share of Marconi's HF company. C&W was finally nationalized after World War II with a charter to operate Britain's international telegraph service, as well as most of the other telecommunications services in its overseas possessions. The colonial European nations built more modest radio links with their colonies.

International Institutions

Two Berlin international radio conferences were convened in 1903 and 1906 for the stated purpose of considering how to regulate maritime radio communications, but also for the unstated purpose of ending the Marconi maritime monopoly. Initial spectrum allocations for maritime radio were agreed on, and a convention was produced by the end of the 1906 Conference that led to the creation of the International Radiotelegraph Union (IRU). The United States participated in both conferences, but did not sign the international agreement primarily because it had the effect of perpetuating the Marconi monopoly by not requiring interconnection between maritime radio systems. The Senate Foreign Relations Committee balked at recommending U.S. ratification of the convention until it was made a prerequisite for participation in a planned London Radio Conference in 1912. Tragically, the conference opened within weeks after the Titanic disaster. Not surprisingly, the British government wisely announced at the beginning of the conference that it would accept a policy of mandatory interconnection without reservation, thereby weakening the Marconi monopoly. A follow-on Washington Conference made further radio allocations and established a Consultative Committee on Radio (CCIR), similar to the Consultative Committees for Telegraph (CCIT) and for Telephone (CCIF) that had been established in the ITU.

The Madrid Conference in 1932 agreed to merge the ITU and IRU. In 1934, the new ITU—the International Telecommunication Union—entered into existence, coincidental with the enactment of the *Communications Act of 1934* in the United States, which merged the wireline regulatory powers of the Interstate Commerce Commission and the wireless regulatory powers of the Federal Radio Commission.

INNOVATION CYCLE OF SUBMARINE TELEPHONE CABLE

Commercial service on submarine telephone cables began with relatively little fanfare in 1956, although the multinational ownership and operation of the new transoceanic telecommunications technology was a significant departure from the previous two eras. The reason for this less than dramatic debut was the cold war, which some have called World War III (see Cohen, 2001).

Geopolitical Dimension

Within months after the founding of the United Nations, confrontations between communist nations and the Western democracies began to mold a bipolarized world that would dictate the *modus operandi* for decades to follow. By 1949, the Soviet Union succeeded in dropping an Iron Curtain to split the European Continent, and Mao Tse Tung finished wresting control of the Chinese Mainland. The West responded with a policy of containment, marked by the formation of a series of regional security organizations, the most significant and enduring being the North Atlantic Treaty Organization (NATO). The UN also sent troops to South Korea to repel communist North Korea's invasion. Yet these developments had the salutary effect of dampening the telecommunications rivalries among the United States, Britain, and Europe that had existed prior to World War II.

Technology

AT&T's developmental work on submarine telephone cable had begun during World War I, and, in 1921 the first "great talking cable" was deployed between Florida and Cuba. Thirty-five years elapsed before the first transoceanic cable— TAT 1—linked the United States and United Kingdom. It was made possible by technical solutions that AT&T and BPO engineers, working independently but in parallel, had devised to overcome two major technical barriers blocking this new costlier, but more efficient, technology. First, underwater repeaters were developed to nullify the loss of voice signal strength and make broader bandwidth available for greater capacity. Second, cheaper lightweight cables were designed to replace the older, bulkier submarine telegraph cables. Also the new cables were not susceptible to ionospheric storms that were seriously disrupting the HF systems (see Brockbank, 1958).

Interaction Among Technology, Industry, and Government

The bipolar geopolitical realignment had the unintended effect of dampening the highly nationalistic competition that had dominated the earlier eras of submarine telegraph cable and HF radio. One by-product was the introduction of a period of normalization in the conduct of commercial relations between the American telecommunications corporations and their PTT counterparts in the North Atlantic, Latin America, and Pacific basin.

This normalization was reflected in the introduction of multinational ownership and the shared use of telecommunications facilities by regional consortia of individual cable systems in contrast to the single country- or single company-owned systems of the submarine telegraph cable and HF radio systems. In the case of TAT 1, the ownership arrangement was comprised initially of AT&T and the BPO, although Canada subsequently purchased minority interest on the North American side. Later cables were built by larger consortia as more European administrations became investors. Cable ownership and maintenance began and ended at the halfway point between cable landing points, with each investor owning outgoing and incoming half circuits. The primacy of the North Atlantic strategic and economic interests was reflected in the deployment of this new technology: TATs 1, 2, and 3 were introduced before the first transpacific cable, Transpac 1, was laid in 1964. Involvement by the U.S. government in this innovative cycle was minimal due to the policymakers' preoccupation with the uncertainties of global politics and, to a lesser extent, the normalization in commercial relations that was taking place.

In 1959, the British convened a Commonwealth Planning Conference to plan a third global communications system, this time employing submarine telephone cable technology. However, political unrest between and among former British colonies comprising the Commonwealth voided realization of the world-girdling undersea project. Interestingly, a technological assessment was made at the time that projected commercial deployment of communications satellites at least 10 to 15 years in the future (see Oslund, 1975, 1977).

International Institutions

During the closing months of World War II and the 4 years thereafter, a handful of IGOs were established to expand the international regulatory framework and, in so doing, facilitate postwar recovery worldwide in the short term, and, implicitly, the globalization process in the longer term.

The International Bank for Reconstruction and Development (IBRD), also called the World Bank, and the International Monetary Fund (IMF) were established at the Bretton Woods Conference in 1944 to aid in postwar reconstruction and to reorganize and restore international trade. In 1946, the UN Educational, Scientific, and Cultural Organization (UNESCO) entered into being. Unfortunately, the vagueness of its charter eventually resulted in ideological collisions between the Western industrialized nations, on the one hand, and the Soviet bloc and nonindustrialized "Third World," on the other, and the consequential withdrawal from membership by the United States in 1984 for nearly two decades. Conventions establishing the International Civil Aviation Organization (ICAO) and Intergovernmental Maritime Consultative Organization (later renamed the International Maritime Organization [IMO]) were negotiated in 1947 and 1948, respectively. Primary purposes of the latter two IGOs were, among other things, to encourage safety of life, facilitate development of navigational technologies, and promote efficient and economic transportation systems. Subsequent agendas of all of these IGOs would include telecommunications-related issues in some form.

Three ITU-related conferences were held in Atlantic City in 1947 to address telecommunications, HF radio, and broader radio issues. In the end, a new ITU structure had been designed that functioned adequately, with some periodic changes, for almost four decades (see Codding & Rutkowski, 1982; Leive, 1970). All of these IGOs eventually became specialized agencies of the UN, which was established in 1945.

INNOVATION CYCLE OF COMMUNICATIONS SATELLITES: THE FIRST DECADE

Communications satellites were a by-product of the cold war. The launching of "Sputnik" in October 1957 heightened East–West political tensions by exporting Soviet–American rivalries into outer space. Communication satellite technology became a weapon in the ensuing space race as it was transformed from an evolutionary telecommunications technology into a more revolutionary tool of American foreign policy. It led to a government-driven acceleration of the development and commercial deployment of the technology and disrupted commercial relationships that were normalizing. It also forced policymakers and regulators to put into place just-in-time domestic and international regulatory regimes.

Geopolitical Dimension

The dramatic launching of Sputnik as the first artificial satellite transformed the dynamics of the cold war. Until then the Eisenhower administration had sought, with considerable success, to keep the growing political rivalries earth-borne. America's response to the Soviets the following year was twofold: The National Aeronautics and Space Administration (NASA) was created to oversee a massive civilian effort for the exploration and exploitation of space, and the Department of Defense (DOD) sped up its own space efforts, punctuated with the Air Force's launching of "Score," the first artificial communications satellite, 1 week before Christmas 1958.

Other geopolitical developments occurred in the late 1950s in the shadows of the cold war. The UN General Assembly established the Committee on the Peaceful Uses of Outer Space (COPUOS) to grapple with this new, unique technology. Regionally, the European Economic Community (EEC) came into being in an experiment to replace nationalism with regionalism in Western Europe. Nationally, a number of new nations started to emerge in their attempt to replace colonialism with nationalism.

Technology

Although the particulars of communications satellite technology are discussed in detail in chapters 2 and 3, certain historical factors need recalling for the purpose of putting this analysis into technological context. First, Sir Arthur Clarke transformed an idea into a concept in the mid-1940s, Dr. John Pierce of Bell Labs spoke of its commercial potential in the mid-1950s, and Early Bird was launched as the first commercial communications satellite in the mid-1960s. Second, the technology was initially projected to evolve from low- and medium-altitude systems (LEOs) to a geosynchronous GEO system. This path appeared to be confirmed in 1962 by the successful launching, by NASA, of the low-altitude elliptical orbiting Bell Labs'-funded Telstar in July and the NASA-funded low-orbit Relay 6 months later, and by the cancellation of DOD's costly and complex multipurpose GEO satellite program, Advent. Third, as discussed in chapter 2, the failure of Advent notwithstanding, a dramatic breakthrough occurred in February 1963 when NASA launched a sole source NASA-funded experimental satellite— Syncom—into GEO orbit. This satellite was built by Hughes Aircraft Company and employed a unique "spinner technology fix" to enable the spin-stabilized satellite to operate while continuously pointing its antenna toward earth. This capability was something that many had regarded as an almost insurmountable technical barrier, but it made Early Bird possible. Last, as discussed in chapter 4, the U.S. government was developing launch vehicle capabilities that could be made available to the military for launching missiles and, at cost, to the commercial sector for launching commercial satellites.

The contrasts between communications satellites and submarine telephone cables were striking. Satellites were multipoint, their scope of operations could cover almost an entire ocean region, and their capacity was markedly greater than that of cables—they could carry video. However, due to limited on-board fuel for station keeping and the lifetime of batteries, spacecraft had to be replaced every 7 to 10 years at the same satellite location regardless of whether their traffic-carrying capacity was being fully utilized. In contrast, submarine cables were point to point with terrestrial extensions at either end, their scope of operations was regional, and their capacity was limited—they could not carry video. Although they had an operational longevity of 25 years, regardless of the years of remaining service, whenever a cable's limited capacity was fully utilized, it was supplemented by another cable, thereby creating a new cable route unlike the case with satellites. These dissimilarities were not appreciated by policymakers at the time, but they eventually proved to be politically problematic.

Interaction Among Technology, Industry, and Government

Communications satellite technology was disruptive to an extreme. As discussed in greater detail in chapter 6, the appearance of the technology brought to the surface in the United States' partisan and ideological differences and slashed across stove-piped national policy sectors—economic, foreign, defense, industrial, national security, trade, space, and regulatory. Internationally, it upset the normalizing international telecommunications relations between the United States and its Western allies.

Policy Partisanship. The Eisenhower administration and the Republican-controlled Federal Communications Commission (FCC), in its regulatory proceedings, approached the proposed commercialization of communications satellites as an evolutionary international common carrier issue. Shortly before the 1960 presidential election, NASA Director Keith Glennan informally unveiled the administration's policy that endorsed industry ownership of communications satellites. On January 1, 1961, a few short weeks before the inauguration of President-elect Kennedy, the Eisenhower White House formally endorsed this statement as a national policy. In May, the FCC concluded that ownership in and operation of an international satellite system by the international common carriers would be in the public interest.

In the months preceding the presidential election, however, Senate Democrats, particularly Lyndon Johnson, Chairman of the Senate Space Committee, had cautioned against overlooking the foreign policy potential of communications satellites, but obviously to no avail. In May 1961, soon after the Soviets' successful orbiting of the first man around the earth and the politically damaging Bay of Pigs fiasco, President Kennedy responded by presenting his "man on the moon" national policy initiative to Congress. The initiative was based primarily on the recommendations of a senior interagency policy group, headed by now-Vice President Johnson, who also was Chairman of the Administration's Space Council (see Logsdon, 1969). Significantly, the initiative also called for "making the most" of growing American leadership in space by accelerating development of space satellites for worldwide communications. Almost overnight, communications satellites were transformed into an instrument of foreign policy, and the establishment of a global satellite system became a major foreign policy goal.

Creation of Comsat. The issue of how communications satellite technology would be projected internationally was debated in Congress from then until August 1962, when the *Communications Satellite Act* was passed to create the Communications Satellite Corporation (Comsat). Established amid controversy and compromise, which is discussed in chapter 6, one writer would call the new corporate entity "the inevitable anomaly" (see Kvan, 1966). Comsat was to be the Nation's chosen instrument, with the twofold mandate to establish "as expeditiously as practicable" a commercial global satellite system and to make satellite technology available to both developed and "economically less developed countries." Thus, four decades after the U.S. government had overseen the creation of RCA as "an effective national instrument" to challenge the British in international terrestrial communications, Comsat was established as the nation's chosen instrument to challenge the Soviets in space communications.

Establishment of Intelsat. The United States sought to open international negotiations to establish a yet-to-be-defined system through bilateral discussions with foreign telecommunications administrations, much along the lines of the approach used in planning submarine cables. Rather than AT&T negotiating on the U.S. "side," senior officials of the Department of State and senior executives of Comsat would be the negotiators. The Europeans delayed discussions until they were able to develop a regional position, and the UK then coordinated this position with its principal Commonwealth partners—namely, Canada and Australia. The Soviet Union declined to participate due to the obvious cold war nature of the U.S. space initiative.

Seven months of hard-nosed international negotiations produced two legal instruments that were signed in August 1964—an *Agreement Establishing Interim Arrangements for a Global Commercial Communications Satellite System*, a political document that was signed by the Parties (or governments); and a *Special Agreement*, a document that addressed commercial issues that was signed by the Signatories (or telecommunications entities designated by their respective governments). The critical word in these documents was *Interim* (i.e., the agreements would be subject to review and revision 5 years hence). The international consortium formally adopted Intelsat as its name in 1965 to become the first international satellite organization (ISO), the interim nature of the agreements notwithstanding.

Interestingly, the framework of the international satellite consortium that was agreed on roughly resembled a submarine cable-type structure that had been worked out by the Ad Hoc Committee of International Common Carriers, which included AT&T and the IRCs (e.g., RCA Global, ITT Worldcom, and Western Union Telegraph), as part of the earlier FCC proceedings on international communications satellites. The international carriers' concept called for the U.S. space segment to be owned by FCC-authorized international common carriers and the foreign space segment to be owned by the foreign PTTs. Ownership in the space segment would be based on projected utilization. Half-circuits between earth stations and satellites would be owned by the respective earth station owners: U.S. earth stations would be owned by FCC-authorized international carriers, and foreign earth stations would be owned by the foreign PTTs.

Departures from the cable consortium approach were dramatic with respect to the international satellite system that emerged. The ownership and operation of the system would be on a global, not a regional, basis. Comsat would own the U.S. portion of the satellite system and, based on its 60% projected utilization of the system, would be the dominant investor. Comsat would also be the system manager due to its operational and technical expertise. Decision making, ranging from personnel to investment to operational matters, would be delegated to an Interim Communications Satellite Committee (ICSC) comprised of the major investors. Wherever possible, decisions would be by consensus that, although cumbersome, would seek to ensure investor support. However, if necessary, decisions could be taken by voting with ICSC members' votes based on their respective investment shares. Developing countries would be assured affordable access to and ownership in the Intelsat system through a 0.05 minimal investment share in the space segment. Further, a policy of nondiscriminatory access ensured universality of service.

The subsequent negotiations of the definitive arrangements, which lasted a little over 2 years (February 1969–May 1971), established a renamed and reconfigured Intelsat as the International Telecommunications Satellite Organization. The *Agreement Relating to the Establishment of the International Telecommunications Satellite Organization—Intelsat* was signed by the Parties, and the *Operating Agreement Relating to the International Telecommunications Satellite Organization—Intelsat* was signed by the Signatories; the instruments entered into force in February 1973. The new structure was generally along the lines of traditional IGOs whose heritage, as noted earlier, was European. First, an intergovernmental plenipotentiary body was established in the form of an Assembly of Parties that would function on a one-nation, one-vote basis in addressing political issues and would meet every 2 years, although there was a provision for extraordinary sessions. Second, a secretariat, headed by a secretary general, was established in the form of an Executive Organ that would be headed by a Director General who would manage day-to-day operations of the satellite system through an international staff. Third, a decision-making council was formed in the form of a Board of Governors with responsibilities that would be similar to those of the former ICSC and would be comprised of Signatories who qualified for membership based on ownership, individually or in groups, and four other Signatories elected by the Assembly to ensure as much international participation as possible. The Board of Governors would meet on a quarterly basis. Because decisions were to be made by consensus, as many as three or four meetings could be required to reach a decision. Only on rare occasions, when industrial policy or political concerns were involved, would decisions be decided by voting, in which case the votes would be based on investment shares held by individual Governors or cumulative shares cast by Governors representing two or more other Signatories. In addition, a Meeting of Signatories, with limited responsibilities, was established to meet annually; its practical purpose was to allow all Signatories to have a participatory role of some kind in the new organization.

As indicated earlier, Intelsat had become the first ISO when it was established in 1964. Its initial membership numbered 14, all industrialized nations, with the exception of the Vatican City (see Table 5.3). However, when the Interim Arrangements were renegotiated and the Definitive Agreements entered into force 9 years later, membership had increased to 83. The majority of the new members were developing nations. Thus, international satellite communications became globalized through this hybrid government/commercial entity. Save for Antarctica, all continents were represented, as were most island nations. Investment was open to any nation who was a member of the ITU. Access to the space segment was nondiscriminatory.

International Institutions

Operation of communications satellites in outer space confronted policymakers with unique controversial political issues, such as who would have access to space and how space would be utilized, and complex technical and operational issues, such as how frequencies would be allocated and how orbital slots would be allotted.

Wishing to capitalize on Sputnik, the Soviet Union proposed the establishment of a UN agency for international cooperation in space research in March 1958, and the United States presented a counterproposal that resulted in the establishment of COPOUS. Among the committee's most significant efforts was a statement of policy adopted unanimously as UN General Assembly Resolution 1721 (XVI) in late 1961. Among other things, the Resolution declared that outer space is free for exploration and use by all states in conformity with international law, and that communications by means of satellite should be available to all nations of the world as soon as practicable on a global and nondiscriminatory basis. The document also urged the establishment of a registry of state-launched subjects "into orbit and beyond."

The ITU's 1959 Geneva Administrative Radio Conference allocated the initial frequencies for "space research purposes." The 1963 Extraordinary Administrative Radio Conference made initial allocations for communications satellites in the International Radio Regulations, and it established regulatory procedures for nations to follow to notify, coordinate, and register their planned satellites. The 1971 World Administrative Conference for Space Telecommunications further defined registration procedures for the use of frequencies as well as for orbital locations of GEO satellites. Frequency allocations and orbital issues have been mainstays on ITU agendas ever since.

TABLE 5.3
Members of Intelsat as Charter Members of Other ISOs

Nation	ISO and Year Established			
	INTELSAT—1964	Arabsat—1976	Eutelsat—1977	Inmarsat—1976
Algeria		x		x
Argentina				x
Australia	x			x
Austria			x	
Bahrain		x		
Belgium			x	x
Brazil				x
Cameroon				x
Canada	x			x
Chile				x
Cyprus			x	
Denmark	x		x	x
Egypt		x		
Finland			x	x
France	x		x	x
FRG	x		x	x
Ghana				x
Greece			x	x
India				x
Indonesia				x
Iran				x
Iraq		x		x
Ireland			x	
Italy	x		x	x
Liechtenstein			x	
Luxembourg			x	
Japan	x			x
Jordan		x		
Kuwait		x		x
Lebanon		x		
Liberia				x
Libya		x		
Malta				x
Mauritania		x		
Monaco			x	
Morocco		x		
Netherlands	x		x	x
New Zealand				x
Norway	x		x	x
Oman		x		
Peru				x
Portugal			x	
Qatar		x		
Republic of San Marino			x	
Saudi Arabia		x		x

(Continued)

TABLE 5.3
(Continued)

Nation	ISO and Year Established			
	INTELSAT—1964	*Arabsat—1976*	*Eutelsat—1977*	*Inmarsat—1976*
Singapore				x
Somalia		x		
Spain	x		x	x
Sudan		x		
Sweden			x	x
Switzerland	x		x	x
Syria		x		
Thailand				x
Tunisia		x		
Turkey			x	x
UAE		x		
United Kingdom	x		x	x
United States	x			x
Vatican City	x		x	
Yemen		x		
Yugoslavia			x	

Note. From Intelsat basic documents, Inmarsat basic documents, and U.S. Congress, Senate Committee on Commerce, Science, and Transportation (December, 1978); *Space Law: Selected Basic Documents, Second Edition*, 95th Congress, 2d Session.

INNOVATION CYCLE OF COMMUNICATIONS SATELLITES: THE SECOND DECADE

The geopolitical agenda broadened during the 1970s. The changes that took place in satellite communications policy were attributable to the introduction of maritime satellite communications and shifts in policy away from the concept of Intelsat as the single global satellite system. Changes in policy at the beginning of the 1970s, primarily domestic, were not that dramatic. In stark contrast was the entry into force in 1979 of an international agreement to establish the International Maritime Satellite Organization (Inmarsat) as the second global satellite system.

Geopolitical Dimension

Soviet–American rivalries remained just below the surface during the 1970s as U.S. foreign policy adapted to changes in other parts of the world. In the early part of the decade, diplomatic relations with the People's Republic of China were opened, beginning with President Nixon's visit in 1972. The Paris Peace Accords, signed the following year, helped bring about this nation's disengagement from the Vietnam War. In the latter part of the decade, President Carter negotiated the transfer of control of the Panama Canal to Panama to signify a major change in policy toward Latin America, and he oversaw the Camp David Accords, which dramatically increased U.S. involvement in the politics of the Middle East. The Soviet Union invaded Afghanistan in 1979.

Technology

The addition of a new frequency capability (L band) for GEO communications satellites to provide maritime satellite communications, as discussed in chapter 3, marked the beginning of the evolution of a new satellite service—Mobile Satellite Service (MSS).

Interaction Among Technology, Industry, and Government

The years 1970 and 1979 proved to be timeline bookends to a watershed decade that witnessed changes in direction in satellite policy.

Intelsat—The Single Global System. When the Intelsat Definitive Agreements were opened for signature in 1971 and entered into force in 1973, they reflected as much as they could the existing U.S. foreign policy objective of maintaining a single global satellite system. However, significant differences had emerged during the negotiations that reflected diverging foreign policies, industrial strategies, and industry structures of the Intelsat member nations.

One of the most contentious issues of the negotiations of the Definitive Agreements involved meshing the single global system concept with member nations' interests in establishing satellite systems separate from Intelsat. The issue was addressed in two articles in the Definitive Agreement. First, Article I drew a distinction between "public telecommunications services" (e.g., telephony, telegraphy, telex, facsimile, data transmission, and transmission of radio and TV programs for further transmission to the public) and "specialized services" (e.g., radio navigation services, broadcasting satellite services for reception by the general public, space research services, meteorological services, and Earth resources services). Second, Article XIV addressed the "Rights and Obligations of Members." A U.S.-proposed "consultative process"—Article XIV(d)—was agreed on that required any Intelsat member planning to establish a satellite system separate from Intelsat for the purpose of carrying "international public telecommunications services" to "consult" in advance with Intelsat's Assembly of Parties through the Intelsat Board of Governors. The Assembly, taking into account the advice of the Board, would then express nonbinding findings intended to ensure that there were technical and operational compatibilities between the systems, and to "avoid significant economic harm" to Intelsat. Article XIV (c) and (e) addressed two other kinds of satellite systems—domestic systems providing public telecommunications services, and domestic or international systems providing specialized services. These would only have to be consulted with the Assembly, through the Board, to ensure technical and operational compatibilities. The West Europeans, in particular, vigorously debated the formulation of this article because they were, at the time, planning to enter the space race with their own satellites and launch vehicles through the European Space Research Organization (ESRO), the forerunner of the European Space Agency (ESA), and the European Launch Development Organization (ELDO). Fearful that they could be blocked from developing their own regional satellite system in the event that they were not able to develop their own launch vehicle capability, the West Europeans elicited a "launch assurance" from the United States to launch a European satellite irrespective of any "significant economic harm" finding by the Assembly. This launch assurance was later extended to other nations as well.

In an effort to minimize industrial policy considerations in the procurement of satellites and launch vehicles, the United States successfully negotiated the "Procurement" article—Article XIII (a)—stating that the bidding process was to be competitive with decisions to be based on "the best combination of quality, price and the most favorable delivery time."

Shifts in U.S. Policy. To the surprise of many foreign governments, given the United States' "commitment" to Intelsat in ongoing international negotiations, the new Nixon administration signaled two changes in the direction of its communications satellite policy. First, in the early months of the administration, an "open skies policy" was proposed by the White House and then adapted and adopted by the FCC. It was designed to encourage and facilitate the development of commercial U.S. domestic satellite communications by any technically and financially qualified company while "off-shore" domestic traffic between the Continental United States and Alaska, Hawaii, and Puerto Rico was to be transferred from the Intelsat system to the domestic satellite systems. A few months later, the administration interrupted international planning for an experimental/evaluation aeronautical satellite (Aerosat) program that was to be jointly owned, operated, and managed by NASA and ESRO. The White House took the position that the program was inconsistent with the Eisenhower administration's communications satellite policy, which supported private ownership, and the recently articulated domestic open skies policy, which encouraged satellite systems to be built in response to the competitive marketplace, rather than to government direction. This was a major step away from the chosen instrument approach that the United States had taken internationally until then. A new NASA/ESRO arrangement was negotiated whereby ESRO was allowed to choose between two competing U.S. systems to provide space segment. Comsat General, a newly organized Comsat subsidiary, was selected as the U.S. industry participant, but an RFP was never issued due to the lack of government funding.

New ISOs.   Further evidence of movement away from the single system concept, this time by other governments, was provided by the formation of three other commercial international satellite systems by the end of the decade. The Arab Telecommunications Satellite Organization (Arabsat) was established in 1976 to carry intraregional traffic. The European Telecommunications Satellite Organization (Eutelsat) was established in 1977 to carry traffic originally planned for an intra-European terrestrial microwave system. The International Maritime Satellite Organization (Inmarsat) came into being in 1979 to provide maritime services. Most members of these new systems were also members of Intelsat (see Table 5.3). All three systems were consulted with Intelsat successfully. This overlapping membership accounts for the fact that these new ISOs were modeled generally along the lines of Intelsat (see Table 5.4). Not unexpectedly, the Soviet Union responded to the Intelsat Definitive Agreements in 1971 with the formation of its own global satellite system comprised of communist countries—Intersputnik.

Contrasts Between Intelsat and Inmarsat.   Inmarsat was the outgrowth of technical and economic feasibility studies that were conducted under the auspices of then-IMCO in the early 1970s. Based on the results and recommendations of these studies, a 2-year International Conference was held in 1975 and 1976 to negotiate the *Convention on the International Maritime Satellite Organization—Inmarsat* and the *Operating Agreement on the International Maritime Satellite Organization—Inmarsat.* The agreements entered into force in mid-1979. The new ISO was headquartered in London and took a West European "tilt" (i.e., the Director General was Olof Lundberg, from Sweden, the Directorate was predominately staffed by Europeans, and ESA was assured a key technical supporting role).

Many charter members of Inmarsat had been original Intelsat investors—namely, the United States, UK, Canada, Australia, Japan, and West European nations such as Italy, Norway, Germany, and France. Yet they chose to establish a separate ISO for a variety of reasons. Some wanted to prevent the U.S.-headquartered and dominated Intelsat from providing maritime satellite services. Other foreign ministries felt that the Soviet Union, a major maritime nation, should be included in any new global maritime ISO, and it was. Many national maritime administrations believed the maritime industry's communications requirements justified the creation of a dedicated system.

There were significant differences between the two ISOs. One involved the competitive procurement process. Article 20 of the *Inmarsat Convention* called for competitive satellite and launch vehicle procurement based on "the best com-

TABLE 5.4

The Rise and Demise of the ISOs

ISO		*Structure*
Intelsat		
Consortium		
	Established 1964	ICSC
	Interim Agreements enter into	(Signatory investors)
	force 1964	Manager (COMSAT)
	Superseded by INTELSAT Organization 1973	
Intersputnik		
	Established 1971	Board
	Agreements enter into force	(all members; not based on investment)
	1971	Committee of Plenipotentiaries
		Directorate
		(DG and Secretariat)
Intelsat		
Organization	Permanent Agreements enter	Board of Governors
	into force 1973	(Signatory investors)
	Privatized 2001	Assembly of Parties
		(Governments)
		Executive Organ
		(DG and Secretariat)
		Meeting of Signatories
		(Signatory investors)
Arabsat		
	Arab Ministers of Information	Board of Directors (Signatory investors)
	and Culture agree in principle	General Assembly (Governments)
	to establish Arab League system 1967	Executive Organ (DG and secretariat)
	Agreements enter into force	
	1976	
Eutelsat		
	Established 1977	Council
	Agreements enter into force	(Signatory investors)
	1985	Assembly of Parties
	Privatized 2001	(Governments)
		General Secretariat
		(SG and Secretariat)
Inmarsat		
	Established 1976	Council
	Agreements enter into force	(Signatory investors)
	1979	Assembly
	Privatized 1999	(Governments)
		Directorate
		(DG and Secretariat)

Note. From Congress, Senate Committee on Commerce, Science, and Transportation (December, 1978). *Space Law: Selected Basic Documents, Second Edition*, 95th Congress, 2d Session; ISO Web sites.

bination of quality, price and the most favorable delivery time," similar to that of Intelsat. As it turned out, the procurement of Inmarsat's first-generation space segment proved to be competitively unique because it included: (a) Comsat General Marisat satellites, discussed in chapters 3, 6, and 7; (b) ESRO experimental MARECS satellites; and (c) "maritime packages" on Intelsat V satellites. There was also a new Article 21—"Inventions and Technical Information." Section (9) of this new article made contractors subject to national export controls that effectively blocked communications satellites with embedded U.S. technology from being launched on the Soviet Union's Proton launch vehicle.

Another major difference was Article 8—"Other Space Segments." This article required Inmarsat members wishing to establish space segments separate from Inmarsat to only "notify" rather than "consult" with Inmarsat's Assembly, through the Inmarsat Council, regarding technical and operational capabilities and "significant economic harm." Article 8 never became the political issue that Article XID(d) became in Intelsat. A major reason for this was Inmarsat's decision not to contest U.S. policy that Inmarsat could provide aeronautical and land mobile satellite services, but on a competitive, rather than an exclusive, basis.

Although Intelsat's basic documents were never amended until the organization was privatized at the turn of the century, the Inmarsat Convention was amended twice—once in 1985 to enable the provision of aeronautical services on a competitive basis, and again in 1989 to enable provision of land mobile services on a competitive basis. This latter amendment process also led to a name change from the International Maritime Satellite Organization to the International Mobile Satellite Organization.

Last, but not least, in 1988, the IMO amended the *1974 International Convention for the Safety of Life at Sea (SOLAS)* to establish an international maritime safety service, the Global Maritime Distress and Safety System (GMDSS). Among the amendments that entered into force in 1992 were provisions requiring Inmarsat to provide the communications backbone for this international public safety service. From the ship management side, GMDSS enabled the introduction of modern satellite terminals on ships that could be serviced by electronic officers as part of their broader on-board maintenance responsibilities. This eliminated the requirement for dedicated HF radio officers whose around-the-clock presence in the HF radio room had been mandated in international treaties resulting from the Titanic disaster decades earlier.

Satellite/Cable Dichotomies. Establishment of Comsat introduced domestic and international structural dichotomies. Domestically, it became a carrier's carrier and leased capacity to the U.S. international carriers, the same international carriers that owned and operated the U.S. portion of submarine telephone cable systems. In contrast, the majority of foreign administrations owned and operated the submarine cables as well as the satellites through their PTTs that also acted in the role of regulators. This was completely opposite of the situation in the United States, in which the FCC, an independent domestic regulatory agency, was charged with ensuring that authorization of telecommunications facilities, including international facilities, was consistent with the public interest.

The traditional *modus operandi* of the U.S. international common carriers in planning a new submarine cable system had been to hold separate bilateral discussions with their foreign correspondents with the tacit understanding that if the terms and conditions of the agreed-on consortium arrangement were within reason, their applications for a Cable Landing License and other regulatory authorizations would be routinely granted. After the *Comsat Act*, the contrasting capacities of analog satellites and submarine cables and their operational lifetimes took on regulatory significance. Mandated to ensure that the most economic, efficient use was made of telecommunications facilities, in the early 1970s the FCC began to regulate how the U.S. international carriers should activate new circuits on cable and satellite facilities. Because the facilities were half-owned by foreign administrations, the domestic regulatory agency was in effect also telling foreign governments how they should operate their international facilities. "Fill formulas" were developed by the FCC to both enable the larger capacity satellites to be efficiently utilized during their relatively shorter operational lifetimes and prevent the smaller submarine cables from saturating many years before the end of their operational longevity. At one point, a "proportional fill" formula required the U.S. international carriers to add five circuits on satellites to one circuit on cables between the United States and a foreign country—a formula that was roughly analogous to the comparative capacities of each technology at the time. A number of foreign ministries submitted formal protests to the Department of State and threatened to stop placing new traffic on U.S.–European routes altogether.

In an effort to avoid further political problems, a bilevel North Atlantic Consultative Process (NACP) for future cable and satellite facilities planning was established in 1975 by the United States, Canada, and the Europeans (collectively working through the Conference of European Post and Telecommunications Administrations [CEPT]). A senior policy level was comprised of West European, Canadian, and U.S. senior government officials and industry executives, and a working level was comprised of senior staff-level subject matter experts on traffic projections, operational planning, and economics from the government and the carriers. During this process, AT&T and the international record carriers were referred to as U.S. International Service Carriers (USISCs). However, another major political problem occurred when the FCC delayed authorizing the entry into service from 1979 to 1983 of the proposed TAT-7, the last analog submarine cable to be laid in the North Atlantic. Unfortunately for the USISCs and their foreign counterparts, the decision was based on mutually agreed-on traffic projections and operational planning principles. A new "balanced loading" methodology emerged, markedly less contentious than the "proportional fill" formula, for adding new circuits on satellite and cable routes between the U.S. and a given foreign country. The FCC originally had recommended delaying the cable until 1985 based on the record of the consultative process, but political pressures, international and domestic, prevailed (see Federal Communications Commission, 1971, 1976, 1977, 1979, 1985; Frieden, 1996, 2001; Oslund, 1983).

International Institutions

The 1970 entry into force of the Convention to establish the World Intellectual Property Organization (WIPO), whose origins can be traced back to the 1883 *Paris Convention for the Protection of Industrial Property* and the 1886 *Berne Convention for the Protection of Literary and Artistic Works*, was another significant step toward globalism. In 1974, the IGO completed negotiating the *Brussels Convention for the Protection of Program-Carrying Signals Transmitted via Satellite*.

INNOVATION CYCLE OF FIBER OPTIC SUBMARINE CABLE

Political changes in the communist world, barely perceptible in the early part of the 1980s, brought about a major realignment in geopolitics at the end of the decade. Telecommunications-wise, the commercial introduction of digital fiber optic submarine cables rearranged the economics and operational planning of transoceanic telecommunications. Communications satellite policies continued to change, and a new policy approach toward the new undersea technology was adopted.

Geopolitical Dimension

Political changes in the Soviet bloc dominated the international stage. The first hint of change occurred in 1980 with the appearance of the Solidarity Movement in Poland. From then on, the Iron Curtain across Eastern Europe gradually rusted, culminating with the symbolic destruction of the Berlin Wall in 1989.

Technology

The advent of submarine fiber optic cable was not unexpected, its having been identified years earlier in the NACP as an emerging technology. Nevertheless, the operational impact of the glass-based technology was stunning. Indeed, when the first fiber optic cable (TAT-8) was "lit up" in 1988, the assumptions that had been used in transatlantic satellite/cable facilities planning in previous years were literally stood on their head as digital fiber optic cable capacities overtook the communications satellites' digitized capacities (see Table 5.5). The economics shifted

TABLE 5.5
Fiber Optics and the Paradigm Shift

Analog Years (half-circuits)			
Cable			Satellite
TAT 1	50	1956	
		1965	INT 1 240
TAT 7	4000	1983	
		1985	INT VA 15,000 + 2 TV channels

Critical Crossover

Introduction of Digital (derived half-circuits)

TAT 8	37.800	1988	
		1989	INT VI 35,000

The Undersea Fiber Bandwidth Explosion (gbs)

Year	Cable	Bandwidth
1956	TAT 1	0.005
1988	TAT 8	0.05
1999	AC-1	80.00
2001	FA-1	2,400.00

Note.   From *Telegeography, FCC Documents*.

to fiber optic cables on the lucrative major international traffic routes, although satellites—with their near ubiquitous geographical overage—retained the operational edge for point-to-multipoint transmissions on major and minor traffic routes alike.

Interaction Among Technology, Industry, and Government

Three significant policy and regulatory shifts in the United States, discussed in chapter 6, affected both transmission media in different ways.

Cable/Satellite Loading Policy.   The FCC's prescribed loading of analog satellite and analog cable facilities had been due, in large part, to critical disparities in their respective capacities. To exploit the disappearance of the disparities in media capacities with the introduction of digital transmission techniques and fiber optic cable, and in an effort to avoid further political problems in transoceanic facilities planning, the FCC moved away from the balanced loading methodology and encouraged Comsat and the USISCs to negotiate long-term leases of satellite capacity in advance of planning any new satellite facilities. There were no more political problems (see Federal Communications Commission, 1988).

New Intelsat Policy. Persistent lobbying by would-be American operators of new international satellite systems and the aerospace industry led to a 1984 "Presidential Determination" by President Reagan that establishment of U.S. international satellite systems separate from Intelsat space segment was "required in the national interest" (see R. Reagan, 1984). This determination came to be known as the "separate systems policy," which is discussed in chapter 6. Contentious but successful Intelsat Article XIV(d) consultations of two U.S. systems, PanAmSat and Orion, discussed in chapter 2, were completed in 1987 and 1989, respectively.

New Fiber Optic Submarine Cable Policy. A separate, but related, policy shift in submarine cable policy occurred the following year when the FCC authorized "Tel-Optik" to link the U.S. and the UK (see Federal Communications Commission, 1985). This cable, which came to be called PTAT-1, was the first private transoceanic cable to be owned by parties other than the traditional international common carriers—hence, the "P" before "TAT."

International Institutions

The existing institutional structures struggled, but were able to cope with these changes, and the institutional framework went unchanged through the decade.

INNOVATION CYCLE OF LEOs . . . AND BEYOND

At the macrolevel, the symmetries of the geopolitical environment, which had been shaped by the dynamics of Soviet–American rivalries, were displaced in the 1990s by new national and non-nation state actors with new asymmetrical threats. In the nonpolitical realm, radical restructuring of the international telecommunications industry was beginning through liberalization, deregulation, and privatization. The debut, or return (depending on one's perspective), of the globe-circling low- and medium-altitude satellites (LEOs) spawned new companies and demanded the attention of governments around the world. Institutionally, an old IGO—the ITU—struggled to avoid being marginalized, and a new IGO—the World Trade Organization (WTO)—was established.

Geopolitical Dimension

The implosion of the Soviet Union, unintentionally accelerated by Gorbachev's experiment with *Perestroika*, marked the end of bipolar politics and of what is seen by many as the end of the cold war—or World War III. New models for policy analysis were needed. Two of the more popular theories bumped up against each other. One was based on the perceived arrival of a universal consensus that embraced liberal democratization (see Fukuyama, 1992). The other foresaw violent collisions with fundamentalist fringes occurring along cultural fault lines (see Huntington, 1996). Both of these theories gave way to real-world asymmetries in the Persian Gulf War in 1991, in the Balkans toward the end of the decade, in New York City and Washington, DC, on September 11, 2001, in Afghanistan in late 2001, and in Iraq in the spring of 2003. Some argue that the Persian Gulf War marked the beginning of World War IV; others contend it began with 9/11 (see Cohen, 2001).

Without much fanfare due to the political events that were unfolding in Eastern Europe in the 1980s, the Soviet Union had signed an MOU with Intelsat and then became a member; Intersputnik also had established formal relations with Intelsat. As a sign of the changing times, and with great fanfare during the Bush–Yeltsin Summit in June 1992, the United States announced it was making an "Inmarsat exception" to its long-standing policy that precluded U.S. satellites from being launched on the Proton. This enabled Inmarsat to consider the Russian launch vehicle as an option for its third generation of satellites—Inmarsat 3. The ISO selected Proton to launch one of its Inmarsat 3 satellites.

Technology

The evolution of communications satellite technologies came full circle in the 1990s with the emergence—or reemergence—of LEO satellites, both Big and Little, as discussed in chapter 3. Recall that in the early years of communications satellites, the concept of lower altitude satellite operations had been technically demonstrated with the successful orbiting of the DOD-funded Score and Courier, the AT&T-funded Telstar, and the NASA-funded Relay. Although the concept of GEO satellites had been shelved with the cancellation of DOD's Advent program before it was ever launched, the unexpected successful launch of Syncom had demonstrated the technical feasibility of GEO satellites after all. Given the GEOs' technical and operational simplicity and lower cost configuration when compared with those of the lower altitude options, they became the technology of choice. However, DOD revived interest in the low-altitude small-satellite constellations in the latter stages of the cold war.

Interaction Among Technology, Industry, and Government

The commercial introduction of the LEOs was facilitated by the United States' aggressive promotional role. However, the announcement of their planned introduction was equaled by the surprise decision to privatize the two global ISOs— Intelsat and Inmarsat.

Commercialization of LEOs. With the demise of the USSR in December 1991, DOD's need to continue to fund the development of technologies for constellations of low- and medium-altitude satellites faded. Outlets for continuing the development of these LEO technologies were explored, including the transfer of the technology to the civilian sector. New American companies, encouraged by the government, formed global consortia with substantial investment by foreign industries, foreign financial institutions, and foreign governmental entities, to finance and operate proposed new LEO and MEO systems. The United States aggressively and successfully lobbied for frequency allocations for mobile Big LEO systems—namely, Iridium and Globalstar—at WARC 1992. At WRC 1995, the United States was similarly successful in obtaining frequency allocations for a fixed LEO system, Teledesic. Inmarsat also sought to enter the mobile Big LEO market based on the U.S. government's condition that the hand-held service would be provided through a fully separate entity, ICO, to which Inmarsat agreed. Subsequent developments are discussed in chapters 6 and 7.

Privatization of ISOs. Possible policy and commercial reasons that senior management of Comsat, the U.S. Signatory to and a primary investor in Intelsat and Inmarsat, called for the privatization of both ISOs, beginning in late 1993, are discussed in chapter 6. Suffice to say here that Comsat was being affected domestically by the previously cited shifts in national policy and regulatory decisions. Internationally, the two global ISOs, structured and acting much like slow-moving IGOs, were facing increasing competition. In the case of Intelsat, the competition was an array of existing and planned submarine fiber optic cable systems and a growing list of potential "separate satellite system" competitors, including a host of systems planning to provide broadband services at higher frequency bands. With respect to Inmarsat, the competitors were the proposed LEO systems.

Separate, and what would prove to be laborious, negotiations to privatize Intelsat and Inmarsat began in 1994. Numerous working groups of Signatories, working groups of Signatories and Parties, and regular and extraordinary meetings of the ISOs' respective Assemblies were convened all over the world through the rest of the century. A high degree of skepticism by many governments and heavy resistance by major foreign investors were encountered initially, but as liberalization and privatization took place at home and the new WTO forged ahead with a new global trading regime, foreign administrations started to relax their opposition. The developing nations, many of whom in the 1960s and 1970s had viewed Intelsat as being an adjunct to America's "cultural imperialism," argued for and achieved assurances that they would continue to have access to the privatized Intelsat, as well as the privatized Inmarsat system.

When Inmarsat was privatized in April 1999, two companies were incorporated under English law—a holding company now called Inmarsat Ventures plc, which has a fiduciary board of directors, and a wholly owned operating subsidiary called Inmarsat Ltd. Business formerly conducted in the International Mobile Satellite Organization was transferred to Inmarsat Ltd. The entities that had been Signatories became ordinary shareholders in the new holding company. A residual IGO also was continued—IMSO (a different acronym for International Mobile Satellite Organization)—which has the same member states as its predecessor and a small Directorate managed by a Director. The Inmarsat Convention was amended to change IMSO's function to monitor the company's compliance with a number of public service obligations. These obligations, which are defined in the Memorandum and Articles of Association for Inmarsat Ventures plc, include continuing the provision of communications satellites services for the GMDSS and giving due consideration to the rural and remote areas of developing countries, and are contained in a public services agreement between Inmarsat Ltd and IMSO. IMSO's director regularly reports to the IMO with respect to new Inmarsat's performance of its GMDSS obligations (see Fig. 5.1).

Intelsat's privatization took effect in July 2001 and generally followed the same approach undertaken for Inmarsat's privatized structure. Two companies were incorporated under Bermuda law—Intelsat, Ltd., a holding company, which has a fiduciary Board of Directors; and Intelsat (Bermuda), Ltd., a wholly owned operating subsidiary. Additionally, Intelsat Global Service Corporation, a wholly owned subsidiary of Intelsat Bermuda and organized as a Delaware corporation, was established to provide contractual technical, marketing, and business support services to Intelsat, Ltd. and its numerous subsidiaries. Business formerly conducted in the International Telecommunications Satellite Organization was transferred to Intelsat (Bermuda) Ltd., and the entities that had been Signatories became shareholders in the new holding company. Nonbusiness activities of Intelsat as an IGO were transferred to a surviving residual IGO—ITSO (a different acronym for International Telecommunications Satellite Organization)—which has the same member states and a small Executive Organ managed by a Director General. The Intelsat Agreement was amended to change ITSO's function to one of

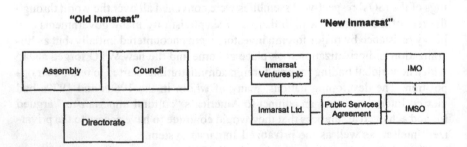

FIG. 5.1.   The old and new Inmarsat.

monitoring the company's compliance with a number of public service obligations. These public service obligations are included in the corporate by-laws for Intelsat Ltd. and are contained in a public services agreement between Intelsat Global Service and ITSO. They include maintaining global connectivity and honoring the lifeline connectivity obligation (LCO) to those customers in poor or underserved countries that have a high degree of dependence on Intelsat due to their economic status or lack of infrastructure necessary to support a local or regional telecommunications system (see Fig. 5.2; see also FCC Report to Congress, 2001). Not surprisingly, Eutelsat, many of whose members belonged to Intelsat and Inmarsat, also privatized in 2001.

International Institutions

The effects of the rapidly restructuring international telecommunications industry and the globalizing marketplace resulted in a major reorganization of the oldest international organization and the establishment of a new one.

ITU's Marginalization. In the late 1980s, ITU was running the risk of becoming marginalized. The IGO was perceived as being unable to respond in a timely, consistent manner to the needs for new regulations, standards, and spectrum allocations that were being generated by new satellite and satellite-related technologies and services, as well as other new technologies and services. Governments and industries were increasingly turning to regional bodies. The CEPT countries spun off the European Telecommunications Standards Institute (ETSI) to harmonize their interests. With strong U.S. encouragement, the countries of the Western Hemisphere reinvigorated the Inter-American Telecommunications Commission (CITEL). Asia-Pacific Economic Cooperation (APEC) emerged as a forum for that far-flung geographic region. Confronted with these developments, the 1989 ITU Plenipotentiary Conference established a high-level committee to

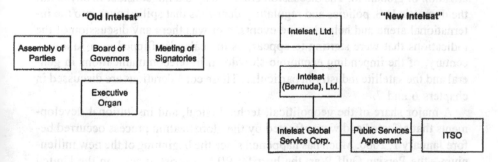

FIG. 5.2.   The old and new Intelsat.

review the IGO's structures and functions and make recommendations as appropriate. In April 1991, the committee reported back with procedures to streamline the organization's ability to act while maintaining its fundamental structure and scope. Significantly, there were no recommendations to expand the charter (see Solomon, 1991). The ITU has since sought to adapt with the formation of three sectors (ITU-T, ITU-R, and ITU-D), and, indicative of globalism, private companies have been allowed to increase their participation in the activities of these sectors, including becoming actual members.

IGO for a Global Economy.   The U.S. campaign to export competition into the international telecommunications marketplace began in earnest in the 1970s, although telecommunications was but one of the service sectors targeted for liberalization and deregulation. The global-wide diplomatic breakthrough was realized in 1994 when the Uruguay Round was completed with two significant diplomatic breakthroughs. First, the long-sought *General Agreement on Trade in Services (GATS)* opened up domestic markets for critical economic services, including telecommunications, finance, maritime, and air transport. Second, the World Trade Organization (WTO) was agreed on—50 years after the United States had rejected the notion of an international organization for trade at the Bretton Woods Conference. In many respects, WTO represents the institutionalization of globalism. Interestingly, the first two international agreements negotiated under the WTO umbrella were the *Information Technology Agreement* in December 1996 and the *Agreement on Basic Telecommunications Services* in February 1997.

CONCLUDING THOUGHTS

This chapter opened by suggesting that Intelsat was an enabler of globalism and ended by opining that the creation of the WTO symbolized the institutionalization of globalism. The discussion consciously did not consider in any detail the domestic U.S. policies and regulatory decisions that spilled over into the international arena and helped shape events. Nor was there any discussion of the indications that were starting to appear, as the calendar flipped over to a new century, of the impending economic slowdown in telecommunications in general and the satellite industry in particular. These considerations are discussed in chapters 6 and 7.

A major share of the geopolitical, technological, and institutional developments that affected and were affected by the globalization process occurred before January 1, 2000. Much has happened since the beginning of the new millennium—the Persian Gulf War, the horrific 9/11 terrorist attacks on the United States, the Afghanistan War, and the Iraqi War. These events have had the effect of making what happened up to and through the 1990s seem to have occurred more than a century—not a few years—ago. The effects of these events on communications satellite technologies and policies are also explored in chapters 6 and 7.

REFERENCES

Archer, G. L., (1938), *History of Radio to 1926*, New York: The American Historical Society.

Bamford, James, (1983), *The Puzzle Palace*, New York: Penguin Books.

Bowett, D. W., (1963), *The Law of International Institutions*, New York: Frederick A. Praegar.

Brockbank, R. A., (1958, November 20), "Overseas Telephone Cables Since the War," *The New Scientist.*

Clark, Keith, (1931), *International Communications: The American Attitude*, New York: Columbia University Press.

Codding, Jr., George A., and Rutkowski, Anthony M., (1982), *The International Telecommunication Union in a Changing World*, Dedham, MA: Artech House.

Cohen, Eliot A., (2001, November 20), "World War IV," *Wall Street Journal*. (Note: Cohen argues: "The cold war was World War III, which reminds us that not all global conflicts entail the movement of multimillion-man armies or conventional front-lines on a map." Key features of World War III included the global nature of conflict that involved a mixture of violent and nonviolent efforts, required mobilization of skill, expertise, and resources—but not necessarily in vast numbers of soldiers, was waged for a long time and had ideological roots. He suggests that World War IV will be different, but will exhibit similar characteristics.)

Federal Communications Commission, (1971), Policy to be Followed in Future Licensing of Facilities for Overseas Communications, 30 FCC 2d 571.

Federal Communications Commission, (1976), Policy to be Followed in Future Licensing of Facilities for Overseas Communications, 62 FCC 2d 451.

Federal Communications Commission, (1977), Policy to be Followed in Future Licensing of Facilities for Overseas Communications, 67 FCC 2d 358.

Federal Communications Commission, (1979), Policy to be Followed in Future Licensing of Facilities for Overseas Communications, 71 FCC 2d 1178.

Federal Communications Commission, (1985), Policies to be Followed in the Authorization of Common Carrier Facilities to Meet North Atlantic Telecommunications Needs During the 1985–1995 Period. 100 FCC 2d 1405.

Federal Communications Commission, (1988), Policies to be Followed in the Authorization of Common Carrier Facilities to Meet North Atlantic Telecommunications Needs During the 1991–2000 Period, 3 FCC Rcd. 3979.

Federal Communications Commission, (1985), Tel-Optik, Ltd., 100 FCC 2d 1033.

Federal Communications Commission, (2001, June 15), Report to Congress as Required by the ORBIT Act (FCC 01-150).

Frieden, Rob, (1996), *International Telecommunications Handbook*, Boston: Artech House. (Note: Documented discussion of contemporary developments in international telecommunications technologies and policies.)

Frieden, Rob, (2001), *Managing Internet-Driven Change in International Telecommunications*, Boston: Artech House. (Note: Updated documented discussion of developments in international telecommunications technologies and policies.)

Fukuyama, Francis, (1992), *The End of History and the Last Man*, New York: The Free Press.

Galloway, Jonathon F., (1972), *The Politics and Technology of Satellite Communications*, Lexington: Lexington Books.

Herring, James M., and Gross, Gerald C., (1936), *Telecommunications: Economics and Regulation*, New York: McGraw-Hill Book Company.

Hudson, Heather E., (1990), *Communication Satellites: Their Development and Impact*, New York: The Free Press.

Huntington, Samuel P., (1996), *The Clash of Civilizations and the Remaking of World Order*, New York: Simon & Schuster.

Kennedy, Paul, (1987), *The Rise and Fall of the Great Powers*, New York: Random House.

Kinsley, Michael, (1976), *Outer Space and Inner Sanctums: Government, Business, and Satellite Communications*, New York: John Wiley & Sons.

Kvan, R. A., (1996), "Comsat: The Inevitable Anomaly," in S. A. Lakoff, Editor, *Knowledge and Power: Essays in Science and Government*, New York: The Free Press.

Leive, David M., (1970), *International Telecommunications and International Law: The Regulation of the Radio Spectrum*, Dobbs Ferry, NY: Oceana Publications.

Logsdon, John, (1969), *The Decision to Go to the Moon: Project Apollo and the National Interest*, Cambridge, MA: The MIT Press.

Maddox, Brenda, (1972), *Beyond Babel: New Directions in Communications*, New York: Simon & Shuster.

Nye, Joseph S., and Donahue, John D., (Editors), (2000), *Governance in a Globalizing World*, Washington, DC: Brookings Institution Press. (**Note:** The authors define *globalism* as "a state of the world involving networks of interdependence at multicontinental distances [that are] linked through flows and influences of capital and goods, information and ideas, people and force, as well as environmentally and biologically relevant substances (such as acid rain and pathogens) . . . When people speak colloquially about globalization, they typically refer to recent increases in globalism. Comments such as 'globalization is fundamentally new' only make sense in this context but are nevertheless misleading. We prefer to speak of globalism as a phenomenon with ancient roots and of globalization as the process of increasing globalism . . ." [pp. 2, 7].)

Oslund, Robert J., (1975), *Communications Satellite Policy: Revolutionary or Evolutionary?*, Washington, DC: The American University. Unpublished dissertation. (**Note:** Documented discussion of historical developments in international telecommunications technologies and policies.)

Oslund, Jack, (1977), " 'Open Shores' to 'Open Skies': Sources and Directions of U.S. Satellite Policy," in Joseph Pelton and Marcellus Snow, Editors, (1977), *Economic and Policy Problems in Satellite Communications*, New York: Praeger Publishers. (**Note:** Documented discussion of historical developments in international telecommunications technologies and policies.)

Oslund, Robert J., (1983, January), "Pacific Facilities Planning Process: An Appropriate Mechanism," Pacific Telecommunications Council.

Oslund, Robert J., (1984), "Reindustrialization: A Telecommunications Perspective," *Telematics and Informatics* Vol. 1. Reprinted in I. B. Singh and V. H. Mishra, Editors, (1987), *Dynamics of Information Management*, Norwood, NJ: Ablex Publishing Corporation.

Oslund, Robert J., (1990), "Restructuring the Domestic Telecommunications Industry: From 'Muddling Through' to 'New Thinking,' " *Telematics and Informatics*, Vol. 7.

Palmowski, Jan, (1997), *A Dictionary of Twentieth-Century World History*, New York: Oxford University Press.

Pelton, Joseph N., (1974), *Global Communications Satellite Policy: Intelsat—Politics and Functionalism*, Mt. Airy, MD: Lomond Systems, Inc. (**Note:** Documented discussion of the establishment of COMSAT and Intelsat.)

Reagan, Ronald, (1984, November 28), Presidential Determination No. 85-2, 40, *Federal Register*.

Solomon, J., (August 1991), "The ITU in A Time of Change," *Telecommunications Policy* Vol. 15.

United Nations, (1995), *Basic Facts About the United Nations*, New York: Author.

U.S. Congress, House, Committee on Interstate and Foreign Commerce, (1921), *Hearings on Cable Landing Licenses*, 67th Congress, 1st Session.

U.S. Congress, Senate, (1929), Committee on Interstate Commerce, (1929), *Hearings on Commission on Communications*, 71st Congress, 1st Session.

U.S. Congress, Senate Committee on Commerce, Science, and Transportation, (1978), *Space Law: Selected Basic Documents, Second Edition*, 95th Congress, 2d Session.

Regulating Communications Satellites on the Way to Globalism

Robert J. Oslund
The George Washington University

> *. . . sometimes you get the feeling that we are imposing on our very complex communication system . . . some rather primitive and patched-up procedures that we have inherited over the last few years.*
>
> —U.S. Senator Gale McGee (D–WY) (1962)

Communications satellite policy that guides the regulation of communications satellites should be viewed within the context of a broader historical continuum. As noted by Senator Gale McGee (D–WY), a former professor of history, during congressional hearings leading up to passage of the *Communications Satellite Act of 1962*, U.S. telecommunications policy has tended to be ad hoc in nature.

> . . . From time to time, we have had occasion in here to wish that we might be able to do over from scratch some of our communications elements. Like Topsy, they just grew, beginning in the 1920s. And sometimes you get the feeling that we are imposing on our very complex communication system now some rather primitive and patched-up procedures that we have inherited over the last few years. And yet as a practical matter, it is very difficult to take the house all apart and put the bricks together again and come out with the right profile. Here we have a chance of putting together what we have learned and the little bit that we know in the satellite realm, to do this perhaps the right way; at least within the limitations of a human being. (see U.S. Congress, 1962)

The preceding chapter explored the historical continuum that has spanned a century-and-a-half transition from imperialism and colonialism toward globalism. This

---

The views expressed herein, unless indicated otherwise, are solely the views of the author.

chapter follows essentially the same continuum, but focuses on the domestic level. Hence, there is some unavoidable overlap. First, policy development is traced through the three eras of terrestrial transoceanic telecommunications technologies—submarine telegraph cable, high-frequency radio, and submarine telephone cable—that led up to the commercial introduction of communications satellites and affected communications satellite policy in varying degrees. Then the evolution of communications satellite policy is traced through presidential administrations. It is demonstrated, from a policy perspective, that despite the chaos of geopolitics and the significant, sometimes unexpected, breakthroughs in international telecommunications technologies, there have been—and there remain—general consistencies in the international telecommunications policies of the United States. For the most part, these policies have been incremental and cumulatively built on precedent. They began with the deployment of the first transoceanic submarine telegraph cable in the 1860s and continued through enactment of the *Radio Act of 1927*, the *Communications Act of 1934*, the *Communications Satellite Act of 1962*, the *Telecommunications Act of 1996*, and up to and into the 21st century (see Brock, 1981; Frieden, 1996, 2001; Galloway, 1972; Hudson, 1990; Kinsley, 1996; Oslund, 1975, 1977, 1984, 1990; Pelton, 1974).

A REGULATORY "TEMPLATE"

Although the focal point of this chapter is on the international dimension of U.S. policies and regulations relating to communications satellites, one must take into account some earlier domestic policies and regulations to which they appear to be related, at least, indirectly. Of particular interest is the manner in which the Federal Communications Commission (FCC) introduced competition in the domestic AT&T-dominated public switched telephone network (PSTN) market between the 1956 *Consent Decree* and the 1982 *Modified Final Judgment* between the Department of Justice and AT&T. This is because it provides a "template," of sorts, that U.S. policymakers, knowingly or unknowingly, followed generally in introducing competition in the Intelsat-dominated international communications satellite market.

Like any governmental institution, the FCC has its own belief system or "bureaucratic ideology" (see Downs, 1967). It is derived from the *Communications Act of 1934* and is reflected to one degree or another in most, if not all, commission decisions. As argued elsewhere, the preponderance of relevant commission policy actions between 1956 and 1982 were based on this belief system: *The promotion of competition, where reasonably feasible and non-harmful to the telephone network, is necessary to stimulate technological development and service innovation, and to allow the public to benefit to the maximum extent possible from both* (see Oslund, 1990).

Alternate Systems Policy

The commission introduced competition to AT&T's universal service monopoly through a phased "alternate systems policy," based to a large extent on the "where reasonably feasible" rationale. At different stages of implementing this policy, an FCC Chairman and a Common Carrier Bureau Chief, in testimony before Congress, explained that each decision was purposely incremental and not intended to do harm to the PSTN. In the end, however, competition was introduced in the PSTN. Indeed it was an inevitable outcome.

Phase 1. In 1959, the FCC concluded a contentious proceeding—*Above 890*—by allocating radio spectrum for intracompany (or private) point-to-point communications systems on the conditions that they would operate independently and not as a cooperative, and that their operations would be subject to continuing review for possible negative impacts on the telephone network. AT&T had vigorously, but unsuccessfully, opposed this frequency allocation by arguing that it would impede the deployment of a commercial international communications satellite system then in the planning stage.

Phase 2. In 1969, by a 4 to 3 decision, the FCC authorized MCI to construct point-to-point non-PSTN common carrier microwave facilities between St. Louis and Chicago in *Microwave Communications, Inc.* In this decision, the regulatory agency explained that this would provide an opportunity to observe the results and consequences of this limited service offering, including whether the system was negatively affecting the PSTN.

Phase 3. Shortly thereafter, the FCC opened a new proceeding—*Specialized Common Carrier*—to investigate whether, as general policy, the public interest would be served by permitting new entrants into the specialized communications field. In 1971, the FCC, to no one's surprise, answered in the affirmative. Again the authorizations were conditioned on avoidance of harm to the PSTN.

Phase 4. Ironically, the momentum generated by these procompetitive policies ended with the unintended early introduction of competition in the PSTN market. This ending was not due to a conscious FCC decision, but was the result of a court interpretation of a commission decision and was made over the objections of both the FCC and AT&T. As an enhancement to its specialized carrier services, MCI had introduced a new offering—"Execunet"—that the FCC authorized, but not as a PSTN service, or so it thought. Eventually, the matter ended up before the D.C. Circuit Court. In confirming MCI's argument that it had been granted entry into the PSTN business, the judges found that in writing the *Execunet* decision, although the FCC did not perhaps intend to open the field of common carrier communications generally . . . its constant stress on the fact that specialized carriers would provide new, innovative and hitherto unheard-of communications services clearly indicates it had no very clear idea of precisely how far or to what services the field should be opened.

The United States moved in similar incremental fashion in introducing competition to Intelsat, which, at the time, was the space-borne component of a global telephone network that was providing universal service. First, it is necessary to trace the evolution of the international policies that led up to the implementation of this policy.

SUBMARINE TELEGRAPH CABLE ERA

Fittingly, the first significant international telecommunications policy action taken by the United States involved trade in telecommunications services. In 1869, President Grant denied a French cable from landing on U.S. shores because, as he explained to Congress 6 years later, the French applicant had already received a long-term exclusive agreement from the French government that precluded U.S. cables from interconnecting with land lines in France. He presented a number of principles that he would follow in considering future applications from foreign companies to land cables in the United States. These included forbidding any foreign-owned cable from landing on U.S. shores unless the respective foreign government allowed U.S. cable companies to land and freely connect and operate through its telegraph systems. Grant further stated that he would continue to pursue this policy in the absence of any congressional advice to the contrary. This executive prerogative was upheld by his successors through President Wilson, with the exception of President Cleveland. As discussed in chapter 5, Congress institutionalized this presidential policy when it passed the *Cable Landing License Act of 1921*. This congressional action blocked Western Union Telegraph's legal challenge of President Wilson's decision to prevent a British cable ship from landing a Western Union Telegraph cable from Bermuda in Florida. Thus, "open shores" became a national policy, and "international comity" became an international principle that remained in the lexicon of international telecommunications policy into the 1990s.

HF RADIO ERA

Generally paralleling Wilson's action against a British cable ship was his intervention to prevent a British businessman, Gugliemo Marconi, from acquiring key HF radio technology developed by a General Electric engineer. This latter action, as discussed in chapter 5, led to the creation of RCA as the nation's de facto chosen instrument for international HF radio and the corporation's ascendancy as the dominant U.S. international radiotelegraph carrier (see Borchardt, 1970). In 1936, Mackay Radio and Telegraph Company filed an application with the FCC seeking to compete against RCA between the United States and Norway, only to have the application rejected by the FCC on the grounds that a competing circuit would not be authorized where adequate service exists; Mackay's appeal was unsuccessful. This policy was waived during World War II at the request of the Board of War Communications to enable the U.S. telegraph carriers (hereafter international record carriers [IRCs]) to provide alternative facilities and services over the same routes, but it was reinstated after the war ended. The State Department then negotiated the *Bermuda Telecommunications Agreement* with the British, which allowed for one telegraph path between the United States and Australia, New Zealand and India, and the opening of one path to the other Commonwealth points. Whereupon the FCC embarked on what one senior staff member called a "cautiously pro-competitive policy," in which circuits were allocated by traffic routes (see Ende, 1969). To enhance their business prestige, RCA and Mackay were authorized "to obtain about as much [Commonwealth] traffic proportionately as they enjoy on a worldwide basis." A dissenting commissioner argued that, to promote competition, the agency had established two monopolies, each to operate exclusively on its assigned routes. Mackay proceeded to challenge what was known as the single-circuit policy when it applied to provide service to and from the Netherlands, Portugal, and Surinam. The FCC found that the volume of traffic on the first two routes could sufficiently support a duplicate circuit and authorized the applications based on the view that there was a national policy of competition. The regulatory action was contested all the way by RCA to the Supreme Court. In the oft-cited 1953 *FCC v RCA Communications* decision, the justices concluded that competition for competition's sake is not a matter of national policy. The commission reopened the docket and reauthorized the competing routes to Mackay by demonstrating the beneficial effects on services and rates that would result from the introduction of competition.

The lesson learned from the prior regulatory episode was that, in the future, policymakers and regulators would have to move in a similar incremental and "cautiously procompetitive" way in introducing competition, especially in a market populated by a government-created chosen instrument—Comsat—and an International Satellite Organization—Intelsat.

SUBMARINE TELEPHONE CABLE ERA

The laying of the first submarine telephone cable between the UK and North America—TAT 1—marked the beginning of a short-lived period of normalization in international communications, as discussed in chapter 5. During this period, AT&T bilaterally negotiated consortium agreements with its foreign partners, the PTTs, for new submarine telephone cables on a regional cable-by-cable basis. The telephone company then filed applications for a Cable Landing License and for authority to construct the underwater facility with the expectation that, if the terms and conditions of the cable agreement were reasonable, the applications would be approved.

The other significant policy development concerned merger of the IRCs. When Congress passed the *Radio Act of 1927*, the lawmakers had sought to promote intermodal competition by prohibiting mergers of the new radio carriers—namely, RCA—and cable companies—a prohibition that was restated in the *Communications Act of 1934*. However, from almost the outset, the federal government had concerns over the economic viability of the IRCs, which, through separate corporate subsidiaries that had been established pursuant to the law, were struggling to keep their structurally separated submarine telegraph cable systems and HF stations up and running while competing against each other in a limited market.

Congress considered but rejected mergers of the struggling IRCs twice in the 1930s and twice in the early 1940s. In 1945, prior to the end of WWII, the Department of the Navy proposed a mandatory merger of all international carriers (i.e., AT&T and the IRCs) into a single telecommunications entity—a "chosen instrument"—although those words were not used at the time. The new international entity would be completely separated from the companies' domestic carrier and equipment manufacturing interests. In congressional hearings on the proposal, the FCC supported the principle of merger, but took no position on whether it should be mandatory or permissive, whereas the Department of State reversed an earlier position and opposed merger altogether.

President Truman's Communications Policy Board revisited the issue again in 1950 and 1951 and concluded that, although a congressional report predicted that the IRC industry, as structured, could not survive, a merger could not be recommended. Among the reasons given were the nation's partial mobilization because of the Korean War and the FCC's being unable to find any compelling arguments in favor of merger. The Board, however, noted that changing conditions could provide compelling reasons for a merger in the future and that future technological developments could determine the fate of these companies.

In 1959, Congress revisited the issue again. The Justice Department opposed any merger, but indicated that the IRCs could lease submarine telephone cable circuits from AT&T to enable them to continue to compete against each other for service while using a common facility. Thus, AT&T became a "carrier's carrier"—a concept that would be clipped and pasted in the formative years of communications satellite policy. In the 1964 *TAT-4* decision, the FCC authorized the IRCs to purchase capacity through circuits on new submarine telephone cables through an indefeasible right of use (IRU) arrangement. This arrangement was continued in subsequent analog submarine cable facilities. Since the arrival of digital submarine fiber optic cable systems, investment has been measured in minimum practical units of ownership (MAOUs) or minimum units of investment for ownership use (MIUs).

COMMUNICATIONS SATELLITES AND THE GLOBALIZATION PROCESS

The following discussion of regulating communications satellites on the way to globalism tracks the development of a national communications satellite policy by successive presidential administrations. It began as a spin-off from a FCC proceeding on frequency spectrum allocation, shifted to the White House during the Eisenhower and Kennedy administrations, and then migrated to Congress where it was ultimately sculptured with passage of the *Communications Satellite Act of 1962* to establish Comsat as a national "chosen instrument." This national policy of a chosen instrument became internationalized when Intelsat was established as the single global satellite system. Policy was adapted the following decade with the creation of Inmarsat to provide global maritime communications to the maritime industry.

Developments in international communications satellite policy from the 1970s through the rest of the century centered on seeking ways to introduce new international competitors and services through accommodating and eventually undoing the concepts of a national chosen instrument and of the single global satellite system concept (i.e., Intelsat in fixed satellite services and Inmarsat in mobile satellite services). Advocates for these changes came from an array of policy sectors— economic, defense, foreign, national security, trade, industrial, space, and regulatory—at the national and international levels as governments responded to geopolitical change and the globalizing economy. In terms of American presidencies, from the Eisenhower administration forward, there were general similarities in the policies of Republican and Democrat administrations, with one notable exception: the abrupt shift during the Kennedy and Johnson administrations that led to the establishment of Comsat and Intelsat. Interestingly, another abrupt shift in the early 1990s occurred when the suggestion to privatize Intelsat and Inmarsat came not from the government, but from Comsat, the U.S. Signatory to and a primary investor in both international satellite organizations.

THE EISENHOWER ADMINISTRATION

The notion of commercial communications satellites was unexpectedly introduced during a domestic regulatory proceeding on frequency spectrum allocations—*Above 890*—when AT&T indicated that it was planning a commercial worldwide low-altitude communications satellite system that would be jointly owned with its foreign partners, the PTTs. The FCC responded by initiating a proceeding in July 1960 in which it asked how space communications could be commercially operated, based on the assumption that a single multiple-satellite system would best serve the public interest. However, the technical, operational, economic, and political magnitude of a global satellite system was such that a contrary view emerged beginning at the Democrat-dominated staff level in NASA along the lines that government should own and operate such a system at least in the beginning. The potentialities of a global satellite system also attracted the attention of Democrats in Congress.

In a preemptive move, Keith Glennan, the NASA administrator who was an Eisenhower appointee, informally coordinated a national policy statement with the White House in the summer of 1960. He announced it in an unlikely venue—a committee meeting of the Oregon State Department of Planning and Development—a few short weeks before the 1960 presidential election much to the displeasure of congressional Democrats. In endorsing private ownership of communications satellites, the policy called for: (a) accelerated development of communications satellites that did not duplicate ongoing industry and military activities; (b) government support for development of the new technology so long as necessary to ensure that the timely development of a commercially feasible system could be completed by private industry; and (c) provision of government launch services on a cost-reimbursable basis.

In December, just weeks after the presidential election, the hawkish Senate Space Committee, chaired by now Vice President-elect Lyndon Johnson, released a report on policy planning for space communications that cautioned against construing space telecommunications too narrowly and, in so doing, overlooking the immediate and rich potential of this technology as an instrument of U.S. foreign policy (see U.S. Congress, 1961). The report emphasized employing satellites to meet the domestic and international communications needs of "the new nations" who were in the middle of the ideological battle between the U.S. and its allies and the Soviet Union and its bloc. On January 1, 1961, less than 3 weeks before President-elect Kennedy's inauguration, the outgoing Eisenhower Cabinet formally announced the earlier informally announced policy of support for private ownership of commercial communications satellites.

THE KENNEDY AND JOHNSON ADMINISTRATIONS

Critics of the Kennedy administration have argued that the aggressive space policy that the president embarked on early in his administration was intended, in large part, to offset the negative political effects of the Soviets' successful orbiting of the first man around the earth and the Bay of Pigs fiasco. Whatever the motivation, on May 25, President Kennedy presented his "man on the moon" message to Congress, which included an initiative to make the most of growing American leadership in space by accelerating development of space satellites for worldwide communications. This had the effect of transforming communications satellite policy from a question of international telecommunications regulation to a national foreign policy goal. The day before the man on the moon message, the FCC released a Report and Order in which it concluded that an international joint venture, with U.S. ownership and operation in an international satellite system limited to the U.S. international common carriers (i.e., AT&T and IRCs), would be in the public interest.

Creation of Comsat

The debate escalated to Congress, where clear battle lines were drawn: Republicans versus Democrats, conservatives versus liberals. Months of contentious congressional debate, including a near summer-long Senate filibuster, ended with a series of compromises that led to passage of the *Communications Satellite Act of 1962* to create Comsat in August 1962. During a temporary suspension of the filibuster in July to enable consideration of the fiscal year budgets for the executive branch agencies, NASA launched AT&T's experimental LEO, Telstar. The timing of the launch was coincidental based on a previous agreement between AT&T and NASA, but its impact was quite dramatic. It energized the supporters of the legislation to muster the votes needed to break the filibuster and pave the way for passage of the legislation.

One of the most significant congressional compromises involved the corporation's stated purpose. During the legislative hearings, the international common carriers had made it clear that no business case could be made for providing satellite services to the less developed countries in the early years of the worldwide system's deployment. However, the White House's National Space Council, the Department of State, NASA, and congressional Democrats countered, with obvious success, that a foreign policy case could be made. Thus, 43 years after RCA had been created as "an effective national instrument" as discussed in chapter 5 to challenge the British in international terrestrial radio communications, Comsat was established as the nation's chosen instrument to challenge the Soviets in international space radio communications. The corporation had a twofold mandate to establish "as expeditiously as practicable" a commercial global satellite system, which led to the establishment of Intelsat, and to make satellite technology available to both developed and "economically less developed countries." It is no coincidence that the wording of the first mandate closely resembled a U.S.-proposed UN General Assembly Resolution that had been adopted the preceding December, *General Assembly Resolution 1721 (XVI)*. The negotiations leading up to the establishment of Intelsat through Interim Arrangements are discussed in chapter 5.

Another significant compromise involved the ownership of the privately owned corporation. Half of the shares would be owned by the international common carriers until such time that they chose to sell their investment, which they eventually did at a significant profit; the other half of the shares would be owned theoretically by individual investors whose taxes had helped finance the development of the satellite and launch vehicle technologies. As shareholders, the international carriers would be entitled to six seats on the Board of Directors, six other members would be elected by the public shareholders, and three members would be appointed by the president subject to confirmation by the Senate. The precedent for government representation on the Board had been set when RCA was established. However, unlike the Navy Department member who sat on RCA's Board as an observer, the three Presidential Directors on Comsat's Board, who were leaders from industry, labor, and academia, could vote.

Compromise language in the Act also required the FCC to resolve two thorny issues, each marked by lengthy, contentious proceedings. The first issue involved Comsat's role as a carriers' carrier, the precedent for which had been established in submarine telephone cables. Under the commission's *Authorized User* decision, only international carriers authorized by the commission would be allowed to access U.S. space segment through Comsat. The second issue involved ownership and operation of the U.S. earth stations accessing the Intelsat space segment. Although the *Comsat Act* had established the new corporation as the sole U.S. investor in the space segment of the new international satellite system, the *Earth Station Ownership* decision halved the ownership of the domestic ground segment between the international carriers and Comsat. The Corporation also became manager of the earth stations and chaired the Earth Station Ownership Committee (ESOC).

The Rostow Report

During the last year of the first full term of the Johnson administration, a number of controversial communications issues posed potential political problems for the White House, including the renegotiation of the then-Interim Intelsat Arrangements, the fragmented international common carrier industry and the continuing plight of the ailing IRCs, the issue of what entity should provide domestic satellite service, the first stirrings of the restructuring of the domestic telecommunications industry, and the need to establish a more coordinated approach to government policymaking. President Johnson's political response was the creation of a "Presidential Task Force for Communications Policy" in August 1967 to study these issues and report back with recommendations within a year, either immediately prior to or shortly after the presidential election. The Task Force came to be known as the "Rostow Task Force," named after its chair, Eugene Rostow, then-Under Secretary of State for Political Affairs. Quite unexpectedly, President Johnson elected not to seek reelection late in the summer of 1968. The Task Force nevertheless finalized its report with recommendations, but the report was not released by the outgoing administration (see *Final Report: President's*, 1968). Under pressure by congressional Democrats, the report was informally released without comment soon after the new Nixon administration took office. The three most significant proposals, from the perspective of this satellite policy analysis, were:

- The basic international transmission functions now performed by Comsat and the international voice and record carriers should be merged in a single entity.
- The United States should remain committed to Intelsat: "Allowing separate U.S. entities to compete with one another by satellite for international traffic would be difficult, if not impossible to reconcile with our commitment to the Intelsat concept of a global satellite system."
- A pilot domestic satellite program should be established. Comsat, as trustee, should own and maintain the space segment, and Comsat, the terrestrial common carriers, and prospective users of wideband services should be eligible to invest in the ground segment.

THE NIXON/FORD ADMINISTRATIONS

The election of a new president from the opposition party always brings with it the potential for change. To the surprise of many, the new Nixon administration signaled changes in communications satellite policy at the same time that the United States was seeking to protect the Intelsat single system concept in negotiating the Intelsat Definitive Agreements. To the surprise of only a few, the integration of the Intelsat system into the traditional terrestrial international telecommunications system proved increasingly problematic. To the surprise of everyone, a commercial global maritime communications satellite system was proposed.

Shifts in Satellite Policy

The changes in policy concerned domestic communications satellites and a step away from the preceding administrations' chosen instrument approach.

Domestic Satellite Policy. Establishment of a domestic satellite system had surfaced at the FCC in the mid-1960s when the American Broadcasting Company (ABC) filed an application to launch and operate a domestic broadcast distribution service via satellite. The regulatory agency's response had been to return the application to ABC "without prejudice" and, owing to the novel legal and technical questions that the application raised, to initiate a formal proceeding in 1966 to determine, "as a matter of law," whether it could even authorize such a system. ABC, the Ford Foundation, AT&T, and Comsat subsequently filed applications. As the proceedings continued, there was speculation that the commission was leaning in the general direction of the Rostow Task Force's recommendation regarding Comsat's role in a domestic satellite program.

In January 1970, an assistant to President Nixon wrote a letter to the FCC Chairman suggesting a dramatically different approach along the lines that government should encourage and facilitate the development of commercial domestic satellite communications to the extent that industry finds them economically and operationally feasible and should not seek to promote uneconomic systems or dictate ownership arrangements (i.e., any financially qualified public or private entity, including government corporations, should be permitted to establish and operate domestic satellite facilities for its own needs). Interestingly, during congressional hearings on legislation to establish Comsat 9 years earlier, a senator had sought the following clarification: "Let me get this straight. If we don't report a bill out of this committee, and the communications satellite situation remains in a status quo, couldn't the Government step in and say that anyone with resources that is fit, willing, and able to pay for the booster can put up a satellite?" The answer was yes (see Oslund, 1975, 1977). Coincidentally, a separate, broader based, high-tech study identified *laissez innover* as an emerging trend in policy, the main thrust of which was that ongoing technological change was increasingly attributable to market forces and skills of particularly capable entrepreneurs, not government policy (see Hetman, 1973).

A few short weeks after receiving the White House letter, the FCC adopted what came to be known as the "open skies policy." It was based on the critical consideration that "what persons with what plans are presently willing to come forward to pioneer the development of domestic communications satellite services according to the dictates of their business judgment, technical ingenuity, and any pertinent public interest requirements laid down by the Commission." Eight applications for domestic satellite systems were filed shortly thereafter. The implication for Comsat and Intelsat was the transfer of U.S. traffic between the continental United States and "off-shore" locations (i.e., Alaska, Hawaii, and Puerto Rico) from the global system to the U.S. domestic systems. Both the White House letter and the FCC decision emphasized that, in establishing domestic satellite systems, the United States would ensure that it would fulfill its Intelsat obligations to go through a consultation pursuant to the Intelsat Agreement.

Aerosat Policy.   Another significant shift occurred in January 1971 when the White House Office of Telecommunications Policy (OTP), newly established with the intention of coordinating national telecommunications policy, issued a policy statement to halt planning that was underway regarding an intergovernmental experimental/evaluation aeronautical satellite (Aerosat) program to be jointly owned, operated, and managed by NASA and European Space Research Organization (ESRO). Arguing that such a program was inconsistent with the Eisenhower administration's policy that communications satellites should be privately, not government, owned, and the Nixon administration's domestic open skies policy that satellite systems should be built in response to the competitive marketplace rather than government direction, OTP directed the FAA, as the lead management agency, to renegotiate the arrangement in accordance with the previously stated policies. Shortly thereafter, as arrangement was agreed on to proceed with a restructured AEROSAT program that allowed the European Space Research Organization (ESRO) to select between competing U.S. companies for the provision of space segment for the program. A formal MOU was agreed in late summer 1974. ESRO selected Comsat General, a newly organized Comsat subsidiary, to provide the space segment. However, a few months later, the FAA canceled the issuance of an RFP due to lack of funding.

Satellite/Cable Dichotomies

Congress created industrial and international structural dichotomies between the United States and foreign administrations when it passed the *Comsat Act*. Domestically, Comsat, as a carrier's carrier, leased capacity to AT&T and the IRCs. These carriers, who came to be called the U.S. international services carriers (USISCs), owned and operated the U.S. portion of submarine telephone cable systems. Because satellite leases could not be capitalized by the USICSs and investments in submarine cables could, there was an economic disincentive to use satellites. Internationally, the general rule was that foreign administrations owned and operated both the submarine cables and satellites through their PTTs. Another international dichotomy was that, besides owning and operating cable and satellite facilities, the PTTs were also the regulators. In sharp contrast, U.S. regulation was carried out by an independent domestic regulatory agency, the FCC, which had the congressional mandate to take into account the nation's "public interest" when authorizing new international facilities.

Prior to the *Comsat Act*, the USISCs bilaterally negotiated cable consortium agreements with their foreign correspondents on a regional cable-by-cable basis and submitted their applications for a Cable Landing License and a regulatory authorization to construct the facility with the expectation that they would be granted so long as the terms and conditions of the agreements were reasonable. After the *Comsat Act*, the aforementioned technological disparities between analog satellite and analog cable capacities and their operational lifetimes took on regulatory significance, as discussed in chapter 5. This is because the FCC was under congressional mandate to ensure, on behalf of the U.S. rate payers, that the most economic and efficient use was made of international facilities. Beginning in the early 1970s, the FCC began to regulate how the USISCs and, consequently, the foreign administrations would activate circuits on cable and satellite facilities. Allocation of traffic was nothing new to the FCC as evidenced by how the agency had allocated HF radio circuits. There was, of course, an obvious difference. In the case of HF radio, the FCC was regulating facilities owned by U.S. companies. In the case of submarine cables and satellites, while telling the USISCs how they should utilize their facilities, the domestic regulatory agency was also telling foreign governments (i.e., the PTTs) how to use their facilities. The FCC's intention was to both seek to enable the larger capacity satellites to be more effectively utilized during their relatively shorter operational lifetimes and prevent the smaller capacity submarine cables from saturating many years before the end of their operational longevity. In so doing, the FCC also was promoting bimodal competition, another adapted "clip-and-paste policy" from earlier times, while it was supporting the continued development of the Intelsat system. The disparity in capacities increased even more with the addition of a second operational satellite route in the Atlantic Ocean Region. At one point, the commission prescribed a "proportional fill" policy that required the USISCs to add five circuits on satellites to one circuit on a cable to and from a given country. This formula was roughly analogous to the comparative capacities of the facilities. The European governments responded with formal protests to the Department of State, and they threatened to stop placing new traffic on U.S.–European routes altogether—a trade war of sorts.

In an effort to prevent a further recurrence, the United States, Canada, and the Conference of European Postal and Telecommunications Administrations (CEPT) agreed, in 1975, to establish a bilevel North Atlantic Consultative Process (NACP) for future submarine cable and satellite facilities planning. The senior level was comprised of CEPT, Canadian, and U.S. government officials and industry chief executives. The working level was comprised of senior staff-level government and industry subject matter experts on traffic projections, operational planning, and economics. This too proved problematic. When AT&T filed an application proposing a 1979 entry into service of TAT-7, which proved to be the last analog submarine cable to be deployed in the North Atlantic, the FCC delayed the cable's entry into service until 1983. To the chagrin of the foreign administrations and the USISCs, this decision to delay was derived from mutually agreed-on traffic projections and operational planning principles, from which there also emerged a "balanced loading" methodology of placing generally the same number of new circuits on each satellite and cable route to and from a given country (see Oslund, 1983).

Prior to the 1975 NACP meeting, OTP had released a study on cable/satellite facilities planning. The study argued that allocation of circuits between transmission media to promote competition represented unnecessary government intervention and proposed that reliance should be placed on "natural market forces." To alleviate the existing economic disincentives, USISCs should be allowed to capitalize long-term satellite leases. The proposal was not accepted. A few years later the FCC, in light of the planned introduction of high-capacity fiber optic cables that would alleviate the glaring disparities between analog cable and satellite facilities, encouraged Comsat and the USISCs to negotiate long-term satellite leases in advance of constructing any future satellites (see chap. 5, Table 5.5).

These leasing arrangements avoided further international political problems and enabled the USISCs to gradually shift major portions of traffic from satellites to fiber optic cables.

Marisat

In response to the U.S. Navy's need for additional satellite capacity for a relatively short period of time, Comsat General developed and, beginning in 1976, operated "Gapsat"—a maritime communications satellite—in the Pacific, Atlantic, and Indian Ocean Regions (see chaps. 3, 5, and 7). The plan was to lease communications satellite services to the Navy so long as the need existed and also to offer commercial services to the maritime shipping and offshore industries. The Comsat subsidiary originally had planned to own and operate the satellite system outright, but the international common carriers argued convincingly before the FCC that the new maritime satellite service would be competing against their ailing maritime coastal radio stations. Consequently, the Marisat joint venture was formed in which Comsat General owned approximately 86% and served as system manager. RCA Global, ITT Worldcom, and Western Union International co-owned the remaining investment shares; AT&T chose not to invest. The United States became the first Intelsat Party to consult with Intelsat pursuant to Article XIV(d) when the application for the Marisat system was filed at the FCC to offer international public communications services.

Preview of Global Market Dynamics

Dramatic changes were starting to take place in the international marketplace, including signs of shifts in trade patterns and balances. This led to the establishment of a White House Council of International Economic Policy early in the Nixon administration. Telecommunications and satellites were identified as two areas that would receive close attention. Across town, the FCC, as part of its domestic-oriented procompetitive policymaking, adopted the "Registration Program" in 1975. The intent of this program was to encourage competition in the provision of terminal equipment by allowing any manufacturer whose equipment met established standards to enter the U.S. domestic market. An unintended consequence was to open the lucrative domestic market to foreign competition before major American equipment suppliers, other than Western Electric, were ready to compete. It was this action, not AT&T's divestiture a few years later, that began an unfavorable shift in the balance of trade in the telecommunications sector and prompted Congress to warn against any further unilateral liberalization in the telecommunications sector when it passed *The Omnibus Trade and Competitiveness Act of 1988* (see Oslund, 1990).

THE CARTER ADMINISTRATION

When the International Maritime Consultative Organization, later renamed the International Maritime Organization (IMO), started to study the feasibility of providing maritime satellite services in 1973, the FCC had initiated a proceeding to look into how the United States could participate in such a system. As the idea of establishing some kind of commercial international maritime satellite system started to gain support, three alternatives emerged: (a) the provision of maritime satellite services through Intelsat, (b) the provision of maritime services through a private system along the lines of Marisat, and (c) the provision of maritime services through a new international satellite organization. Complicating the commission proceeding was the existence of two competing applications submitted by Comsat: one from the Intelsat Signatory division within the corporation, and the other from Comsat General, the principal owner and operator of the Marisat system.

An early major initiative of the Carter administration was the creation of the National Telecommunications and Information Administration (NTIA), the successor to OTP. NTIA undertook a series of consultative meetings with foreign administrations to exchange views on maritime satellite-related issues. Following these meetings, there was a discernible shift toward the alternative that involved the establishment of a new international satellite organization that ultimately resulted in the creation of Inmarsat. Comsat was designated as the U.S. Signatory when Congress passed the *Inmarsat Act* in 1978. The FCC rulemaking, practically speaking, was overtaken by the international events. The establishment of Inmarsat is discussed in chapter 5.

THE REAGAN ADMINISTRATION

Telecommunications and trade policy converged in the 1980s. According to one senior policymaker, trade officials were being "prodded and pulled into communications issues which previously were the exclusive domain of communications engineers and regulatory officials," and telecommunications technologies and applications were among the "hottest new topics" (see Feketekuty & Hauser, 1984). Efforts to liberalize rules and regulations in foreign markets and, later, to privatize PTTs headed the agendas of almost each and every U.S. delegation going into bilateral and multilateral discussions.

Policy "Spins"

U.S. policymakers announced two generally similar spins to further promote competition in both the communications satellite and fiber optic submarine cable realms.

New Intelsat Policy. Three international satellite systems (i.e., Inmarsat and two regional systems, Arabsat and Eutelsat) were consulted with Intelsat under Article XIV(d) of the Intelsat Agreement without controversy. Strongly encouraged by the U.S. aerospace manufacturers, six U.S. companies expressed interest in establishing international communications satellite systems. Most notable of the potential entrants were Orion Corporation and PanAmSat. In a move specifically linked to the notion of the FCC's earlier discussed domestic *Above 890* decision, Orion, in its application to the FCC, argued that as a privately owned system, it was planning to sell or lease transponders to major business users on both sides of the Atlantic Ocean. Further, because no common carrier services were to be offered, the system would not be subject to the Article XIV(d) consultation with Intelsat concerning "significant economic harm." Instead, Orion should only be subject to coordinating with Intelsat for technical compatibility. This argument did not pass muster with the State Department's Legal Adviser, who determined that Article XIV{d) consultation would be required of any U.S. system proposing to provide international public communications services (see Legal Adviser, Department of State).

In November 1984, President Reagan announced a widely anticipated "Presidential Determination" that the establishment of "separate international communications satellite systems are required in the national interest" (see Reagan, 1984). The Departments of State and Commerce were directed to inform the FCC of criteria necessary for the U.S. Party to "meet its international obligations and to further its telecommunications and foreign policy interests" in implementing what later became known as the "separate systems policy" (see Secretary of Commerce, 1984). Two criteria were transmitted to the FCC chairman in a letter from the secretaries of the two departments. The first criterion appeared, in part, to have been "clipped and pasted" from the earlier FCC domestic "Alternate Systems" decisions. Each system would be restricted to providing services through the sale or long-term lease of transponders or space segment, but none of the services was to be interconnected to the PSTN, thereby avoiding any significant harm. The second criterion was that at least one more "foreign authority" would need to join the United States in initiating the Article XIV(d) consultation. Panamsat became the first such system to be consulted successfully in early 1987. Orion followed in 1989. However, both consultations were time-consuming and contentious.

New Submarine Cable Policy. Another paradigm shift occurred in 1985 when the FCC authorized "Tel-Optik" (*aka* PTAT-1) to link the United States and the UK. In setting a new "general policy direction" in *Tel-Optik*, the commission stated that private alternative cable systems should be allowed to introduce "more meaningful competition" in the North Atlantic. This system was to be jointly owned by a private American corporation and Cable & Wireless, in contrast to the existing submarine cable systems that were jointly owned by the USISCs and foreign PTTs.

THE BUSH "'41" ADMINISTRATION

The end of the cold war and the accompanying change in the geopolitical realignment brought about an adjustment in the nation's launch vehicle policy and the transfer of DOD-funded technologies to the commercial sector.

Shift in Launch Policy

After the Soviet Union ceased to exist, the United States reversed its "*nyet*-Proton" launch policy during the June 1992 America/Russian Summit. As discussed in chapters 4, 5, and 7, the reversal involved the "Inmarsat exception" that enabled Inmarsat to consider the Proton as an alternative to launch the new Inmarsat 3 series of satellites. The long-standing cold war-driven launch policy had come under increasing scrutiny as Soviet policies had visibly been liberalized by the communist party in the late 1980s, *albeit* by political necessity, not by choice. Also, opponents of the policy had argued that it was at odds with the aggressive stance being taken internationally by U.S. negotiators to promote competition and remove barriers to trade. This change in policy not only gave communications satellite system operators a new launch option, which some argued was more reliable and cost-efficient than existing Western launch vehicles, but it also set the stage for U.S. launch vehicle manufacturers to start "thinking about the unthinkable" and to explore joint ventures with their Russian and Ukrainian counterparts.

Commercializing LEO Technologies

As part of its evolving war-fighting strategy, the Department of Defense (DOD) began to seriously investigate the deployment of small low- and medium-altitude satellite technologies in the 1980s. After the Soviet government fell and changed the geopolitical equation, funding for these programs could no longer be justified. Nonmilitary applications for the emerging LEO technology were explored, including the transfer of the technology to the private sector. As it turned out, unlike the earlier initial transfer of the DOD-funded development of Internet to the civilian side of government before it was transferred to the commercial sector, the decision was made to transfer at least some of the new technologies directly to the commercial sector. Iridium became the de facto U.S. chosen instrument for mobile LEOS, and the government aggressively lobbied for and achieved badly needed MSS frequency allocations during WARC 92. Globalstar and Inmarsat's ICO were two other serious competitors. Comsat was allowed to participate in Inmarsat's entry into the Big LEO market, but on the condition that the new handheld service would be provided by Inmarsat through a fully separate entity, ICO, to which Inmarsat agreed. Inmarsat later sold off its ownership in ICO. Teledesic, a fixed LEO system, was the result of further technology transfer. The Clinton administration made Teledesic the de facto chosen instrument for fixed LEOs when the United States aggressively lobbied for and achieved badly needed frequency allocations during WRC 95.

THE CLINTON ADMINISTRATION

An acceleration in the globalization process occurred in the early years of the Clinton administration. A suggestion was made that the time had come to privatize Intelsat and the other global ISO, Inmarsat. However, it came not from the government, but from senior executives of Comsat, the U.S. Signatory to and a primary investor in both ISOs. The Uruguay Round ended successfully with agreements to open telecommunications markets to competition through the *General Agreement on Trade in Services (GATS)* and the establishment of the World Trade Organization (WTO).

Privatization of the ISOs

In 1993, Comsat senior management suggested the full privatization of Intelsat and Inmarsat. This is because their governance "didn't allow [them] to be quick of foot in a heavily competitive environment" (see Gifford, 1998). Indeed, in the midst of the globalizing economy, efforts by the Intelsat and Inmarsat owners to adapt were hampered by their institutional frameworks. The consensual form of decision making was proving to be much too cumbersome. For example, 12 to 16 months were normally required for a decision to be taken by either the Intelsat Board of Governors or the Inmarsat Council on a major space segment procurement or a major new service offering. Further, procedures to amend the ISOs' governing instruments to enable them to enter new markets were likewise cumbersome. In the case of Inmarsat, it took the Parties 5 years to complete the process of amending the convention and thereby allow the organization to provide aeronautical satellite services on a competitive basis.

As discussed in chapter 5, Intelsat was facing competition from an increasing number of fiber optic cable systems and new separate satellite system competitors at existing and new frequencies. Domestically, the FCC had taken a number of incremental, procompetitive decisions affecting Comsat directly and Intelsat indirectly. Examples included *Transborder Satellite Video Services* (see FCC, 1981) that authorized transborder use of U.S. domestic satellite systems whose footprints spilled over into neighboring countries, although provision of PSTN services was prohibited, and *Ownership and Operation of U.S. Earth Stations* (see FCC, 1984) that required the unbundling of the ground and earth segments at the U.S. earth stations and eventually led to the dissolution of ESOC and AT&T's assumption of ownership of the major continental U.S. earth stations. After years of putting on hold a growing number of petitions to allow U.S. carriers and other users to bypass Comsat altogether, and to directly access the Intelsat space segment economically and operationally, in 1990 the commission had initiated the *Direct Access* proceeding. Such direct access to Intelsat and Inmarsat was eventually granted, but was conditioned on the privatization of the ISOs. Further, at the urging of the United States, Intelsat had started to relax the Article XIV(d) significant economic harm test by setting and then gradually raising allowable caps on PSTN and non-PSTN traffic. The eventual elimination of the significant economic harm test altogether was a foregone conclusion. With respect to Inmarsat, on the mobile satellite side, while the ISO had started to provide aeronautical and land mobile services on a competitive basis, a number of proposed Big LEO systems were seeking financing based on business plans targeting Inmarsat's maritime, aeronautical, and land mobile markets.

THE BUSH "'43" ADMINISTRATION

Efforts to further facilitate and adapt to the globalization process have continued during the Bush "43" administration, but they have been overshadowed by the horrific events of September 11 and the military operations in the Middle East. The most dramatic developments of the administration's communications satellite policy, from this author's perspective, have taken place in the realm of national strategies to achieve redefined national interests. They are discussed in chapter 7.

THINKING GLOBALLY

From the early 1990s forward, the globalization process increasingly generated its own momentum, helped along the way by continuing negotiations in multiple venues to break open national markets in general, and telecommunications markets in particular. Among the worldwide venues were the Uruguay Round of the GATT negotiations, the International Telecommunication Union, and the Organization for Economic Development and Cooperation (OECD). Regional forums included the European Commission (EC), Asia-Pacific Economic Cooperation (APEC), the Organization of American States (OAS), and the Inter-American Telecommunications Commission (CITEL). At the same time, bilaterals were held with foreign economic and trade ministries. Of course the most far-reaching breakthrough, as discussed in chapter 5, was the successful completion of the Uruguay Round in 1994 that produced the *GATS* to start opening national markets in telecommunications, finance, and air and maritime transport services, and the establishment of the WTO. The *Information Technology Agreement* and the *Basic*

Telecommunications Agreement were the first two trade pacts to be negotiated within the WTO framework.

Thus, global became the scale at which incumbents, competitors, policymakers, and regulators have since had to scramble to survive and adapt. The sampling that follows reveals the scope and breadth of the resulting changes, although the fluidity of events defies presenting them chronologically or in order of importance.

Global Linkages

Shortly after the WTO came into being to oversee the implementation of *GATS*, the FCC acted in the *Foreign Carrier Entry Order* (see FCC, 1995) to formally link the national regulatory framework to the new global trade regime. A new regulatory standard—effective competitive opportunities (ECO)—was adopted to govern foreign carrier entry into the U.S. market. Reminiscent of the conditions that President Grant had applied to allow foreign cables to land on U.S. shores some 125 years earlier, the "ECO test" was intended to ensure that governments of foreign applicants seeking to provide terrestrial or radio common carrier services in the United States were affording effective competitive opportunities to U.S. companies in their markets. This litmus test was carefully portrayed as not being trade-related reciprocity, but had the effect of replacing *international comity* as the term of art that had been used in international telecommunications since the Grant declaration of "open shores." With respect to communications satellites, the *DISCO I* decision (see FCC, 1996), which was described at the time as a response to "the trend toward a globalized market," relaxed restrictions of the earlier transborder and separate satellite systems policies and eliminated distinctions between U.S.-licensed domestic and international satellite systems. Coincidental with the entry into force of the WTO's *Basic Telecommunications Agreement*, the FCC replaced the ECO test, in the *Foreign Participation Order*, with "an open entry standard." This standard was based on "a rebuttable presumption in favor of entry" for foreign carriers from other WTO members seeking to provide services in the United States, to land submarine cables in the United States, or, as discussed in chapter 7, to exceed ownership restrictions of radio licenses unless it could be demonstrated that an application could pose a high risk to competition in the U.S. market and FCC safeguards and conditions would be ineffective. At the same time, *DISCO II* (see FCC, 1997) established a uniform licensing framework for foreign-licensed satellite systems wishing to provide services in the United States, including an open entry standard similar to the *Foreign Participation Order* (see FCC, 1997). In this instance, the standard was based on "a presumption that entry by WTO member satellite systems will promote competition in the U.S. satellite services market" unless demonstrated otherwise.

Global ISO Restructuring

Over 5 years of international negotiations led to Inmarsat's privatization, effective mid-April 1999, while Intelsat privatization took effect in July 2001. Late in the privatization process, Congress sought to apply additional pressure on Intelsat with highly publicized hearings and passage of the pro-privatization Open-Market for the Betterment of International Telecommunications Act (ORBIT Act) in March 2000. This action was resented by the foreign Parties and Signatories in light of the progress being made. In the end, the privatized Intelsat and Inmarsat entities were generally consistent with the guidelines of the legislation. The Orbit Act enabled the Lockheed Martin Corporation, after 2 years of deliberations before the FCC and Congress, to finalize a merger with Comsat. Almost simultaneously, Comsat was folded into a new Lockheed Martin subsidiary, Lockheed Martin Global Telecommunications (LMGT). Since then Lockheed Martin has sold off its business enterprises in the two ISOs and has disbanded LMGT. Overall, this was a curious and unexpected final chapter of the history of Comsat—"the inevitable anomaly" (see Kvan, 1966).

Global GEOs

Domestic systems have merged, changed into regional systems, and, in turn, been subsumed into global systems. Five fixed global GEO systems were operating at the turn of the century. Intelsat remained the largest in terms of investors and countries served, although SES Global and Panamsat had more satellites in orbit. The fourth system was New Skies, a spinoff from Intelsat as part of the privatization process. Last, Intersputnik continued its operations. Speculation continues regarding further consolidation of these systems. Based on optimistic assumptions regarding the dot.com explosion and the continued growth in the telecommunications industry and markets, multiples of Ka satellites and systems for broadband services were planned; some were even registered in the ITU. However, the sequential implosions of the dot.coms and telecommunications service and manufacturing sectors have resulted in these systems being shelved, some for the moment, others forever.

Global LEOs

Within the reorganized ITU, proponents of Big LEO systems negotiated a "Memorandum of Understanding to Facilitate Arrangements for Global Mobile Communications by Satellites" in 1997 to facilitate the mutual recognition and cross-border movement of the hand-held terminals, thereby allowing users of the proposed LEO systems to roam across political frontiers. The relatively informal low-key manner in which the arrangements were negotiated and agreed on allowed operators of the new systems to bypass formal elongated ministerial negotiations that could have resulted in the erection of barriers to this new form of trade in services.

Entry into service of two of the Big LEOS—Iridium and Globalstar—arrived after much fanfare, but later and costlier than originally planned. A third, ICO, launched one satellite. All three have since sought Chapter 11 protection. The Big LEOs' financial plight has been attributed primarily to the faster than expected ubiquitous build-out of cellular and PCS systems. Orbcomm, a Little LEO, was also deployed and had to apply for Chapter 11 protection. Suffice to say, plans for the handful of other mobile LEO systems have either been placed in the inactive file or abandoned entirely. The proposed fixed LEO system, Teledesic, an "internet in the sky," has not launched any satellites and may never do so.

Global Fiber Optic Submarine Cable Systems

Tel-Optik (*aka* PTAT-1), jointly and privately owned to link the United States and UK, became the first private system to be authorized by the FCC in 1985. As noted earlier, *Tel-optik* articulated a new "general policy direction" through which private alternative cable systems would introduce "more meaningful competition" in the North Atlantic. A later authorization, the *Japan–U.S. Order* (see FCC, 1999), had the effect of completing the transfer of the separate systems policy from satellites to fiber optic submarine cables. In the proceedings leading up to the authorization of the Japan–U.S. cable system, the incumbent cable system owners and operators were referred to by their would-be competitors as the "cable club," comprised of U.S. common carriers and their foreign partners, the Telecommunications Organizations who formerly were part of PTT structures before being split off and, in some cases, privatized. Plans for and actual deployment of other noncommon carrier-owned fiber optic cable systems followed in the Atlantic, Pacific, and Indian Ocean regions and the Caribbean Sea. Those that were deployed produced a "giga-glut" in fiber optic capacity that was exacerbated by the collapse of the dot.coms and the bursting of the telecommunications bubble. Two ambitious hybrid-owned global systems, Fiberoptic Link Around the Globe (FLAG) and Global Crossing, were constructed, although both sought Chapter 11 protection. Two other ambitious global systems were proposed, but were not constructed—TYCO's Global Network and OXYGEN. The latter was intended to be "a living network" that was to have linked approximately 100 landing points in 78 countries and locations.

Global Alliances

AT&T Divestiture in the early 1980s and continued changes in domestic regulations and the marketplace set into motion a restructuring of the U.S. domestic and international carrier industries that is still taking place. MCI and Sprint became long-distance competitors to AT&T in the long-distance telephone markets, domestically and internationally, while the IRCs (i.e., ITT World Com, RCA

Global, and WUI) disappeared through becoming parts of other companies or simply being dissolved. The three major long-distance carriers turned to ambitious international alliance building with foreign correspondents: AT&T's "World Partners," MCI's "Concert" with British Telecommunications, and Sprint's "Global One" partnership with France Telecom and Deutsche Telekom. Subsequently, the MCI/BT "Concert" was interrupted by Worldcom's acquisition of MCI, AT&T "World Partners" was disbanded when AT&T and BT forged a new "Concert" arrangement that has since been canceled, and Sprint sold its investment in "Global One," which in turn has ceased to operate as an alliance of any sort.

Global Geopolitics Revisited

The global marketplace developed in new dramatic ways, not only due to globalization, but also to decommunization. Consider, for example, the evolution of the bipolar relations between the United States and the Soviet Union, now Russia. Although it was still the USSR, the Soviet Union, as discussed in chapter 5, quietly signed a Memorandum of Understanding with, and, subsequently, became a member of Intelsat, and Intersputnik established formal relations with Intelsat. Although export controls remained "a" not "the" sticky problem in bilateral relations, "an exception" to U.S. launch policy was announced during the Bush–Yeltsin Summit, which allowed Inmarsat to consider and eventually select the now-Russian Proton to launch one of its new third-generation satellites. Further, Lockheed Martin Intersputnik constructed LM-1 that is part of the Intersputnik space segment. As discussed in chapter 4, Lockheed Martin formed a joint venture—Lockheed–Khrunichev–Energia International (LKEI)—in 1992 and, in turn, a joint venture with LKEI—International Launch Services (ILS)—in 1995 to market commercial Atlas and Proton launch services. Last, Boeing entered into an international partnership—Sea Launch—that included companies and employed and integrated technologies from Russia, the Ukraine, and Norway. The latter's contribution was in the form of a converted oil rig.

CONCLUDING THOUGHTS

As this chapter argued, amid the chaos of geopolitics and significant breakthroughs in international telecommunications technologies, there has been a general consistency in U.S. telecommunications and communications satellite policies and related policies. For the most part, the policies have been incremental, "clipping and pasting" precedents along the way, beginning soon after the Civil War and continuing through enactment of the *Radio Act of 1927*, the *Communications Act of 1934*, the *Communications Satellite Act of 1962*, the *Telecommunications Act of 1996*, and up to and through the turn of the 21st century. The Kennedy administration's cold war-related intervention concerning how communications satellites were to be commercially deployed in the form of a single global satellite system for not-too-subtle political reasons stands in stark contrast to the policy approaches of the immediately preceding Eisenhower administration and subsequent administrations of both parties. One could easily argue that the establishment of Comsat and Intelsat during the Kennedy and Johnson administrations created dilemmas that U.S. policymakers and regulators spent the rest of the century seeking, initially, to accommodate and, ultimately, to undo. Thus, the focus on Comsat-, Intelsat-, and Inmarsat-related policies in discussing overall U.S. communications satellite policy has not been accidental either in the preceding chapter or in this one.

One of the most striking things about the two decades leading up to the Millennium was the simultaneous ending of the cold war (or World War III) that erased politically imposed physical boundaries and the acceleration of the globalization process that pushed aside artificially imposed trade barriers. Although separate, the gradual end of World War III and the continued ascendancy of globalism nevertheless were related; they had to have mutually reinforced each other to some degree. Clear lines had been easy to draw in earlier time periods to delineate geopolitical alignments and strategic weaponry, to say nothing of international telecommunications policies and regulations, domestic and international telecommunications markets, and/or transoceanic telecommunications systems. By the turn of the 21st century, they had become virtually nonexistent. Asymmetries in political relationships and weaponry, on the one hand, and the increasingly uncertain global marketplace and churning in the domestic and international telecommunications and communications satellite industries, on the other, are the obvious consequences, geopolitically and economically, respectively, in the transition to globalism.

REFERENCES

Borchardt, Kurt, (1970), *Structure and Performance of the U.S. Communications Industry*, Boston: Graduate School of Business Administration, Harvard University.

Brock, G. W., (1981), *The Telecommunications Industry: The Dynamics of Market Structure*, Cambridge: Harvard University Press.

Downs, Anthony, (1967), *Inside Bureaucracy*, Boston: Little, Brown.

Ende, Asher H, (1969, Spring), "International Telecommunications: Dynamics of Regulation of a Rapidly Expanding Service," *Law and Contemporary Problems*, 34.

Federal Communications Commission v. RCA Communications, Inc., 346 U.S. 86 (1953).

Federal Communications Commission, TAT 4, 37 FCC 1151 (1964).

Federal Communications Commission, Transborder Satellite Video Services, 88 FCC 2d 258 (1981).

Federal Communications Commission, Modification of Policy on Ownership and Operation of U.S. Earth Stations That Operate with the INTELST Global Communications Satellite System, 100 FCC 2d 250 (1984) (*Ownership and Operation of U.S. Earth Stations*).

Federal Communications Commission, Tel-Optik, Ltd., 100 FCC 2d 1033 (1985).

Federal Communications Commission, Market Entry and Regulation of Foreign-Affiliated Entities 11 FCC Rcd 3873 (1995) (*Foreign Carrier Entry Order*).

Federal Communications Commission, Amendment to the FCC's Regulatory Policies Governing Domestic Fixed Satellites and Separate International Satellite Systems et al., Report and Order, 61 FR 9946 (1996) (*Disco I*).

Federal Communications Commission, Rules and Policies on Foreign Participation in the U.S. Telecommunications Market, Report and Order and Order on Reconsideration, 12 FCC Rcd 23,891 (1997) (*Foreign Participation Order*).

Federal Communications Commission, Amendment to the Commission's Regulatory Policies to Allow Non-U.S.-Licensed Space Stations to Provide Domestic and International Satellite Service in the United States, Report and Order, 12 FCC Rcd 24,094 (1997) (*Disco II*).

Federal Communications Commission, AT&T Corp. et al. Joint Application for a License to Land and Operate a Submarine Cable Network Between the United States and Japan, SCL-LIC-19981117-00025, Cable Landing License, 14 FCC Rcd 13066 (1999) (*Japan-U.S. Order*).

Feketekuty, Geza, and Hauser, Kathyrn, (1984), "A Trade Perspective of International Telecommunications Issues," *Telematics and Informatics*, Vol. 1.

Final Report: President's Task Force on Communications Policy, (1968, December 7), Rostow Task Force, Washington, DC: Government Printing Office.

Frieden, Rob, (1996), *International Telecommunications Handbook*, Boston: Artech House. (**Note:** Documented discussion of contemporary developments in international telecommunications technologies and policies.)

Frieden, Rob, (2001), *Managing Internet-Driven Change in International Telecommunications*, Boston: Artech House. (**Note:** Updated documented discussion of developments in international telecommunications technologies and policies.)

Galloway, Jonathon F., (1972), *The Politics and Technology of Satellite Communications*, Lexington: Lexington Books.

Gifford, James M., (1998, May 1), "And in this corner . . . Betty Alewine, Chief executive officer of Comsat Corp," *Satellite Broadband*.

Hetman, Francois, (1973), *Society and the Assessment of Technology*, Paris: OECD.

Hudson, Heather, (1990), *Communications Satellites: Their Development and Impact*, New York: The Free Press.

Kinsley, Michael, (1976), *Outer Space & Inner Sanctums: Government, Business, & Satellite Communications*, New York: John Wiley & Sons.

Kvan, R. A., (1966), "Comsat: The Inevitable Anomaly," in S. A. Lakoff, Editor, *Knowledge and Power: Essays in Science and Government*, New York: The Free Press. (Cited in Oslund, 1975, 1977)

Legal Adviser, U.S. Department of State, "Memorandum of Law in the Matter of The Orion Satellite Corporation and International Satellite, Inc., Applications for International Satellite Communication Facilities," dated November 21, 1983.

MCI Telecommunications Corp. v. Federal Communications Commission 561 Fd 356 [D.C. Cir., 1977].

Open Market Reorganization for the Betterment of International Telecommunications Act, Public Law No. 106 (enacted March 17, 2000). (Orbit Act)

Oslund, Jack, (1977), " 'Open Shores' to 'Open Skies': Sources and Directions of U.S. Satellite Policy," in J. Pelton & M. Snow, Editors, (1977), *Economic and Policy Problems in Satellite Communications*, New York: Praeger Press. (**Note:** Documented discussion of historical developments in international telecommunications technologies and policies.)

Oslund, Robert J., (1975), *Communications Satellite Policy: Revolutionary or Evolutionary?*, Washington, DC: The American University, 1975. Unpublished dissertation. (**Note:** Documented discussion of historical developments in international telecommunications technologies and policies.)

Oslund, Robert J., (1983, January), "Pacific Facilities Planning Process: An Appropriate Mechanism," Pacific Telecommunications Council.

Oslund, Robert J., (1984), "Reindustrialization: A Telecommunications Perspective," *Telematics and Informatics* Vol. 1. Reprinted in I. B. Singh & V. H. Mishra, Editors, (1987), *Dynamics of Information Management*, Norwood, NJ: Ablex Publishing Corporation.

Oslund, Robert J., (1990), "Restructuring the Domestic Telecommunications Industry: From 'Muddling Through' to 'New Thinking,' " *Telematics and Informatics*, Vol. 7.

Pelton, Joseph N., (1974), *Global Communications Satellite Policy: Intelsat—Politics and Functionalism*, Mt. Airy, MD: Lomond Systems, Inc. (**Note:** Documented discussion of the establishment of COMSAT and Intelsat.)

Reagan, Ronald, (1984, November 28), Presidential Determination No. 85-2, 40 *Federal Register*.

Secretary of Commerce and Secretary of State letter to FCC Chairman, dated November 28, 1984.

U.S. Congress, Senate, Committee on Aeronautical and Space Sciences, (1961), *Staff Report on Policy Planning for Space Communications*, 86th Congress, 2d Session (*1961 Staff Report on Policy Planning*).

U.S. Congress, Senate, Committee on Commerce, (1962), *Hearings on Communications Satellite Legislation*, 87th Congress, 2d Session (*1962 Senate Commerce Committee Hearings*).

CHAPTER SEVEN

Dual Use Challenge and Response: Commercial and Military Uses of Space Communications

Robert J. Oslund
The George Washington University

> *Let no one think that the expenditure of vast sums for weapons and advanced technological systems of defense can guarantee absolute safety. . . .*
> —Dwight D. Eisenhower

Different adjectives have been used to describe three wars in the Middle East within a 12-year span. The Persian Gulf War in 1991 was a Space War, indeed the first Space War. The Afghanistan War in 2001 was an Information War. The Iraqi War in 2003 was a Digital War. This is because a full complement of global commercial and military space communications systems—communications satellites, remote sensing satellites, and position locating satellites—were utilized with increasing effectiveness in each armed conflict. This progression of leveraging information technology is further evidence of the ongoing Revolution in Military Affairs (RMA; see Berkowitz, 2003; Owen, 2000). What is striking is that, although the primary use of this troika of space-borne technology systems was for military purposes in the three theaters of military operations, these same facilities were used simultaneously for nonmilitary purposes within these regions and throughout the rest of the world. Hence, the notion of *dual use*.

In the case of commercial communications satellite systems, Intelsat and Inmarsat were most frequently mentioned in the Persian Gulf War. PanAmSat and Iridium were added to the list in Afghanistan, and Eutelsat was added in Iraq. In the case of commercial worldwide remote sensing satellites, the Department of Defense (DOD) was the exclusive wartime customer of the French "Spot" and

---

The views expressed herein, unless indicated otherwise, are solely the views of the author.

U.S. "Ikonos" systems for images from the Persian Gulf and Afghanistan. However, this exclusivity was relaxed for the "Spot," "Ikonos," and newer U.S. "Quickbird" systems during the Iraqi War. Operational use of DOD's Global Positioning System (GPS), the sole such system, was simultaneously employed by Coalition forces in all three wars and by civil and commercial users worldwide.

Supporting these systems in space has been the earth-borne development of international treaties and agreements and of domestic regulatory mechanisms. This terrestrial legal regime is intended to accomplish a number of goals: (a) promotion of international competition in goods and services; (b) encouragement of technological innovation; (c) creation and maintenance of high-tech jobs; (d) accessibility to reliable, state-of-the-art services by nonmilitary and military users; and (e) avoidance of the transfer of militarily sensitive technologies to nations and interests not friendly to the United States and its Allies, lest the former use these technologies against the latter. Therein is found the dual use challenge and response of space communications—balancing the peaceful uses of these technologies with their not-so-peaceful uses, and national economic concerns with national security interests.

Dual use of space communications is "an iceberg issue" (i.e., there are more underlying issues than those that are immediately apparent on the surface). It should come as no surprise that these issues cut across national policy sectors—economic, defense, foreign, national security, trade, industrial, space, and regulatory—with advocates and constituencies from each sector competing to impose their own perspectives in defining and resolving space communications-related issues. Adding more complexities are the adjustments from established relationships between parties to the diminished contest between East and West cold war ideologies and symmetrical nuclear threats, to the fluctuating asymmetrical relationships arising from clashes with fundamentalist fringes of civilizations and newer kinds of cyber and physical threats.

This chapter, written from a U.S. policy perspective, is a "first cut" at the iceberg issue. Due to the breadth and depth of the underlying issues, it is limited to an overview of four of the most significant contemporary dual use issues associated with commercial communications satellites and, given their symbiotic relationship, commercial launch vehicles. The chapter begins with a summary of the evolution of the international consensus regarding peaceful use of outer space and ends by tracing the emergence of a relatively new U.S. space policy and a new DOD joint doctrine for military operations in outer space. In between are discussions of two issues that directly affect the availability of communications satellites and their services and are directly affected by the dynamics of geopolitical change and globalization—export controls and foreign ownership. Given the unpredictability of these dynamics, the national and international applications of laws and regulations governing technology transfer and non-national proprietorship at different times have produced both complementary and contradictory results. There is no guarantee that the consequences of their application will be any different in the future.

PEACEFUL USE OF OUTER SPACE

The Soviet Union's successful launching of Sputnik in October 1957 was a dramatic cold war coup. One month earlier, Secretary of State John Foster Dulles had proposed to the United Nations (UN) General Assembly that an international study be undertaken on the use of outer space "for peaceful and not for military purposes," in an effort to engage the Soviets in negotiations on nuclear disarmament. Within weeks after Sputnik, the Soviet delegation proposed the establishment of an agency in the UN for the purpose of international cooperation in space research, conditioned on the negotiation of an agreement to ban the use of outer space for "military purposes." The proposal received little support outside the Soviet bloc. The United States countered successfully with a proposal to establish "appropriate international machinery" to foster international cooperation in "the potential uses of outer space for peaceful purposes" in areas such as science and engineering, meteorology, transportation, and communications. Since then, a UN consensus has solidified around the formulation that "peaceful use" means "nonaggressive" rather than "nonmilitary" (see Petras, 2002).

The General Assembly established the Committee on the Peaceful Uses of Outer Space (COPOUS) in 1959. Included among the many resolutions that COPOUS would draft and the General Assembly would adopt in subsequent years—to promote the peaceful use of and international cooperation in space—was the December 1961 Resolution 1721 (XVI)—"International Cooperation in the Peaceful Uses of Outer Space." Among other things, the resolution expressed the belief that communications by means of satellites should be available to the nations of the world as soon as practicable on a global and nondiscriminatory basis. The resolution also underlined the need to prepare the way internationally for means to enable the effective operation of satellite communications.

Eight months after Resolution 1721 (XVI) was adopted, the *Communications Satellite Act* was signed into law. This created and, in turn, directed Comsat to follow generally the guidelines of the resolution by establishing, in conjunction and cooperation with other countries as expeditiously as practicable, a commercial global communications satellite system, and to provide such services to economically less developed countries and areas as well as more highly developed ones (see chaps. 5 and 6). Two years to the month after the *Comsat Act* was approved, the *Agreement Establishing Interim Arrangements for a Global Commercial Communications Satellite System* was finalized internationally and entered into force. The Preamble opened with reference to Resolution 1721 (XVI). Subsequent definitive institutional instruments of the two global International Satellite

Organizations (ISOs)—*Intelsat Agreement* and *Inmarsat Convention*—retained the reference.

Military Use of Outer Space

U.S. military interest in using satellites for communications has a long history. The Army Signal Corps bounced radio signals off a satellite for the first time in 1945, but the satellite was a natural one—the moon. At almost the same time, the Army Air Corps funded Project RAND that, among other things, was to study satellite communications. The U.S. Air Force successfully launched and orbited the first artificial communications satellite, the low altitude Score in 1958, followed 2 years later by the DOD's low-altitude Courier. As discussed in chapter 5, the DOD was unsuccessful in developing a multipurpose geosynchronous satellite, Advent. It then entered into discussions with the new Comsat Corporation concerning the development of a shared system arrangement; at approximately the same time, Comsat and the Department of State were negotiating the *Interim Agreement* for what would become Intelsat. After a Military Subcommittee of the House Committee on Government Operations held a number of hearings that revealed the cost and complexity of such a civil/military undertaking, the discussions were canceled (see U.S. Congress, 1964). Another form of system sharing occurred in the mid-1970s when Comsat General, a Comsat subsidiary, developed, funded, and operated Gapsat for the Navy, the initial principal customer; as discussed in chapters 3, 5, and 6. This system, of course, also became the platform for the commercial Marisat system.

Since then, defense ministries of nations around the world, besides using dedicated military satellite systems, have used commercial satellites in varying ways and degrees for military purposes. Traditionally, the commercial satellites were used for general purpose communications, and military satellites were used for critical command and control communications. However, commercial satellites since have been increasingly used to directly support regional military operations. Besides the Persian Gulf, Afghanistan, and Iraqi Wars, they have been used in the Falklands (Malvinas) War in the 1980s and military operations in Somalia, Bangladesh, Bosnia, and Croatia.

ISOs and the Peaceful Purposes Issue

Intelsat and Inmarsat, the two global ISOs, handled military use of their space segments in contrasting ways prior to their privatization (see Morgan, 1994).

Intelsat. Military use of Intelsat was addressed in the early months of negotiating the organization's Definitive Agreements. In 1969, during a meeting of the Preparatory Committee of the Plenipotentiary Conference, the Algerian delegation proposed language to prohibit Intelsat from providing separate satellites for military needs. The intent of this proposal was subsequently reflected in the definition of "specialized services" in Article III (e) of the *Intelsat Agreement*, as discussed in chapter 5, in which no reference was made to "military" (see U.S. State Department, 1971). Algeria also took the position that there should be no objection for a country to use channels "in a regular Intelsat satellite for security purposes." That view was not seriously contested and thus became the ISO's standing policy. In a post-Persian Gulf policy review, Intelsat's General Counsel reaffirmed that the global system could legally provide public telecommunications to defense ministries. However, the issue disappeared with Intelsat's privatization in 2001.

Inmarsat. The Soviet Union raised the military use issue during exploratory discussions of "an international maritime satellite service" by a Panel of Experts (POE) of the International Maritime Consultative Organization (later renamed the International Maritime Organization [IMO]) in 1972. In a paper intended to provide a basis for discussion, the Soviet delegation proposed a number of provisional principles that included draft articles stating that the activities of such a service should be in compliance with GA Resolution 1721 and applicable treaties, and that the proposed system "shall act exclusively for peaceful purposes" and "should not permit the military use either directly or indirectly of the technical means which will be at its disposal." The POE's final report included the first draft of a Convention that referred to GA Resolution 1721 and the Outer Space Treaty and incorporated the language relating to "peaceful purposes." The *Inmarsat Convention*, which ultimately was adopted by the International Conference on the Establishment of an International Maritime Satellite System in September 1976, opened with reference to the General Assembly Resolution and the Outer Space Treaty. Further, Article 3(3)—"Purpose"—stated: "The Organization shall act exclusively for peaceful purposes." No reference was made to *military use* in the finalized text (see Inmarsat, 1989).

Inmarsat was used extensively by the United Kingdom during the Falklands (Malvinas) War. Indeed a number of books on the military action were published in the UK after the war that prominently featured pictures of that nation's naval vessels as well as converted commercial passenger ships equipped with Inmarsat ship earth stations (shipboard terminals). This prompted some members of the organization to privately raise concerns. These concerns, no doubt, were reflected in a 1987 summary legal opinion prepared by Inmarsat's Legal Adviser that narrowly interpreted Article 3 of the Convention to the effect that use of Inmarsat shipboard terminals on ships engaged in armed conflict, even in situations in which they would be acting in self-defense, would not be considered a use for peaceful purposes.

When pictures appeared in the press showing U.S. forces using Inmarsat terminals during operations related to the Persian Gulf War in 1991, Inmarsat's Director General raised concerns with the U.S. government and called attention to the 1987 opinion by the organization's Legal Adviser. In preparing a response to

Inmarsat, the Department of State—in coordinating an opinion for internal use that had been prepared by the Navy's Judge Advocate General—stated the broader view that military activities in space are not excluded if they are consistent with the UN Charter, and noted that Inmarsat had long approved the installation of shipboard terminals aboard warships. In its formal response to Inmarsat, the Department of State gave assurances that "appropriate steps have been taken to avoid recurrence of such publicity." Inmarsat's since retired Legal Adviser would later characterize these assurances as "a classical diplomatic response." Further, in revisiting the issue of peaceful uses, he concluded that use of Inmarsat by UN peacekeeping or peacemaking forces acting under the auspices of the UN in implementing a UN Security Council decision, and use of Inmarsat for self-defense, are not prohibited (see von Noorden, 1995). However, this became a nonissue when Inmarsat was privatized in 1999.

The Hot Line

Often overlooked is the separate, but related, significant peacekeeping role that communications satellites—namely, Intelsat and the Soviet Union's government-owned Molniya system—played in providing the hot line during the cold war. This electronic linkage through space was intended to enable American and Soviet civilian leaders to communicate directly to avoid any misunderstanding that could have led to a major military confrontation and possible escalation to a nuclear war. Previously, the cold war connection had been routed through transatlantic submarine telephone cable systems and coaxial land-line extensions from the European cableheads through Europe to Moscow. However, the submarine cables had proved vulnerable to cuts accidentally by fishermen and sometimes intentionally by Soviet trawlers, and the coaxial cables were reported to have been accidentally cut by farmers plowing their fields.

GLOBALISM AND EXPORT CONTROLS

Not unexpectedly, contemporary export controls have a cold war heritage. Domestic and international controls were established almost concurrently to prevent the transfer of militarily sensitive technologies to the Soviet Union and the Warsaw Pact countries, as well as China.

The *Export Control Act of 1949* delineated three basic purposes for export controls: further national security, further foreign policy, and protect commodities in short supply domestically. As Soviet–American relations evolved, the Nixon administration started to link foreign policy issues, such as human rights and trade, to issues involving technology transfer. Related to this linkage policy was the *Export Administration Act of 1969*, which called for a relaxation of controls of those technologies that were already being made available from foreign sources and were deemed to be of marginal military value. Seven years later, the *Arms Export Control Act* established the International Traffic in Arms Regulations (ITAR) to formalize the procedures for the Department of State to follow in licensing technologies listed on the Munitions List (i.e., those technologies that are inherently military in character or have a predominately military application). The *Export Administration Act (EAA) of 1979* further sharpened the distinction between military and dual use technologies and their regulation. The Department of Commerce was charged with developing the Commodity Control List for guidance in licensing dual use technologies. At the same time, the Department of State was directed to establish a Militarily Critical Technologies List with a view toward further trimming the number of items on the Munitions List. The intention of this law was to further refine the balance between promoting trade in high-tech goods and protecting critical military technologies.

Internationally, the Coordinating Committee for Multilateral Export Controls (COCOM) was established in 1949 as an informal nontreaty organization headquartered in Paris. With a membership comprised of NATO countries, minus Iceland, plus Japan, its purpose was to coordinate national efforts to control the transfer of sensitive military and dual use technologies. Communications satellites (and related telemetry, tracking, and control facilities) and launch vehicles were added to the domestic and international export control lists as they were developed.

Changes in Geopolitics and Regulatory Policies and Practices

In the early years, the uncertainty that existed internationally in export controls was minimal given the common threat and response to that threat which was shared by the United States and its Western allies. However, change—being the constant that it is in international politics, economics, and technology—caused these lines to start blurring. The appearance of a realignment of relations between the Soviet Union and Eastern Europe and between the Soviet Union and China led to subtle shifts in the individual Western nations' bilateral relations with the Soviet Union, Eastern Europe, and China. Breeches in the COCOM consensus accompanied these shifts. The tearing down of the Berlin Wall at the end of the 1980s and the disintegration of the USSR at the beginning of the 1990s further diluted the consensus as national perceptions of the nature of the threat diverged, in some cases significantly. Additionally, from the 1970s forward, international trade became increasingly important. Liberalization within the global marketplace gained momentum, and the balance (or imbalance) of trade on national economies assumed greater importance. Technologically, U.S. leadership gradually eroded as America's Allies and cold war adversaries developed their own technological capabilities. This international diffusion of technological capabilities had a number of consequences. Demands grew for greater commercial access to the once tightly controlled technologies.

COCOM members lobbied for greater latitude in exporting a growing number of dual use technologies. As foreign companies were granted greater marketing latitude by their governments, differences between the U.S. government and industry became more public. Increasingly, questions were raised at the national and international levels regarding the relevancy of, and political motivations behind, the retention of certain items on the lists of the export control regimes (see National Academy of Sciences, 1987, 1991).

"Waasenaar" for COCOM

When the Soviet Union became Russia on December 31, 1991, the bipolarism that had dominated East–West relations formally ended, and not all NATO countries shared the same perceptions with respect to the impact of these events on their own national interests as well as the collective interest of NATO. On April 1, 1994, the inevitable occurred when COCOM became NOCOM. Two years passed before a new, more flexible regulatory regime appeared in the form of the "Waasenaar Arrangement of Export Controls for Conventional Arms and Dual-Use Goods and Technologies." Its purpose was not to prevent, but to facilitate, the exchange of information by subscribing states to promote cooperation in their national and international export control regimes with an eye toward controlling exports of militarily sensitive technology to the "rogue nations." Russia, the Ukraine, and the nations of Eastern Europe formally subscribed to the arrangement. To date this multilateral arrangement has had "but lukewarm success" due to the lack of a single major threat and lack of agreement on what constitutes a smaller serious threat from the rogue nations (see Defense Science Board, 1999).

In anticipation of the geopolitical realignment and the growing threat of the rogue states, in 1987 the Western allies had fashioned the Missile Technology Control Regime (MTCR) as part of a concerted broader nuclear nonproliferation effort to restrict the flow of technologies that could be used as missile delivery systems for weapons of mass destruction. Russia and the Ukraine became members of the MTCR due, in part, to their desire to get into the commercial launch vehicle market. Although not a formal member, China has agreed to abide by the regime on each occasion that the United States has applied sanctions in response to Chinese violations of this nonproliferation regime.

Bilateral "Rules of the Road" for Launch Vehicles

In addition to the multilateral regime for missile technology, commercial bilateral "rules of the road" agreements were negotiated by the U.S. Trade Representative with foreign competitors entering the worldwide launch service marketplace. The first attempt at an agreement was related to Europe's Arianespace, then China's Long March, and, following the demise of the USSR, Russia's Proton and the Ukraine's Zenit. China, Russia, and the Ukraine shared one commonality—they were nonmarket economies (NMEs) whose launch prices, based on non-Western accounting practices, were markedly below those of American launch companies and of Arianespace. The bilateral agreements between the United States and the NMEs resolved the launch cost issue through negotiated "par pricing," which held that so long as the proposed cost of the NMEs' launches fell within a preset percentage of those of the West, they would not be subjected to more detailed analyses. Separate bilateral agreements between these same governments were also negotiated to ensure that MTCR-sensitive technology was not diverted to third countries, in particular the rogue nations.

Interestingly, the first formal "rules of the road" agreement was not reached with the Europeans, who had been competing with U.S. providers the longest. Some informal U.S./ESA discussions were held in 1990, but they were followed by only one inconclusive formal session in early 1991. The first bilateral was negotiated and agreed with China. The 6-year bilateral trade agreement, which included par pricing and capped the number of allowable Chinese launches for GEOs, was signed in 1989, renegotiated in 1995, and amended in 1997. It was allowed to expire on New Year's Eve 2001, with no public indication on either country's part of whether a new agreement would be sought. However, the agreement regarding MTCR-sensitive technology remained in place (see Smith, March 2002, April 2002).

The reversal of the U.S. "*nyet*-Proton" policy was modified during the June 1992 American–Russian Summit with the "Inmarsat exception," as discussed here and elsewhere, and paved the way for a bilateral agreement in mid-1993 on allowable Proton launches. This agreement was amended in 1996, but was essentially phased out at the end of 2000. A cap agreement on the number of launches was reached with the Ukraine in early 1996, with an expiration date of the end of 2001, but the agreement was terminated by the U.S. in mid-2000 to recognize the new nation's strong commitment to international nonproliferation norms. However, the launch vehicle technology safeguard agreements with Russia and the Ukraine were allowed to remain in force.

As stated earlier, export controls were established for three purposes: further foreign policy goals, further national security goals, and protect commodities in short supply domestically. The prior discussion has been related to the second of the three (i.e., furthering national security through seeking to protect the export of critical military technologies as part of the broader strategy of the nonproliferation of weapons of mass destruction). Yet technology transfer agreements have also been suspended and sometimes resumed to further foreign policy goals, such as the promotion of human rights and political reform.

Thus, 'it should be apparent that controlling exports, especially of the high-tech, dual use variety, is not an objective scientific endeavor, but rather is a subjective response to shifting externalities. This is because the laws and regulations to implement them have, by design, given the policymakers wide discretion to act. Consequently, how, when, and why controls have been employed has, in some instances, entailed elements of unpredictability and, in others, has had the appearance of inconsistency. This is attributable, many argue, to the ambiguous international milieu and changing national interests that present policymakers with challenges that require responses. Instances in which these responses allow or disallow export controls have been described as "lightswitch diplomacy" (see Bertsch, 1998), "an imperfect panacea" (see Defense Science Board, 1999), or "a particularly troublesome corner of federal policy" (see National Research Academy of Sciences, 1991). Whatever words one chooses to describe these actions, the fact remains that they have had—and continue to have—significant effects on the commercial entities involved in the manufacturing of satellites and launch vehicles, as well as the provisioning of satellite communications and launch services, not only domestically but internationally (see Meredith & Fleming, 1999).

Churning Domestic Policies

Beginning in 1992, responsibilities for export licensing for commercial communications satellites started to be shifted from the Department of State to the Department of Commerce in response to geopolitical change. In the fall of 1996, Commerce assumed primary responsibility. However, subsequent developments, primarily involving China, led to a reversal of responsibility that illustrates the symbiotic relationship between communications satellites and launch vehicles (see Smith, March 2002, April 2002).

In June 1989, following the suppression of the student-led Tiananmen Square demonstrations, all pending satellite export licenses were suspended by President Bush for foreign policy reasons, along with all other military exports. Congress followed suit by conditioning further licenses to China on a presidential finding either that political and human rights reforms have been instituted or the export of satellite technology is in the U.S. national interest. Based on a national interest-based presidential finding, the suspension subsequently was lifted. Further sanctions on China have since been imposed and then lifted for violations of the MTCR regime on at least two other occasions.

The troublesome aspects of export controls is dramatically demonstrated by developments that took place after failures at launch of Long March launch vehicles carrying APStar-2 in January 1995 and Intelsat 708 in February 1996. Postevent discussions concerning the possible causes of the launch failures were held between Chinese officials and representatives of U.S. satellite manufacturers, including Space Systems/Loral, a division of Loral Space & Communications, the manufacturer of the Intelsat satellite, and Hughes Electronics, a subsidiary of General Motors, the manufacturer of APStar-2. In 1998, press reports stated that these two companies were under investigation by the U.S. government for allegedly providing technical data to China that improved the reliability of Chinese nuclear missiles. In January 2002, the Department of State announced that Loral had reached a settlement, and in March 2003, the Department of State announced that a consent agreement had been signed with Hughes Electronics Corporation (the General Motors subsidiary that has since become part of Boeing Satellite Systems) and Boeing Satellite Systems (see U.S. State Department, 2003). In a separate case in June 2000, the State Department announced that it had reached a settlement with Lockheed Martin involving the transfer of technical data to China without an export license.

The controversy surrounding the Loral and Hughes issues led to the establishment of the "Select Committee on U.S. National Security and Military/Commercial Concerns with the People's Republic of China" ("The Cox Committee"). Based on the report's findings, in 1998 Congress tightened Executive Branch oversight of all facets of communications satellite exports and reinstated the Department of State as the lead agency. These actions disrupted agreements, especially with the Western allies, and added further delays in the licensing process. These effects were magnified even more by a number of other factors that were negatively affecting the satellite manufacturing industry: consolidation, overcapacity in satellite production (due to shrinking commercial and military markets, stiffer competition from Europe and consequential narrowing profit margins, and longer lived satellites already in orbit), declining employment, and overall national budget cuts. Many of these maladies were and are also affecting the U.S. launch vehicle sector. One privately funded research organization has concluded that budget cuts in the early 1990s "gravely harmed" the industrial base for the military space programs, and changes in export controls "are completing the damage by eroding the commercial industrial base through loss of sales." Further, these changes in export controls appear to have "overstated the 'uniqueness' (and thus risks to national security)" of U.S. satellite and launch technology (see CSIS Satellite Commission, 2002).

An Unexpected Reversal

An almost forgotten—and ironic—example of how a sudden reversal in dual use policy can negatively affect the commercial launch vehicle industry came in 1972. This was the federal government's decision to end reliance on expendable launch vehicles (ELVs) and, instead, target the Space Shuttle as the primary launch vehicle for military, civil, and commercial launches. As a result, ELV development and production facilities in the United States were scaled back, and launch sites were neither updated nor expanded. However, the tragic Challenger accident in 1986 led to the White House's unexpected suspension of commercial use of the Space Shuttle. Almost overnight, the domestic launch vehicle industry found itself thrust back into the international marketplace, well behind Europe and, in turn, China and Russia, who had continued to develop their expendable initiatives. Some disparity remains, although with the geopolitical realignment some members of the domestic launch vehicle industry have entered into joint ventures with Russia and the Ukraine to increase their launch service product lines (see chaps. 4, 5, and 6, and Vice President's Space Policy Advisory Board, 1992a, 1992b).

GLOBALISM AND FOREIGN OWNERSHIP

In contrast to the restrictive nature of domestic and international export controls, foreign ownership is part of a cluster of domestic permissive regulations intended to implement the broad provisions of the 1994 *General Agreement on Trades in Services (GATS)* and the sector specific provisions of the 1997 WTO *Agreement on Basic Telecommunications Services*, which include commercial communications satellites. However, provisos pertaining to national security have been built into the international and national regulatory structures.

When *GATS* was negotiated to fold services, including telecommunications, into an international trading framework, the national security concerns of the participating parties were addressed specifically in the "Security Exemptions" of Article XIV of the agreement. Under these provisions, no member is required to disclose any information considered contrary to its essential security interests, nor is any member to be prevented from taking any action considered necessary to protect its essential security interests relating to the supply of services, directly or indirectly, to provision a military establishment, or from taking any action in time or war or other emergency in international relations. Further, no member is to be prevented from taking "any action in pursuit of its obligations under the UN Charter for the maintenance of international peace and security." These security exemptions, in turn, enable states to take national security concerns into account when drafting regulations to implement the *Agreement on Basic Telecommunications Services* or any other service-related arrangements.

Historical Perspective

From a historical perspective, as discussed in chapter 5, the national security implications of foreign-owned telecommunications facilities were raised decades ago during the eras of submarine telegraph cable and HF radio.

Based on operational experiences with radio during the Spanish–American War in 1898 and observations of the use of radio during the Russo–Japanese War a few years later, the Navy strongly, but unsuccessfully, advocated inclusion in the *Radio Act of 1912* a provision to prohibit non-U.S. citizens from being granted radio licenses. Also, before World War I, the Department of Commerce unsuccessfully opposed the granting of a radio license being sought by a company, indirectly financed by German capital, to operate a facility on Long Island. However, Commerce contested the license application not for reasons of national security, but because the German government was not granting American companies radio station licenses in Germany. During the early months of World War I, the German-controlled radiotelegraph facility warned German ships sailing off the East Coast to seek cover. This ceased late in the summer of 1914, when President Wilson, consistent with U.S. neutrality in the European war, issued successive executive orders (EOs). The first EO prohibited all radio stations under U.S. jurisdiction from sending or receiving messages of an "unneutral nature" and from rendering an "unneutral service" to any of the belligerents. The second one placed all stations capable of transoceanic communications under government control as long as the hostilities lasted. All radio stations were denationalized by an EO in February 1920 (see Herring & Gross, 1930).

The *Radio Act of 1927* limited ownership by foreign companies of U.S. licenses to 20% in an effort to prevent a recurrence of espionage. Congress increased foreign ownership restrictions even more in Section 310 of the *Communications Act of 1934*, but maintained the 20% investment cap. However, in 1974, when the *Communications Act of 1934* was amended, Section 310 (b) (4) established a 25% cap for ownership by foreign individuals, corporations, or governments of a U.S. common carrier radio license, but gave the FCC the discretion to allow up to 100% foreign ownership on determination that such ownership is not inconsistent with the public interest (see National Security Telecommunications Advisory Committee, 2000).

A Paradigm Shift

These ownership restrictions became a point of contention in bilateral trade discussions and in the WTO negotiations in the early 1990s as the Executive Branch and the FCC argued for greater deregulation of international communications; they were also debated in the lengthy congressional deliberations concerning the proposed *Telecommunications Act of 1996*. With respect to the latter deliberations, FCC Chairman Reed Hundt called on Congress to "shift the focus of Section 310 from its original national security rationale to an approach which better accommodates global trends." Section 310 (b) (4) was untouched in the final version of the telecommunications reform act.

When the Commission in 1977 implemented the WTO's *Basic Telecommunications Agreement* in two decisions—the satellite-specific *DISCO II* (see FCC, 1997) and the nonsatellite-specific *Foreign Participation Order* (see FCC, 1997), it recognized that situations could arise involving national security and/or law enforcement concerns. Significantly, however, 9 months after establishment of the WTO in 1994, the regulatory agency, in adopting the *Foreign Carrier Entry Order* (see FCC, 1995), already stated that there are areas involving matters—"additional public interest factors"—that fall outside the agency's traditional public interest analyses and are within the purview of the Executive Branch (i.e., national security, law enforcement, foreign policy, and trade). To that end, the FCC said that it would take those concerns into account on a case-by-case basis in processing applications involving ownership, control, and/or operation of international facilities and services. Beginning with the proposed MCI/BT merger that was approved by the FCC, but was then preempted by Worldcom's acquisition of MCI, the FCC has imposed certain conditions on commercial transactions involving foreign carriers. These conditions are based on agreements reached between the commercial entities and, as appropriate, DOD, the Department of Justice, the FBI, and, more recently, the Department of Homeland Security. For example, in situations involving foreign satellite systems providing services to and from the United States, system operators have been required to construct a gateway earth station in the United States, if one does not already exist, to enable law enforcement authorities to conduct electronic surveillance. Other conditions have included: (a) limitations on foreign access to certain information; (b) citizenship requirements for certain personnel; (c) disclosure of personal data regarding personnel who occupy sensitive network positions; and (d) establishment of formalized reporting requirements.

Executive branch oversight of proposed foreign investments in national security-related domestic industries is also conducted by the Committee on Foreign Investments in the United States (CFIUS), a Cabinet-level interagency committee established by President Gerald Ford in 1975. This oversight was strengthened by the "Exon-Florio Amendment," which was included in the *Omnibus Trade and Competiveness Act of 1988*. The amendment stipulates that in the event of an acquisition that involves "an entity controlled or acting on behalf of a foreign government" that could affect U.S. national security, a CFIUS review becomes mandatory. The committee confidentially reports its findings and recommendations to the president, who then decides to either allow or disallow the transaction. There is no provision for judicial review.

In an examination of the foreign ownership issue in 2000, the President's National Security Telecommunications Advisory Committee (NSTAC) found that this regulatory structure, comprised of FCC regulations and the CFIUS review, has thus far satisfied the diverse interests of all parties in government and industry (see National Security Telecommunications Advisory Committee, 2000).

NEW PERSPECTIVES ON DUAL USE OF COMMUNICATIONS SATELLITES

In seeking to adapt to the post-cold war environment and the emerging asymmetrical international relationships and threats, U.S. policymakers and military strategists undertook a number of strategic policy planning activities that started in the mid-1990s. They culminated in 2002 with the integration of commercial communications satellites into the critical infrastructures supporting both the nation's Homeland Security strategy and the Pentagon's global operational planning. The activities that led up to this integration proceeded along two tracks. The first track was a series of three senior-level commissions dealing with Critical Infrastructure

Protection (CIP), a redefinition of national security in the 21st century, and national security-related operations in space The second track, under the guidance of the Joint Chiefs of Staff, involved medium and long-term Joint Visions that led up to the formulation of a joint doctrine for military operations in outer space.

The framework for all these activities, save for the CIP undertaking, was the White House's first post-cold war "National Space Policy" that was officially released in September 1996 (see White House Fact Sheet, 1996). Of most relevance to this chapter were the declarations that "access to and use of space is central for preserving peace and protecting U.S. national security as well as civil and commercial interests," and that space systems of any nation are considered to be national property with the right of passage through and operations in space without interference. Complementing these were guidelines intended to promote international cooperation in space:

- Defense—maintain the capability to execute the mission areas of space support, force enhancement, space control, and force application; and
- Commercial Space—support and enhance U.S. competitiveness in space activities while protecting U.S. national security and foreign policy interests; purchase by government of "commercially available space goods and services to the fullest extent feasible"; transition from negotiated trade in launch services "towards a trade environment characterized by the free and open action of market economies"; and ensure that government's telecommunications policies support a competitive international environment for space-based telecommunications.

Commissions

Three bipartisan, senior-level commissions were convened to reassess this nation's national security posture and its multiple national interests in the light of the changing geopolitical environment and globalizing economy:

- The President's Commission on Critical Infrastructure Protection (PCCIP);
- The U.S. Commission on National Security/21st Century (Hart–Rudman Commission); and
- The Commission to Assess U.S. National Security Space Management and Organization (Rumsfeld Commission).

The PCCIP, in its September 1997 report, addressed national security in terms of the nation's economy as well as its defense posture (see President's Commission, 1997). Attention was drawn to the interdependencies of the national critical infrastructures, such as finance and banking, transportation, energy, and government services at the national, state, and local levels, and to the significant role that converging telecommunications and computer systems were playing in supporting and interlinking these infrastructures, to say nothing of providing national security/emergency preparedness (NS/EP) communications. Because industry owns and operates 85% of the infrastructures on which the civil and military agencies depend, one of the report's primary recommendations called for the establishment of a government/industry partnership in the CIP process to address the new asymmetrical cyber and physical threats. However, commercial communications satellites were not mentioned in the commission's recommendations. *Presidential Directive 63—Critical Infrastructure Protection—*operationalized many of the commission's findings in May 1998, but again commercial communications satellites were not referenced (see Oslund, 2002). This oversight was finally corrected in the *Homeland Security Act of 2002*, which directed the Secretary of Homeland Security to develop "a comprehensive National Plan" for securing this nation's key resources and critical infrastructures, and included "satellites" as part of the Information and Communications Infrastructure. Additionally, in late spring 2003, the Media Security and Reliability Council, a federal advisory committee established by the FCC as a result of September 11, reconfirmed that communications satellites continue to be the predominant national means for distributing emergency and disaster-related communications.

Chartered by the Secretary of Defense to conduct a thorough study of national security processes and structures, the Hart–Rudman Commission issued a series of three broad-sweeping reports between September 1999 and March 2001 (see U.S. Commission, 1999, 2000, 2001). The commission's approach was built, in part, on the findings of an earlier report by a privately organized and funded Commission on America's National Interests (see Commission, 1996). The Hart–Rudman Commission examined in depth the challenges of the new international environment and attendant asymmetrical threats—not only in terms of mass destruction, but also of mass disruption. Reflecting strategic studies that were underway in DOD, the report stressed that continued use of space for peaceful purposes remained a major priority, but observed that space—outer space as well as cyberspace—could become "warfare environments" alongside land, sea, and air. The commission further found that as use of space for military purposes by the United States and other countries expands, other countries will also learn to exploit space for commercial and military purposes, thereby making the risk of deployment of weapons in space a distinct possibility.

The last in the trio of the commissions—the Rumsfeld Commission—issued its Final Report on January 11, 2001 (see Commission, 2001), less than 2 weeks before George W. Bush was inaugurated as president. Prompted in part by the first two reports of the Hart–Rudman Commission, this Commission on national security operations in space identified a threefold national interest: (a) the promotion of the peaceful uses of outer space; (b) the use of the nation's potential in space to support its domestic, economic, diplomatic, and national security objectives; and (c) the development and deployment of the means to deter and defend against hostile acts directed at U.S. space assets and against the uses of space hostile to U.S. interests. The Commission reiterated that the United States and most other nations interpret "peaceful" to mean "nonaggressive." Such an interpretation was considered to be consistent with customary international law that allows for "routine military activities" in outer space, the same way it does on the high seas and in international airspace. However, the Report opined that, save for nuclear weapons, "there is no blanket prohibition in international law on placing or using weapons in space, applying force from space to earth, or conducting military operations in and through space." Based on these factors, the Commission recommended a number of actions. These included: working to favorably shape the amalgam of treaties and international and domestic laws and regulations that govern outer space, leveraging "the commercial space sector" (i.e., communications satellites and launch vehicles), and fostering multinational alliances to help maintain the nation's position as a leader in the global space market. Significantly, a Commission staff study addressed the ever-present need to balance national security with national economic policy. Among other things, the study first called attention to the potential effects of foreign ownership and foreign facilities limitations on national security, but then pointed out the need to establish regulatory policies that would encourage, rather than restrict, the availability of space products worldwide so as to maintain the U.S. technological lead (see Haller & Sakazaki, 2001).

"Visions" and a Doctrine

The Pentagon's "Joint Visions" were part of the broader Revolution in Military Affairs to establish joint operations as a *modus operandi* for the Army, Air Force, Navy, and Marines in the decades ahead. As indicated earlier, they also provided the foundation for deriving a joint doctrine for space operations.

The first to be issued by the Joint Chiefs of Staff—*Joint Vision 2010*—was intended to provide a "conceptual template" for joint military operations based on "full-spectrum dominance" of future theaters of operations (see Joint Chiefs of Staff, 1996). Central to the template was leveraging existing and planned information technologies to support four new operational concepts—dominant maneuver, precision engagement, full-dimensional projection, and focused logistics—and enable more effective information operations and information superiority.

Shortly thereafter, the existing concepts of the *national information infrastructure* (NII) and the *global information infrastructure* (GII) were redefined in military terms as the *defense information infrastructure* (DII). The DII includes command and control, tactical, intelligence, and commercial information systems used to transmit DOD information to all levels of command (see *DOD Dictionary of Military Terms*).

The then U.S. Space Command joined other major service commands in projecting their own conceptual planning out another 10 years. *USSPACECOM Vision for 2020* called for the exploitation of the "troika" of space communications cited at the beginning of this chapter—communications satellites, reconnaissance satellites, and GPS (see U.S. Space Command, 1997). The authors of this Vision conceded that DOD-owned and operated space capabilities could no longer solely support military operations, but assumed that the projected growth in non-military civil space systems, as well as commercial and international space systems, was such that these systems could be leveraged to take some of the burden off the military civil systems through a strategy of forming global partnerships. Such partnerships would be of mutual advantage to all parties through sharing costs and risks, increasing information connectivity, strengthening alliances, and building confidence in coalition warfighting.

The multiple Visions were synthesized in mid-2000 by the JCS in *Joint Vision 2020* (see Joint Chiefs of Staff, 2000). The Vision made clear that achieving transformation of the Armed Forces and gaining full-spectrum dominance, as called for in *Joint Vision 2010*, will be dependent on creation of a "global information grid" (see *DOD Dictionary of Military Terms*); access and freedom to operate in all domains (space, sea, land, air, and information [cyberspace]); and realization of the potential of the information revolution, including communications satellites. The military planners implicitly recognized the impact of the globalization process by noting that, although communications and information technologies will continue to evolve, the nation's potential adversaries will have access to the same global commercial industrial base and much of the same technology as the U.S. military. Further, the "increased availability of commercial satellites, digital communications, and the public internet all give adversaries new capabilities at a relatively low cost." Consequently, new information operations strategies and concepts will be required both to defend against information attacks by nontraditional adversaries engaging in nontraditional conflict, and to degrade the enemy's information capabilities.

A few months after the release of *Joint Vision 2020*, the JCS issued *Joint Doctrine for Space Operations* (see Joint Chiefs of Staff, 2002), predictably incorporating the previous Visions. One of the basic thrusts of the doctrine as it relates to this chapter is that as the military, civil, and communications sectors become more heavily dependent on satellite communications, these resources increasingly will be viewed by adversaries as potential vulnerabilities. Thus, the United States "must be able to protect its space assets (and when practical and appropriate, those of its allies) and deny the use of space assets by its adversaries." Further, because both the United States and its potential adversaries will have "almost equivalent access" to the same advanced satellite technologies and innovative commercial services, "sustaining military advantage may largely rest on U.S. ability to integrate those technologies and commercial services into its force structure faster and more effectively than the adversary can."

OBSERVATIONS AND CONCLUSIONS

Ten years ago, peaceful uses of outer space, export controls, and foreign owner-ship were perceived as being three separate, distantly related, public policy issues. Today, due to the rearranged geopolitical map, a globalizing economy, rapid tech-nological change and diffusion, and the vulnerability of the United States, espe-cially post-9/11, to possible asymmetrical physical and cyberattacks, they are more than related; they are now intertwined. Outer space is projected to become a fourth medium for warfare; cyberspace is the fifth.

Space commerce today is a critical economic component, nationally and inter-nationally, whose economic woes, ironically, may be reminding policymakers, ci-vilian and military, of just how critical this sector is. Due to the globalizing nature of space commerce, a change in export controls in one country today can else-where disrupt the "space technology food chain" in the manufacturing of com-mercial communications satellites or launch vehicles, in the launching of satel-lites, and/or in the ready availability of space segment for commercial and military use. As regards the latter, the *Joint Doctrine for Space Operations* pointed out that adversaries can now bypass the development of their own space capabilities simply by purchasing them in the global marketplace. A shift in how a nation abides by its obligations pursuant to the WTO's *Basic Telecommunications Agreement* can upset broader international commercial relationships and the bal-ance sheet in trade. A downturn in one industry sector in one country can bring about cascading implosions in other industry sectors and in other countries, such has been the case with the successive collapses of the dot.coms, the telecommuni-cations sector, and the commercial space sector.

Today, the military needs commercial communications satellites to carry out many of its expanding and increasingly far-flung missions as much as the sag-ging commercial communications satellite industry needs the military as a customer (see Nelson, 2003; Johnson, 2003; Noguchi, 2003). Indeed, the satel-lite companies are revising their business plans to meet not only the Department of Defense's growing global communications requirements, but also the new Department of Homeland Security's growing domestic communications re-quirements.

Government studies estimate that during Desert Storm and Desert Shield, 45% of all communications between the United States and the Persian Gulf were via commercial communications satellites (see General Accounting Office, 2002). Yet at that time, military use of commercial satellites was predominately for gen-eral—not mission-critical—communications (see General Accounting Office, 1992). Reports in the press indicate that military use of commercial satellite ca-pacity increased tenfold during the Iraqi War (see Stern, 2003). This suggests that military use of these satellites for mission-critical communications has increased dramatically, especially in light of the broadband requirements of new weapons systems.

An industry study was recently reported to have found that many of the international satellites being utilized by the military are of foreign origin, and that by 2010, 80% of commercial communications satellite services could be provided by foreign-owned companies. At the same time, Pentagon planners are projecting heavy reliance on the commercial communications satellites until 2020, at which time they will be used for surge capacity (see General Accounting Office, 2002).

A decade ago, commercial telecommunications and satellite industry analysts and corporate planners were projecting near exponential growth in demand, primarily driven by strong national economies, expanding international markets, and a surge in Internet traffic. Geopolitically, potential adversaries were attracting increasing attention by the national security policymakers, and the notion of asymmetrical warfare was gaining support among the military strategists and foreign policy experts. As the previously cited commissions carried out their mandates and the military planners started formalizing their Visions, these new threat perceptions appear to have been heavily weighted in the preparation of their final work products. What appears not to have been foreseen or too lightly weighted was a montage of other looming externalities that affected the assumptions on which the Visions had been based. These externalities, sometimes colliding, included: weakening of national economies in general; impending implosion of the dot.coms, the bursting of the telecommunications industry bubble, and the accompanying drop in customer demand for satellites and launches; shifts in export control policies; liberalization of telecommunications policies, such as foreign ownership of facilities and services; and privatization of Intelsat, Inmarsat, and Eutelsat. These externalities, no doubt, demanded attention when the former U.S. Space Command conducted "Schriever 2001" (see Singer, 2001) and "Schriever 2003" (see McMullin, 2003), the first two war games to focus on space as a medium for conflict.

In the coming months, government policymakers, military planners, and satellite communications industry executives, domestic and foreign, will have to be nimble, proactive, innovative, and prudent in their responses to unforeseen challenges. At times it may become harder to maintain the balance between national security and economic security that they have achieved thus far. There may be missteps by government or industry as there have been in the past. However, so long as they are reversible and incremental, the dual dilemma that space communications poses will likely remain manageable.

REFERENCES

Berkowitz, Bruce, (2003), *The New Face of War: How War Will be Fought in the 21st Century*, New York: The Free Press.

Bertsch, Gary K., (Editor), (1998), *Controlling East-West Trade and Technology Transfer*, Durham: Duke University Press.

Commission on America's National Interests, (1996, July), *America's National Interests*.

Commission to Assess United States National Security Space Management and Organization, (January 11, 2001), *Report of the Commission to Assess United States National Security Space Management and Organization.* (Rumsfeld Commission)

CSIS Satellite Commission, (2002, April), *Preserving America's Strength in Satellite Technology,* Washington, DC: The CSIS Press.

Defense Science Board, (1999, December), *Final Report of the Task Force on Globalization and Security.*

Department of Defense, *DOD Dictionary of Military and Associated Terms,* www.au.af.mil/au/ database.

Federal Communications Commission, (1995), Market Entry and Regulation of Foreign-Affiliated Entities, 11 FCC Red. 3873 (Foreign Carrier Entry Order).

Federal Communications Commission, Rules and Policies on Foreign Participation in the U.S. Telecommunications Market, (1997), 12 FCC Red. 23891 (Foreign Participation Order).

Federal Communications Commission, (1997), Amendment of the Commission's Regulatory Policies to Allow Non-Licensed Space Stations to Provide Domestic and International Service in the United States, 12 FCC Red. 24094 (DISCO II).

General Accounting Office, (1992, May 22), *Military Satellite Communications: Potential for Greater Use of Commercial Satellite Capabilities,* Washington, DC.

General Accounting Office, (2002, August), *Critical Infrastructure Protection: Commercial Satellite Security Should Be More Fully Addressed,* Washington, DC.

Haller, Linda L., and Sakazaki, Melvin S., (2001, January 11), *Appendices: Staff Background Paper— Commercial Space and United States National Security,* prepared for Commission to Assess United States National Security Space Management and Organization.

Herring, James M., and Gross, Gerald C., (1936), *Telecommunications: Economics and Regulation,* New York: McGraw-Hill.

Inmarsat, (1989), *Basic Documents, Fourth Edition.*

Johnson, Nicholas, (2003, March 13), "Digital Capital: Satellite Phones Go To War," *Washington Post.*

Joint Chiefs of Staff, (1996), *Joint Vision 2010,* Washington, DC: Department of Defense.

Joint Chiefs of Staff, (2000, June), *Joint Vision 2020,* Washington, DC: Department of Defense.

Joint Chiefs of Staff, (2002, August 9), *Joint Doctrine for Space Operations,* Washington, DC: Department of Defense.

McMullin, Jenna, (2003, March 7), "War Games Test Space Assets," www.spacedaily.com.news.

Meredith, Pamela L., and Fleming, Sean P., (1999), "U.S. Space Technology Exports: The Current Political Climate," *Journal of Space Law, 27*(No. 1).

Morgan, Richard A., (1994, September–October), "Military Use of Commercial Communications Satellites: A New Look at the Outer Space Treaty and 'Peaceful Purposes,' " *Journal of Air Law and Commerce, 60.* (Note: This documented legislative history of the "military use" deliberations in Intelsat and Inmarsat was used as a primary information source.)

National Academy of Sciences, (1987), *Balancing the National Interest: U.S. National Security Export Controls and Global Economic Competition,* Washington, DC: National Academy Press.

National Academy of Sciences, (1991), *Finding Common Ground: U.S. Export Controls in a Changed Global Environment,* Washington, DC: National Academy Press.

National Security Telecommunications Advisory Committee (NSTAC), (2000, May), *Globalization Task Force Report.*

Nelson, Emily, (2003, March 12), "The All-Digital War," *Wall Street Journal.*

Noguchi, Yuki, (2003, April 17), "Digital Capital: With War, Satellite Industry Is Born Again," *Washington Post.*

Oslund, Jack, (2002, May), "National Security: The Definitions Landscape," *Compendium of Papers—Security in the Information Age: New Challenges, New Strategies,* U.S. Congress, Joint Economic Committee, 107th Congress, 2d Session.

Owen, Bill, (Adm. USN Ret), (2000), *Lifting the Fog of War,* New York: Farrar, Straus and Giroux.

Petras, Christopher M., (2002), " 'Space Force Alpha': Military Use of the International Space Station and the Concept of 'Peaceful Purposes,' " *Air Force Law Review, 53.*

President's Commission on Critical Infrastructure Protection, (1997, October), *Critical Foundations: Protecting America's Infrastructures.* (PCCIP)

Romero, Simon, (2003, March 10), "Military Now Often Enlists Commercial Technology," *The New York Times.*

Singer, Jeremy, (2001, January 28), "Air Force War Games Highlight Space Deterrent," www.space.com/news.

Smith, Marcia S., (2002, March 22), *Space Launch Vehicles: Government Activities, Commercial Competition, and Satellite Exports/CRS Issue Brief for Congress,* Congressional Research Service. (**Note:** This brief was used as a primary information source.)

Smith, Marcia S., (2002, April 9), *U.S. Space Programs: Civilian, Military and Commercial/CRS Issue Brief for Congress,* Congressional Research Service. (**Note:** This brief was used as a primary information source.)

Stern, Christopher, (2003, March 20), "Pentagon Scrambles for Satellites," *The Washington Post.*

U.S. Commission on National Security/21st Century, *New World Coming: American Security in the 21st Century—The Phase I Report* (September 15, 1999) (Hart–Rudman Report).

U.S. Commission on National Security/21st Century, *Seeking a National Strategy: A Concert for Preserving Security and Promoting Freedom—The Phase II Report* (April 15, 2000) (Hart–Rudman Report).

U.S. Commission on National Security/21st Century, *Road Map for National Security: Imperative for Change—The Phase III Report* (March 15, 2001) (Hart–Rudman Report).

U.S. Congress, House of Representatives, Military Operations Subcommittee of the Committee on Government Operations, (1964, October), *Report: Satellite Communications (Military-Civil Roles and Relationships),* 88th Congress, 2d Session.

U.S. State Department T.I.A.S. Series 7532, (1971), *International Telecommunications Satellite Organization (Intelsat), Agreement Between the United States of America and Other Governments and Operating Agreement.*

U.S. State Department, (2003, March 5), "Media Note: U.S. Department of States Reaches Settlement with Boeing and Hughes."

U.S. Space Command, (1997, February), *Vision for 2020.*

Vice President's Space Policy Advisory Board, (1992a, November), *A Task Group Report: The Future of the U.S. Space Industrial Base.*

Vice President's Space Policy Advisory Board, (1992b, November), *A Task Group Report: The Future of the U.S. Space Launch Capabilities.*

von Noorden, Wolf D., (1995), "Inmarsat Use by Armed Forces: A Question of Treaty Interpretation," *Journal of Space Law, 23* (No. 1).

White House, (1996, September 19), "Fact Sheet: National Space Policy," Washington, DC.

BUSINESS, MEDIA, AND ECONOMICS

CHAPTER EIGHT

Satellites, Internet, and IP Networking

Indu B. Singh
Robinder N. Sachdev
American University

Electricity . . . ended sequence by making things instant.
—Marshall McLuhan (2001)

Satellite-based Internet is now changing both the developed and developing world. Its impact is not only large, but is now growing at tremendous speed. This is because these interlinked technologies are global, instantaneous, and cost-effective. Internet networks have the potential to be increasingly pervasive in every country in the world.

Although the satellite communication industry has created viable business models in media, broadcast, and voice and data communications, it is yet to come up with a similarly effective model in the Internet-related sector of the satellite industry. This is partly due to the fact that the technology curves in the media and broadcast industries have reached maturity levels, whereas satellite Internet services are still at an early stage of development.

Despite this still being an early stage, it is already an important part of the overall business. As of 1997, there was modest Internet traffic on the major satellite systems such as Intelsat and Panamsat, but today Internet traffic represents the fastest growing category of fixed satellite service revenues and represents some 10% to 20% of total revenues on most global systems.

This Internet satellite traffic, however, is largely backbone or trunk traffic between countries. Only a small portion of this satellite traffic is broadband Internet traffic that is connected directly to business or end users. It is the potential of direct-to-the-user broadband Internet traffic that represents the greatest opportunity for the future, but also the largest market disappointment to date. It is here (at the level of broadband service directly to the end user) where viable business models are still being actively sought and business risks are actually quite large. It is in the broadband direct Internet service where satellites and Internet could ultimately create a powerful new synergy that takes the world of satellites to a new level of importance in global society. It will be a decade from now before the impact of satellite Internet services on the 21st century can be fully recognized and measured in a social, economic, and political sense.

SATELLITE-BASED INTERNET SERVICES
AND THE DEVELOPING WORLD

The use of satellites to connect Internet users around the world, of course, depends on the degree of Internet penetration. Satellite Internet thus depends on the diffusion of basic Internet and telecommunications technologies. This is true not only in the "developing world," but in the "developed world" as well. Internet diffusion is fundamentally tied to the growth of the basic building blocks (i.e., the core infrastructure of information and communications). In most developing countries, this development is still incomplete and, in fact, is just starting. The same can still be true in developed countries, especially in rural and remote areas, in the inner cities, and other "islands of economic and social isolation."

This "information gap" condition (namely, the lack of telecommunications, information technology, and educational development) for developing and newly industrializing countries is summarized in various statistical data provided in Table 8.1.

Network growth for Internet strongly correlates to the growth of telephone density, literacy, and personal computer population. Without these basic building

TABLE 8.1
Connectivity and PC and Internet Penetration in Some Developing Countries

Country	Population in Millions	Urban Population (%)	GNP per Capita US$	Illiteracy Rate/100 Male/ Female	Teledensity Lines/100 People	PCs per 1,000 People
India	997.5	28.1	450	32.2/55.5	2.2*	2.7*
Indonesia	207.0	39.8	580	8.5/18.7	2.7*	8.2*
China	1,200	31.6	780	8.8/24.5	6.96*	8.9*
Sri Lanka	19	23.3	820	5.7/11.4	2.84*	1.1**
Malaysia	22.7	53.7	3400	8.9/17.2	19.76*	58.6*
Philippines	76.8	57.7	1020	4.7/5.1	3.7*	15.1*
Brazil	168.1	78.4	4420	15.2/15.1	12.05*	30.1*
Chile	15.0	84.4	4740	4.2/4.6	20.55*	48.2*

blocks and without relaxation of governmental regulations that restrict free market growth, the development of the Internet will likely continue to be slow.

The nature of and relationship between development and technology constantly challenge intellectuals, policymakers, and practitioners in both the public and private spheres. Even in highly developed economies, many factors are needed to stimulate Internet growth, and these include not only basic telecommunication and information infrastructure and levels of education, but also business systems, market conditions, public regulatory policy, telecommunications pricing, and several other factors.

The public sphere is also impacted by the infrastructures of communication, and the emergence of the Internet has vastly affected the lines and patterns of societal interconnection. The billions of binary digits flowing by means of many different media and at a wide range of speeds (i.e., copper lines, fiber optic cables, microwaves, satellites, and terrestrial wireless networks) are giving a unique new dimension to the communication of thought and the implementation of ideas worldwide. Access to new information, trading opportunities, and political thought was greatly expanded by satellites in the 1960s, 1970s, and 1980s, and the combination of satellite service and Internet even more dramatically expanded global flows of information and knowledge.

Clearly the patterns of societal and economic interconnection within various countries and along international routes vary widely. Today, the pattern shows the likelihood of increasing gaps in communications and connectivity rather than a lessening of these factors. Even so, the potential of access to data and information via Internet and satellite multicasting has expanded exponentially in the past decade. It is this very potential for fundamental shifts in the flow of information globally that is at once so very intriguing, but also so very challenging.

Figure 8.1 depicts the connectivity between regions in terms of volume of Internet traffic as measured in megabits/second. This figure clearly illustrates the great differences in information flow via the Internet today. If one were to show the patterns of Internet traffic within urban areas, and from urban to urban areas, as opposed to rural to rural and rural to urban, the patterns would be even more manifest.

This having been said, the rapid growth of the Internet is creating enormous new opportunities for multimedia communications, e-commerce, and entertainment on a global scale. The underpinning technology, known as Internet Protocol (IP), makes it simpler and more economic to offer these services under one network. Within the current open-access tariff policies that exist in most countries today, even the poorest communities can participate if they can gain access.

According to Forrester Research, the worldwide e-business market is expected to continue to grow rapidly—into the trillions of dollars by 2005. Many industry observers believe that the existing legacy communication infrastructure will be inadequate to support this explosive growth engendered by Internet traffic alone. Emerging broadband multimedia service providers that have the capability to

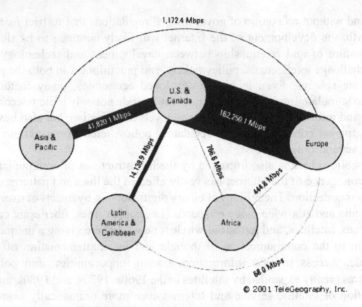

FIG. 8.1. The asymmetrical information flows via Internet globally.

blend voice, data, and video seamlessly over a single network appear well positioned to capture a significant share of this growth. This trend will create a great deal of turmoil among classical telecommunications companies and new Internet-based service providers that offer not only e-mail, but increasingly new voiceover IP, videoconferencing, and multimedia over Internet networks at a small fraction of the cost of the dominant carriers such as AT&T, Sprint, KDD of Japan, British Telecom, France Telecom, and so on.

The Internet is changing from a simple point-to-point exchange medium, where each end user connects from a personal computer to relay text-based content, to a system that utilizes peer connection via flexible networks as well as point-to-multipoint (known as multicasting) architecture. In this new environment, the Internet will support multimedia Web content, streaming audio and video, and sophisticated Internet applications. In this new Internet architecture, information can be distributed concurrently from one source to multiple end users. This reflects the convergence of traditional voice communications with data and broadcast media, which will further strain the Internet, especially in locations lacking the high bandwidth infrastructure necessary to support such applications. Satellite networks are particularly well suited to provide this type of dynamic connectivity.

According to a study by Frost and Sullivan (2001), total revenues from Internet Protocol telephony (or voiceover IP) in Latin America alone are projected to grow from $18 million in 1999 to as much as $11 billion in 2006. The majority of this traffic is most likely to flow via digital broadcast-based networks that transmit over satellite communications networks to support backbone Internet traffic. There are similar studies again showing dramatic growth of Internet and e-commerce business in Latin America, Africa, Asia, and the Arab world.

SATELLITES AND THE RISING IMPORTANCE OF IP TRAFFIC

One of the factors that will fuel the growth of Internet-related revenues is the now almost continuous introduction of value-added telecommunications services. These include international roaming, universal messaging, and bandwidth intensive applications such as video conferencing, virtual private networks, and call conferencing. For example, the current market for international and domestic long distance, worldwide, is in the order of a half trillion dollars. Voiceover IP is projected to account for about 10% of that global telecommunications market by 2005.

Satellite systems compete with other media to meet customer telecommunications needs for long-distance and international service. End users or telecommunications customers are for the most part what might be called *technology agnostic*. This is to say that most telecommunications customers base their decisions on product attributes such as price, reliability, branding, ease of use, and so on just like any other service. Few, if any, users say they would like to use terrestrial wireless, fiber, coax, or satellites to meet their telecommunications needs; they just want what is most reliable, lowest in cost, and most accessible.

The use of satellites is often the result of complementing other terrestrial networks in terms of backing up these networks, and providing connectivity where fiber or coax networks are not available or where broadcasting or multicasting networks are needed. In this context, satellite service is ideally suited to regions of the world that lack the basic terrestrial networks or in cases where satellites are necessary to accommodate low-, medium-, or high-speed data transmission that simply cannot be accommodated by traditional copper lines. In many cases, this can actually apply even in more developed markets, especially in rural and remote areas, or even in urban, suburban, or exurban areas that have not been equipped with fiber optic cable, cable modem service, or DSL service. This case of where broadband service is not available, for instance, includes perhaps some 15% of the U.S. market.

The key advantage offered by satellite communications is its broad area coverage. A single satellite can cover up to one third of the earth's surface, and it can also focus capacity on high-demand regions using spot beams or contouring the satellite's footprint. Satellites can flexibly redeploy capacity where needed in accordance with "peak load" requirements from time zone to time zone, and they also provide needed capacity when congestion occurs in terrestrial networks. Thus, satellites often complement terrestrial networks rather than compete with them.

The coverage capability of satellites allows for "instant communications infrastructure" in countries or regions that compress the time frame for implementation of services once a go decision has been made, and the distance insensitivity of satellites makes them well suited to provide thin route services, whereby the needs of rural and remote regions can be combined with the high-capacity geographical markets.

Internet via satellite started to appear as late as 1997, when satellite traffic consisted of a few point-to-point links on Intelsat and other new private competitive satellite systems such as Panamsat, New Skies, and other global systems. It was only after 1997 that Internet via satellite started to become a significant market opportunity for satellite operators and satellite service providers.

Along with the growing opportunity, the structure of the demand has also been changing. Today, satellite systems are providing not only backbone interconnectivity between countries (its primary revenue generator), but also a small but increasing amount of direct Internet access. This direct-to-the-business user and mass consumer business is the greatest area for potential growth. These direct satellite services can now go to end customers; content providers, advertisers, and broadcasters; tier 1 Internet Service Providers (ISPs); tier 2 regional and smaller ISPs; large multinational corporations; SMEs; residential users; and other segments (institutions, governments, etc.). Revenues from satellite-based broadband delivery services to the end users are projected to increase from $750 million in 2001 to several billions of dollars by 2005. Some studies project these numbers to be as high as $14.5 billion in 2005, with particularly strong demand in Asia and Latin America in the next 2 to 3 years. Others studies (see Table 8.2) project that growth to be less—namely, around $4.8 billion by year end 2005.

Against the background of Table 8.2 and similar forecasts, there are several challenges that continue to face the satellite Internet industry. Some challenges are generic to the industry, such as the deflation of telecom and Internet markets and the unpredictability of demand and analyst reports, including numbers in the prior table. Anecdotal evidence suggests that many of the satellite links are short-term measures intended for use only until fiber comes along. Accompanying these

TABLE 8.2
Broadband IP Satellite Markets Worldwide (Combined Wholesale
and Retail Revenues in $ Billion)

Broadband Satellite Service Markets/Year	Year End 2001	Year End 2002	Year End 2003	Year End 2004	Year End 2005	Year End 2006
Broadband Access	$0.35	$0.80	$1.70	$3.30	$4.80	$6.6
Multicasting	$0.25	$0.55	$1.05	$1.85	$2.90	$3.9
IP Trunking/Broadband Transmission	$0.50	$0.95	$1.65	$2.70	$3.60	$4.7
Total Estimated Revenues	$1.1	$2.3	$4.4	$7.85	$11.30	$15.2

Note. From *The New Satellite Industry*, International Engineering Consortium, 2002.

economic and forecasting difficulties are certain technical features that need to be resolved. A key technical attribute—latency or the time delay in the transmission of a communication—is higher in satellite communications compared with terrestrial alternatives. However, recent developments in transmission protocols that are optimized for satellite IP transmission may be able to significantly improve transmission efficiency, as well as address problems with security by accommodating IP Sec (IP Security) without major loss of effectiveness.

The 1990s provided certain learning points with which the industry is grappling to come to terms. Several satellite ventures were conceived and sold to corporations and investors in this period, yet many of these failed or offered service at higher than projected costs and at prices higher than customers anticipated.

Estimates of product development and product life cycles, both in the satellite and competing terrestrial systems, played havoc with several of these ventures. The idea that markets could change so rapidly in only a few years time so as to invalidate market projections was likely the key oversight of several satellite networks.

As one industry insider recently put it at a satellite conference,

> ... five independent consultants typically offer six independent projections of the future. Sometimes, you can even see correlation. But no one can effectively predict fundamental changes from trend lines ... where we really need help, and we have no good methods of attack in identifying beforehand major departures from trend lines; changes that in a sense are measurable on the Richter scale. (Personal interview with Ravi Subbarayan, Intelsat)

THE BUSINESS CASE—AND THE ATTENDANT RISKS

The terrain of political economy for satellite Internet, mired in technological, international, market, and development issues, depicts a struggle to reduce gaps between ideas and viable business models. The front loading of setup costs and the length of the deployment periods, coupled with the currently rigid nature of service offerings, significantly impact the business case. Perhaps the most significant risk involves the fact that implementation of a new satellite system inevitably takes many years, and much can change between the time the initial service concept is approved and the time the service is actually introduced. This is accompanied with the challenge of effectively marketing and selling sometimes new and untried technologies.

Most of the business cases for new satellite services factor these varying levels of risk and try to address them in some way. These factors tend to create plans based on both largeness of scale and long lead-time implementation periods. Unfortunately, during these long-term start-up periods, the target market can change dramatically, and competition can make unforeseen advances. New, global satellite projects require heavy investment just to reach the market. This level of expenditure requires complex financing arrangements, along with the need to attract large numbers of subscribers to the new system quickly. A quite solid economic case for the projects, such as those envisioned by Bill Gates and Craig McCaw's Teledesic, Lockheed Martin, TRW's Astrolink, or Hughes' Spaceway systems, has found that market demand for direct broadband access Internet services has radically altered from the time these projects were first conceived until the time they were ready to provide customer services.

In the current depressed state of equity markets and investment banking, any attempt to finance a new broadband Internet-based satellite service would seem an uphill task—at least until such time as one or more of these new types of satellite networks actually succeeds in the marketplace. Further, long-term, high-technology satellite networks must clearly place a high reliance on debt. If key milestones are not met, this too can limit the project's flexibility and lead to a downward spiral. The recent troubles of the telecom industry bear witness to that, where most problems can be traced to insurmountable debt and interest payments that cripple the enterprise.

In light of the multiyear period of system implementation for new satellite networks, it has not been uncommon that the target markets morph in a manner not incorporated in the product design or may even fail to grow as projected at product planning stages. Advances by competition further color this picture. During the long period of system and product development, competition has the opportunity to react and prepare.

The mobile satellite communication segment as represented by Iridium, New ICO, Orbcomm, and Globalstar represent the most clear-cut examples for this type of market failure. In this case, alternative terrestrial mobile cellular and digital mobile Global System for Mobile Communications (GSM) technology filled the market void that mobile satellites were planning to fill. The rapid and successful proliferation of terrestrial GSM cellular phones during the 1990s and early 2000s (i.e., the period that mobile satellite networks were being planned and built) largely served to crush the hopes of satellite phones for a mass market. The terrestrial mobile networks were increasingly capable of providing not only voice, but also data and Internet services to consumers at competitive prices in those areas where satellites planned to serve. As terrestrial mobile networks were extended along Interstate and other major highways and into smaller and smaller communities, the potential market for mobile satellites continued to shrink. The satellite phone industry is now struggling to identify niche segments (i.e., such as broadband Internet) where it meets specific customer needs—especially for businesses—while conceding the mobile telephony business to the digital cellular providers.

Thus, the economic case for satellite Internet seems to be evolving in differing patterns in the developed and developing countries. In the developed countries, satellite Internet seems to be addressing the end customer—particularly in areas without DSL or cable modem access. In the developing countries, satellite Internet

TABLE 8.3
Internet Services Directly Relying on Satellite Networks for Direct Access

Region	Number of ISPs (Approx. Total)	Number of ISPs (With Direct Satellite Links)	Number of ISPs (With Indirect Links)
West Europe	3,857	31	8
Africa	308	88	57
Asia	1,766	96	30
Latin America	1,173	135	642

Note. From DTT Consulting, 2000.

seems to be addressing the needs of other service providers such as regional and smaller ISPs. It is providing them with connectivity to the Internet backbone, thereby either overcoming structural constraints of the incumbent provider or competing with the few incumbents in these markets for backbone connectivity. Table 8.3 gives a picture of ISPs using satellite connectivity as of end 2000.

SATELLITE NETWORKS, INTERNET SERVICES, AND THE COMMODIZATION OF DIGITAL SERVICES

A key trend shaping the telecommunication industry is particularly affecting the satellite and some telephony operators in an adverse manner. This trend is what might be called the *commodification of bandwidth*, or simply the reduction of telecommunications and especially Internet services to that of a simple commodity. The first wave of commodification struck in the early 1990s when operators realized that networks designed for pure voice had to transition to become data networks that could flexibly handle digital voice, data, e-mail, video streaming, or multimedia. This shift also led to the digitization of telecommunications networks, required newer investments, and skewed the pricing models.

The second wave of commodification struck in late 1990s, when large-scale investments were made at least in the developed world. These huge investments were made in new fiber, wireless, and satellite infrastructures. This in turn led to a bandwidth glut in the industry for backbone or trunking systems. Ironically, this glut of heavy-duty telecommunications systems to provide thick routes of traffic (or backbone service) still did not provide any answers to the bedeviling problem of last-mile connectivity (i.e., linking the terminal equipment on the desks of end-user consumers at their homes and offices). Over 90% of the total system length that one might seek to measure in national telecommunications networks is in that famous *last mile* that directly links consumers' terminals in their workplaces or their homes.

Meanwhile telecommunications planners in the developing world were also following the path of rapid telecommunications planning that might be called the *commodization road.* Thus, they believed the analysts who had helped create the "dot com bubble" and pursued testosterone-driven forecasts. In short, they were adopting aggressive strategies to augment their communication infrastructures rapidly. In this period, there was still hope that differing media would be able to differentiate their offerings and perhaps avoid a head-on collision in the rising demand for simple bandwidth.

The third wave of bandwidth commodification appears set to level the ground for all technologies and reduce the issue to simple economics of bandwidth and availability. This wave will create a tiered commodification of bandwidth, whereby the tiers will be fiber, copper, wireless, and satellite. Each tier will fit into the value chain at specific points, thus providing a hybrid model to meet customer needs.

If bandwidth ultimately becomes a simple utility or commodity, much like water on tap, then the buyer is indifferent to the faucet. As a corollary, the consumer is almost completely sensitive to the pricing. The market for bandwidth (including satellite bandwidth) is elastic, and price competition is a dominant strategy in the industry. At least temporarily, satellite Internet can make some inroads in the markets of developing countries precisely because of pricing issues. The near monopoly conditions occupied by the incumbent telephone carriers in many of the developing countries have created possibilities for satellite providers to by-pass the structural constraints. Archaic interconnect and other rules, while applying to undersea cables, can be avoided when the backbone access is not through the traditional channels. This is advantageous for local service providers because they can buy bandwidth at lower rates.

However, this segment has a tendency to shift over to buying bandwidth from fiber access once an operator's total volume of service goes beyond 10 to 20 megabits per second. This is because the service provider has now successfully penetrated the market and is now able to afford, if need be, larger bandwidths from the fiber provider without being encumbered by latency. In this regard, satellite and fiber technology still often complement each other in the extension of Internet services, especially in the developing countries.

SOCIAL, ECONOMIC, AND POLITICAL FACTORS

The political economy of satellite Internet is ultimately based on a geopolitical view of the world. Harold Innis (1950), the Canadian economist and social scientist, described all media as possessing a bias—either a bias of space or a bias of time. Railways, rivers, and such media that enabled central governments to extend control over territories that Innis termed *empires* reflect a bias of space. In contrast, media reflecting a bias of time were institutions with qualities that enabled them to endure over long periods of time.

For example, as Herbert Schiller (1996) noted, after World War II, the United States intended to dominate satellite communication at the expense of Great Britain, which at that time held substantial share of the world market through its undersea cable systems. Testifying before Congress in 1966, McGeorge Bundy, former chief aide to President Kennedy and afterward president of the Ford Foundation, recollected that he was part of the executive branch during the period that led up to the establishment of Comsat (the U.S. Communication Satellite Corporation). He stated: "I do clearly remember what the record fully confirms—that Comsat was established for the purpose of taking and holding a position of leadership for the United States in the field of international global commercial satellite services."

Moving beyond Innis, the bias of space exhibited by the developed countries is concurrently affecting the bias of time in developing countries. Extension of the empires is creating a grid through which flow content and cultural mores, thereby impacting local cultures. Among other lenses, this issue can be viewed through the perspective of the German philosopher, Heidegger (2000), who argued that technology alienates the language and being of man. Heidegger traced human history and pointed out that mankind has a state of being in each period, and that technology disrupts that cosmic rhythm to the detriment of human potential. It can be argued that the Heideggerian view is analogous to the Marxian critique of capitalism, where Marx posited that capitalism alienates the worker from his means of production. This type of philosophic inquiry as to the meaning of digital technology, satellites, and the Internet is explored in greater detail by Professor Hamid Mowlana later in this book (chap. 13).

Thus, if technology and capitalism together could be considered to alienate both the being of man and the relationships to his production, then empires with a bias for space will realize the need for temporal domination to augment their spatial advantages. Temporal domination can be achieved through institutions and media that are long lasting in time. To maintain or grow their dominance, the spatial empires will then emphasize the notions and institutions of freedom, democracy, justice, markets, and communications. The end of history, as proposed by Francis Fukuyama at the end of the cold war, may well apply to the evolving characteristics of empires. Spatial empires are incorporating temporal aspects, whereas temporal empires are building spatial capabilities. In this clash, the global digital communications infrastructure has become a crucial piece because it delivers capabilities for spatial, temporal, and cultural dominance. Certainly the instantaneous global nature of satellite and the Internet poses a host of new issues and ways to look at our 21st century that was anticipated by McLuhan (1994) in his book *Understanding Media*.

SATELLITE COMMUNICATIONS AND EMERGING INTERNET APPLICATIONS

If there is one aspect that can depict the progress of the satellite Internet industry, it is product development and the subsequent diffusion of the innovation in the intended markets. In the 1990s, the use of the Internet in the daily life of people and

TABLE 8.4
Satellite Internet Service Evolution

Year	Suite of Service Offerings
1998	ISP links to Backbone-Hybrid Access Services
1999	1998 Offerings + Caching and Usenet Feeds
2000	1999 Offerings + two-way access services
2001	2000 Offerings + Voiceover IP (Trunking/VSAT)

Note. From DTT Consulting.

businesses around the world exploded. At the same time, terrestrial GSM-based mobile communications took off on a sharp curve of exponential growth, and fiber optic networks increasingly became the predominant long-haul transport of choice, although satellites retained their primacy for broadcast applications and for connectivity to developing countries. Microprocessor speeds continued to adhere to the predictions of Moore's Law given to us by Gordon Moore, Intel's founder (i.e., a doubling in capacity every 18 months). Finally, circuit-switched voice communications began to be overtaken by packet-based data communications. The recent advances in Transmission Control Protocol–Satellite (TCP–Sat), which extend the TCP/Internet Protocol (TCP/IP) capabilities to satellite-based Internet services, are expected to considerably enhance the competitiveness of satellite Internet solutions in coming years. The recent evolution of satellite Internet services over the last 4 years is summarized in Table 8.4.

Specific technological issues that impact satellites services, which are the main distinguishable characteristics of a satellite link compared to a terrestrial network that affect the performance of TCP, are:

- Effect of Satellite Link Errors
- Satellite Propagation Delay
- Bandwidth and Path Asymmetry
- Channel Access and Network Interactions
- Size of Window Required to Correctly Detect Network Congestion
- Efficient Methods for Applying IP Sec to Network Traffic

The result of all this is that there has been some skepticism concerning how well TCP will operate over a satellite link. In fact this doubt may be mostly unfounded. TCP was designed to work over satellites. The original goal of the research project that created TCP was to link a pilot satellite network (SATNET) with the terrestrial Internet (ARPANET). TCP–Sat could also serve a replacement for IP Multicast in terrestrial networks because it addresses the fundamental limitations in the transport protocol. Nevertheless, specific methodology involving simulation, spoofing, and network adaptation are needed to ensure efficient operation of IP traffic over geosynchronous satellite networks.

Since the late 1990s, satellite Internet services can be grouped into two parts: international broadband and Internet services, and voiceover IP services. The first of these, international broadband and Internet services, is comprised of Internet & Broadband Access (T-1, ISDN, ADSL), Wholesale Internet Access (Core access), Digital Video Broadcasting (DVB), Virtual Private Networks (VPNs), and Video Conferencing. Internet access solutions are designed to meet the bandwidth and network requirements of a broad range of customers. This service falls into two broad categories: connectivity services and network management services.

Connectivity services provide satellite-based Internet access to telecommunications companies, ISPs, media companies, and multinational corporations. Network management services allow customers to manage data traffic flow and optimize the efficiency of their networks. These services are evolving rapidly, particularly in Latin America, Asia, and Africa. A typical satellite Internet operation such as the one shown in Fig. 8.2 depicts a service between the United States and South America.

In those countries with no reliable terrestrial links to the Internet, broadband duplex service allows an end-to-end solution that supports high-speed, two-way Internet connectivity over satellite network. In other countries where copper lines or other relatively low-speed terrestrial links to the Internet are available, receive-only bandwidth services allow customers in these markets to improve their Internet performance.

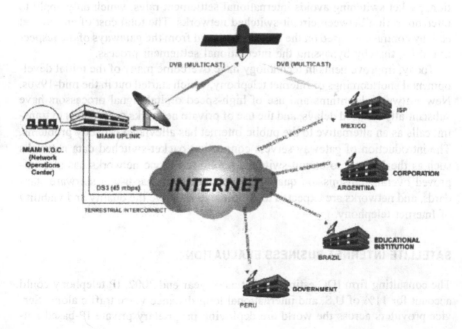

FIG. 8.2. A satellite Internet operation.

Network management services allow users to manage and optimize data traffic flow. Using these services, users can request specified amounts of guaranteed bandwidth and have the ability to monitor bandwidth use in real time over the Internet and distribute capacity to and among multiple Internet points of presence on an as-needed basis.

Satellite broadcast and multicasting of data and video services are targeted at organizations that need to transmit large amounts of data or video files to many locations at the same time through the use of satellite medium. Satellite video broadcasting provides corporate video broadcast and takes advantage of multicasting technology, wherein a customer can avail of the "one-to-many" capabilities of multicast technologies. A camera at a central site can broadcast encoded IP data to an unlimited number of satellite receiving stations in the service provider's satellite footprint area. These data can also be reflected to the customer's local area network, allowing all users in the network to view the original feed from their desktop workstations.

Voiceover IP and other value-added services comprises Internet telephony and fax, universal messaging and voice mail, one number portability, call conferencing, toll-free international access, prepaid services, and so on. A significant attribute of Internet telephony is the ability to reduce overall transmission costs versus traditional circuit-switched networks. Packet-switched networks are substantially less expensive to operate than circuit-switched networks because carriers can compress voice traffic and place more calls on a single line. In addition, packet switching avoids international settlement rates, which only apply to interconnection between circuit-switched networks. The total cost of an Internet call, by contrast, is based on the local calls to and from the gateways of the respective ISPs, thereby bypassing the international settlement process.

Today, improvements in technology have overcome many of the initial developmental shortcomings of Internet telephony, which started out in the mid-1990s. New software algorithms and use of high-speed digital signal processors have substantially reduced delays, and the use of private networks or Intranets to transmit calls as an alternative to the public Internet has alleviated capacity problems. The introduction of gateway servers connecting packet-switched data networks such as the Internet to circuit-switched public telephone networks has also improved overall transmission quality. Developments in hardware, software standards, and networks are expected to continue to improve the quality and viability of Internet telephony.

SATELLITE INTERNET BUSINESS EVALUATION

The consulting firm IDC estimates that as of year end 2002, IP telephony could account for 11% of U.S. and international long-distance voice traffic alone. Service providers across the world are deploying proprietary private IP-based networks that will offer high-quality voice and fax transmission at a substantially lower cost than over circuit-switched networks, particularly in emerging markets. In the developing world, satellite Internet services will continue to represent a substantial portion of the market. Even in the U.S. market, where cable modems and DSL lead the way, satellite-based IP services are still expected to be between 1.5% and 2% of the market as of year end 2003.

Another key aspect of product development and business implementation in the satellite Internet industry is the fallacy of market forecasts. Compared with other industries, the consequences of inaccurate trend analysis can be fatal in this industry primarily due to the long lead times of satellite projects and the associated difficulties in changing business models once a project has started. The most commonly accepted approach for mitigating this risk is to conduct extensive market research. This has become the primary means for defining product attributes, building a business case, and quantifying the return on investment. For a new satellite system to achieve market penetration, it must be highly competitive with terrestrial alternatives because the niche segments, where satellite Internet truly offers a value proposition (i.e., where terrestrial links are absent, such as remote areas, oil drilling rigs, scientific expeditions, etc.), are not large enough to justify investor interests.

The satellite Internet industry has learned some key lessons from the world of mobile satellite communications. These include the realization that market research for a 5- to 10-year forecast is a false beacon given the pace of development of competing technologies and changes in customer expectations regarding product attributes. Thus, the forecasts provided in this chapter have come from the more conservative IEC consulting study as opposed to the much more optimistic Pioneer Consulting market assessments.

Satellite Internet forecasts should probably be evaluated using Michael Porter's classic management model of the forces impacting a firm. According to Porter (1998), the forces affecting a firm at any given time include the threat of substitutes. In the case of mobile satellite telephony, it was the substitute cellular products that sounded the death knell of those ventures.

Risk-mitigation strategies include incorporating high levels of flexibility into new satellite systems at inception. Geostationary satellites using bent-pipe transponders lend more flexibility because the system management and complexity is resident in the ground segment of the system, as opposed to globally integrated systems with their complexity built into the space segment. Thus, it is easier to make changes in system specifications at the ground segment than where the control systems are integrated in the space segment. However, the trade-off here is that the greater flexibility obtained through a geostationary system compromises its ability to serve a targeted application. The upside of such an approach is that if the market demand does not develop as anticipated for the planned service offering, a differing or modified service can be introduced with comparatively less resources, and thus the returns on investment can be reasonably assured.

SATELLITE NETWORKS, INTERNET, AND NATIONAL DEVELOPMENT

Satellite Internet telecommunications service issues that impact development actually are quite numerous. Economic issues include sanctions policy, World Trade Organization (WTO) compliance, intellectual property rights, foreign aid for information development, global information development gaps and socioeconomic development, technical standards, and bilateral agreements and joint technology ventures. Political issues include cyberwarfare, viruses, info-bombs, cyberterrorism, international jurisdiction in contract dispute settlement and remedies, global information infrastructure, access, and use. Social issues cover collaborative research, social issues related to a digital development gap, and international contention for satellite bandwidth allocation and orbital slots, especially as related to the North–South gap.

In light of the rapid spread of the free-market system approach to the provision of satellite services and the largely unregulated nature of the Internet, the dominant prescription for most development policy seems to be much more economic than political in nature. However, Joseph Schumpeter (1991), the famous Austrian economist, contended that sociology can accomplish much that economists cannot do, and thus proceeded to make this philosophy the cornerstone for what he termed *economic sociology*. He argued that social institutions, property, inheritance, family, and such are partly economic and partly noneconomic in nature. Therefore, social institutions cannot be analyzed with conventional economic theory—pure economic theory is only applicable to topics such as value, price, and money. According to Schumpeter, the modern state is dependent on the existence of a healthy private economy; if no income is produced, there is nothing to tax. Taxation of entrepreneurial profit should be lenient, whereas the profit of cartels can be taxed more harshly. Schumpeter went on to assert that if people demand more and more public services, the state will eventually collapse.

However, even within Schumpetrian thought, the issue remains as to what are the public services that a people can demand without precipitating the collapse of the state from economic reasons? Much of the developing world has dense population, rural life and economic deprivation, illiteracy, extensive dependence on natural processes for food security, widespread occurrence of natural disasters, need for hazard mitigation, rehabilitation, and a lack of modern infrastructure. Several, if not all, of these have a pressing need for information to be obtained at the right place, time, and speed, while the act of information gathering and delivery occupies all strata of society—hospitals, civic administration, rural telephony, literacy, and so on.

This background is prompting policy planners and entrepreneurs to adopt multiple modes of communication technologies—telephones, Internet, satellite broadcast, videoconferencing, wireless, and so on. The challenges in this task are manifold. One great challenge is that of teledensity. This can be as low as 10 lines per 1,000 people for telephone and as low as 1 in 1,000 people for Internet, compared with 1 in 2 or 1 in 3 in the United States and Europe. There are many other indicators of low teledensity. These include poor communication infrastructures with low call-completion rates, poor transmission quality, limited repair capabilities, lack of locally relevant Web content, lack of regional interconnectivity, and traffic being routed through longer routes. For example, some international calls within Africa may still be routed through Europe. This can add substantially to the continent's annual phone bill. Further problems include low corporate investment in R&D, poor investment climate, and of course high costs for Internet access where it is actually available.

Several practitioners, including the World Bank, are now advocating South–South cooperation as a strategy to alleviate several of the previously mentioned concerns. Cooperation and knowledge sharing among developing countries provide for effective transfer of knowledge, diffusion of innovations, and creation of viable business models driven by local and regional investment psychology, as opposed to the unbridled expectations of Wall Street.

Areas of cooperation among the developing countries include communication infrastructures that reduce the carriage on international and longer routes, and regional initiatives that pool the traffic to be carried to the United States and Europe. According to one estimate in 2000, there is an average of 20 router hops and 4 network hops between content origin and end user worldwide. Communication infrastructures facilitate the development of markets for intermediate goods—a feature that serves as impetus for the growth of small market economies. For example, small firms in Africa can utilize high-quality software from India at lower costs than if the goods were packaged in the North and then sold back to them. The Indian software firms of course benefit from a large number of small clients. Heavy investments in the communications infrastructure require that the relevant technologies be used, particularly for the rural interiors, where majority of the developing world's population resides. India's experience with VSAT systems for servicing rural farmers, primary health centers in villages, fishermen, and so on, can be of value to similarly situated states. Cultural and technological exchange programs facilitate knowledge sharing, as well as sow the seeds for economic cooperation. Key aspects of the communication infrastructures and traffic management, which has evolved well in the developed countries, are the Internet exchange points. Internet exchange points are locations where carriers—terrestrial, wireless, and satellite—aggregate their traffic in metropolitan or high-density areas and then can hand off or pick up the traffic from their peers. Such exchange points are in their infancy in the developing world and could go a long way in facilitating the penetration of Internet access.

CONCLUSION

When Joseph Pelton (1974) wrote that "... economists, political scientists, and sociologists among others have just begun studying the impacts of space technology on society," the world of satellite communications was an emerging field, and the Internet network was perhaps a seed in the minds of the military-industrial complex. By the early part of the 21st century, the complexities and variables have grown enormously.

The truly revolutionary aspect of satellites and Internet—especially when combined to provide e-mail and Web access as well as to provide value-added, IP-based telecommunications services—is the instantaneously global reach of these digital systems. Radio and TV, the telephone, Internet, e-mail, and the Web create a way to move us beyond the global village into an integrated planetary consciousness. The power of Internet is in its ability to integrate other technologies and media seamlessly together and on a global scale. The fact that it can also do this at increasingly lower cost only serves to leverage its impact. Satellite Internet is truly changing both the developed and developing worlds and seemingly at an ever-increasing pace.

REFERENCES

Frost, & Sullivan, (2001, July), *World Caching, Streaming, and Multicasting via Satellite Markets*.

Heidegger, Martin, (2000), *Being and Time*, New York: Blackwell.

Innis, Harold, (1930), *Empire and Communications*, Toronto: University of Toronto Press.

McLuhan, Marshall, (1994), *Understanding Media: The Extensions of Man*, Cambridge, MA: MIT Press.

Pelton, Joseph, (1974), *Global Communications Satellite Policy: INTELSAT, Politics and Functionalism*, Mt. Airy, MD: Lomond Systems, Inc.

Porter, Michael E., (1998), *Competitive Strategy: Techniques for Analyzing Industries and Competitors*, New York: The Free Press.

Schiller, Herbert I., (1996), *Information Inequality: The Deepening Social Crisis in America*, New York: Routledge.

Swedberg, Richard, (Editor), (1991), *Joseph A. Schumpeter: The Economics and Sociology of Capitalism*, Princeton, NJ: Princeton University Press.

BACKGROUND READINGS

Acland, Charles R., and Buxton, William J., (Editors), (1999), *Harold Innis in the New Century: Reflections and Refractions*, Montreal, Canada: McGill Queen's University Press.

Communicating by Satellite: Report of the Twentieth Century Fund Task Force on International Satellite Communications, (1975), Millwood, NY: Twentieth Century Fund.

Galloway, Jonathan F., (1972), *The Politics and Technology of Satellite Communications*, Lexington, MA: D.C. Heath and Company.

Internet via Satellite 2000, (2000), Hampshire, United Kingdom: DTT Consulting.

Nagler, Gerald M., and Baxter, Cedric, (2000), *Investing in New Satellite Systems—Lessons Learned from the Recent Past*, Comsat Mobile Communications.

O'Brien, R., (1998), *The Political Economy of Communications and the Commercialization of the Internet*.

Patterson, Graeme, (1990), *History and Communications*, Toronto, Canada: University of Toronto Press.

Pelton, Joseph N., (2002), International Engineering Consortium, *The New Satellite Industry. Revenue Opportunities and Strategies for Success*, Chicago, IL: International Engineering Consortium.

Ploman, Edward W., (1984), *Space, Earth and Communication*, Westport, CT: Quorum Books.

Rorty, Richard, (1991), *Essays on Heidegger and Others: Philosophical Papers*, Cambridge: Cambridge University Press.

Sachdev, Robinder, et al., (1999), *Activity Based Costing in the Telecommunications Industry*, American University MBA Research Paper.

Schiller, Herbert, (1997), *The United States Fifty Years On: Fighting for Communication Control*, Le Monde Diplomatique.

Spar, Deborah L., (2001), *Ruling the Waves: Cycles of Discovery, Chaos, and Wealth*, New York: Harcourt.

Staples, Gregory, (2001), *Telegeography 2001*, Washington, DC: Telegeography.

Subbarayan, Ravi, and Oweis, Loay, (2000), *Mediacasting: Internet Streaming Media Distribution*, Lockheed Martin Global Telecommunications: World Systems Broadband Center.

Williams, Michael S., (2000), *What's Happening to My Business Plan*, Lockheed Martin Global Telecommunications.

CHAPTER NINE

Benefits of Satellite Telecommunications

Dr. Norman C. Lerner
Transcomm, Inc.

Where There Is No Vision, The People Perish.
—Old Testament, Proverbs, XXIX, 18

Satellite telecommunications systems and services have provided substantial socioeconomic benefits worldwide for nearly four decades, and this impact likely will only continue to grow. However, the distribution of these benefits are not spread evenly throughout the world, nor is there any assurance that they have necessarily gone to those most "deserving." Moreover, there is little agreement to date on how to categorize or even accurately measure those benefits.

The following analysis seeks to identify the incidence and magnitude of these socioeconomic benefits, as well as consider the associated problems that these gains have created. To accomplish these objectives, the following analysis presents a structure for categorizing and evaluating the benefits derived from satellite communications. The approach utilized here develops a hierarchy of beneficiaries related to representative satellite systems, specific services and their economic contribution at different levels of commercial enterprise, and different geographic categories within this structure.

This analytic process allows the benefits to be quantified by using relevant proxy economic and financial data. This approach seeks to separate, as far as possible, the specific gains that satellite communications services and products have produced, as distinct from those gains associated with the benefits of telecommunications in general.

The existence of net economic benefits appears to be strongly demonstrable. This can be shown in an economic sense by the many instances where customers have purchased services available from satellite networks as they were built and deployed. Clearly, in such instances, there had to be perceived economic benefits in that consumers purchased the service or product in a global market where other alternatives competed to be purchased for their dollar.

Alternatively, economic benefit was not universally perceived. Some satellite networks were built and space telecommunications services offered, but customers "didn't come." These satellite networks failed when consumers did not perceive economic benefit. These included Comsat General's Direct Broadcast System, Satellite Business System (SBS), the American Mobile Satellite Company (AMSC), Astrolink, Teledesic, Globalstar, Iridium, Orbcomm, and more. In a certain small percentage of the cases involving new satellite systems, the expected economic benefits did not materialize, or at least they did not materialize quickly enough to make the systems economically viable on their first market introduction.

BENEFITS: INCIDENCE AND STRUCTURE

It is easiest for benefits to be discussed in economic terms and thus be precisely specified in dollar terms. Yet we have tried to speak more broadly as well. Thus, benefits in this chapter have been identified as not only financial impacts or effects, but also socioeconomic effects too—both qualitatively and quantitatively. Social benefits that are not easily quantifiable also have value. Both financial and socioeconomic benefits generally accrue—sometimes cumulatively—from the bottom–up in the telecommunications "food chain" and other times from the top–down. The telecommunications and satellite economic activities later in this chapter are thus characterized as a hierarchy of industry and economic sectors. This starts with users and builds its way up the economic structure to carriers, operators, suppliers, manufacturers, investors, and other stakeholder groups. The hierarchy and associated benefits can also be categorized geographically for analytical purposes into "Global–International" and "Regional–Local," each of the consequent categories evaluated with respect to the benefits derived from the effects of various services available from specific satellite systems. This evaluation "matrix" allows benefits to be quantified and compared.

Measures of the benefits—both qualitative and quantitative—can be derived from market-related factors specific to each type of satellite system and representative applications. These factors are benefit proxy variables, which are measures of commercial activity and are represented by various financial-economic variables (e.g., values and patterns of financial and economic growth at different levels of disaggregation or economic separation). The satellite activities to which the benefits are attached can be identified as either specific services offered through dedicated, single-purpose satellites or as part of multipurpose satellite networks, which include a large number of the satellite systems since operational inception in 1965.

It is important to keep in mind that the magnitude and scope of benefits can differ among elements of the matrix hierarchy for any specific satellite system and applications. This is a result of the framework within which any benefits may occur, particularly as contrasted among developed and developing countries. Thus, for example, the availability of services, extent of competition, pricing flexibility, and general regulatory–administrative–economic conditions will either enhance or repress potential benefits for categories that otherwise have similar technical characteristics.

SATELLITE SYSTEMS EVOLUTION

Generally, the benefits of satellite communications are proportional to the development of the support systems or infrastructure that they either create or that were necessary for their own creation. Most all satellite systems, applications, and services are provided through the facilities of public and private organizations, and most of these have been geostationary satellites (GEOs), gradually being supplemented as time progresses, but now medium earth orbit (MEO) and low earth orbit (LEO) systems are also being deployed. No economic benefit is instantaneous, and in the satellite field research costs and development periods must be considered as part of the startup costs. For instance, it took two decades to get from the conception of these satellites, through early U.S. military experimentation (Star, Tackle, and Decree) and then the early civilian development stage that included Courier 1B, SCORE, Echo, Telstar, Relay, and Syncom 2 and 3 to reach the first commercial operation in April 1965. (Note: Research and development programs continue in the satellite field around the world, and these costs must continue to be part of the economic equation.)

The first of these satellites—namely, the Early Bird satellite with 240 voice channels or one black-and-white video channel—started a global communications revolution on virtually a preoperational basis. As of 1965, Early Bird was the only method of transmitting live TV across the North Atlantic, and it also increased the number of telephone voice channels over then existing submarine cables by about two thirds. The satellite was also used for the first time for service restoration when transatlantic cable service failed. This modest beginning was the start of a major shift in the international telecommunications business. In light of the fact that the satellite network was reasonably well booked from the beginning suggests almost instant economic benefits.

As noted earlier, President Kennedy merged business, social, and political objectives when he insisted on sharing space technology in a practical way to link the world through a global satellite system. Consequently, these operations were

integrated under the auspices of what was first established as the International
Telecommunications Satellite Consortium (Intelsat) in 1964. Intelsat was the first
of several international organizations dedicated to the same objectives in the
1970s. However, due to global economic shifts and regulatory reforms, organizations such as Intelsat, Inmarsat, and Eutelsat have been privatized, in whole or in
part, into privately held corporations without official governmental status. They
now have been restructured and their ownership transformed so as to provide the
most economic and competitive satellite services, rather than pursuing the original social, political, and peaceful-oriented goals first identified several decades
ago. Because these systems continue to be used for tele-education, tele-health, rural and remote services, economic development, and price equalization, one must
conclude that social benefits remain even though the organizational structure and
formal goals of these organizations have shifted over time.

The benefits of eventual global connectivity, beyond the initial North Atlantic region, were first provided through a global cooperative oriented toward providing space segment capacity for international public telecommunications services. The benefits were so manifest that several other international and regional
organizations were similarly established in the 1970s, but then commercial satellite organizations and domestic satellite systems also sprouted in the 1970s
and 1980s, with Canada and the Soviet Union being among the first to have their
national satellite systems. However, as shown in Figs. 9.1 and 9.2, there has
been a worldwide trend to pursue a shift to privatization and deregulation. This
shift has taken place in the belief that private sector management, control, and
innovations—coupled with increased competition—will lead to greater benefits
more rapidly than those accruing through the original public sector, government-owned structures.

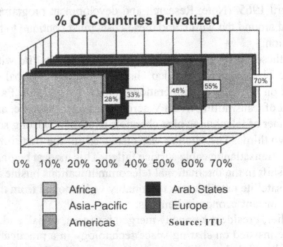

FIG. 9.1. Percentage of countries privatized.

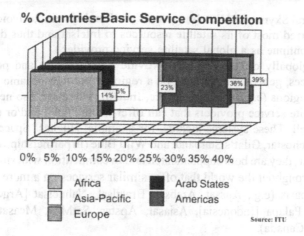

FIG. 9.2. Percentage of countries with basic service competition.

CHANGING INDUSTRY STRUCTURE

The services provided by these and subsequent expanded facilities and organizations are varied as are the associated effects and benefits. For example, the satellite services include:

- fixed voice and data
- land mobile voice and data
- maritime mobile voice and data
- video
- broadband data
- radionavigation and radiodetermination
- remote sensing and imaging

These services are made available through various global or regional coverage satellite systems, generally classified by altitude and number of satellites, as:

- geosynchronous (GEO); 22,237 nautical miles
- medium earth orbit (MEO); 6,000–12,000 nautical miles
- low earth orbit (LEO); 600–1,000 nautical miles

Although Intelsat capacity, as the world's first commercial satellite system, has clearly fostered international telecommunications between and among countries, the value associated with its early monopoly is now being shared among numerous private sector competitors, including a privatized Intelsat spinoff—New Skies—that joins the ranks of other competitive players such as Panamsat, SES

Global, Loral Skynet, and Eutelsat. (Note: The Loral system, to avoid bankruptcy, has transferred most of its satellite resources to Intelsat and thus does not seem likely to continue as a global satellite service provider.)

These globally oriented systems provide point-to-point and point-to-multipoint services, generally competing on a regional and transoceanic basis defined by ocean regions (e.g., Atlantic, Pacific, Indian). There are also new broadband fixed satellite service providers that can offer international and/or domestic services as well. These systems include, for example, Hughes' Spaceway satellite network, Echostar, Gilat's Starband, and Wild Blue (in partnership with Intelsat).

However, they are both supplemented and in competition with smaller regional systems throughout the world that offer similar services on a more focused geographical basis (e.g., Aussat, Arabsat, Brasilsat, Nahuelsat [Argentina, South America], Palapa [Indonesia], Asiasat, Apstar, SatMex, Measat [Malaysia], Telesat of Canada).

The most predominant trend at this time seems to be the formation of new global satellite organizations through merger and acquisition. In this regard, SES-Astra, which started out as only a European regional carrier, has now grown through acquisitions (particularly with GE Americom) to be the largest integrated satellite system provider in the world, with full or partial ownership of over 40 spacecraft. Thus, its now enlarged fleet, built by acquisition, has outstripped both Intelsat and Panamsat (each with about 20 satellites in operation, although Intelsat's latest acquisitions from Loral will boost its numbers to 26). All of these carriers furnish facilities and services that have provided, and will continue to provide, the numerous socioeconomic and financial benefits briefly summarized in this chapter.

SOCIOECONOMIC AND FINANCIAL BENEFITS

This brief summary of satellite institutional structure provides an overview of the framework within which the benefit evaluation matrix structure seeks to identify the socioeconomic and financial effects of representative satellite systems worldwide. The benefits are defined in terms of changes in selected socioeconomic and financial conditions derived from various satellite systems and services applications within each of the geographic categories—Global–International and Regional–Local—as depicted in Fig. 9.3.

There is also a need for a qualitative assessment for all of the socioeconomic effects. These are discussed and quantified, as far as possible, with proxy variables. The inherent overlap between and among the various elements of the matrix is addressed by segregating the effects and disaggregating the categories as far as possible. Nevertheless, the matrix framework, although not exact or completely discrete, is still useful for defining and approximating the benefits of satellite communications.

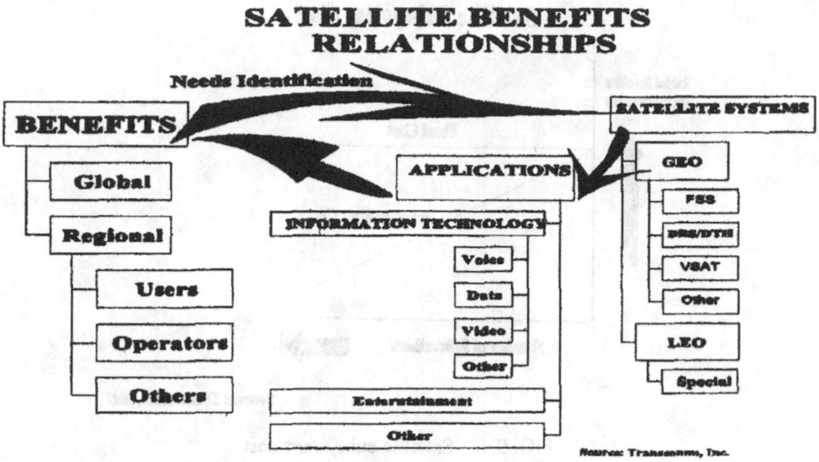

SATELLITE BENEFITS
RELATIONSHIPS

FIG. 9.3.   Satellite benefits relationships.

STRUCTURE OF SATELLITE BENEFITS

The major advantage to satellite communications, and the characteristic from which most beneficial effects are derived, is that it provides equal capability and access within the geographic coverage area of the satellite on the earth's surface, also called the satellite "footprint." For GEO satellites, the latter is defined by a specific location in earth orbit and the satellite antenna radiation pattern.

In general, the satellite systems are characterized by high initial costs to establish the space subsystem. However, once the initial fixed investment cost for the satellite is established, the costs of providing services within the coverage area are driven by the incremental cost of additional earth stations, not the satellite, providing the basis for the economies of scale. Consequently, for the life of the satellite, the average satellite system cost is, on average per unit, reduced with each such additional access (i.e., per earth station or per user circuit access; see Fig. 9.4).

Both satellite and terrestrial capacity costs are sensitive to cross-sectional density (i.e., number of equivalent voice channels carried in each facility). However, the distance insensitivity and consequent lower incremental costs for satellites contrasts with the terrestrial, distance-sensitive costs within the same coverage area. In the latter instance, any additional coverage requires a multihop expansion of facilities, be they copper, fiber-optic cable, or point-to-point microwave facilities.

The contrast is even more extreme when coverage for sparsely populated "thin routes" or rural areas is required. In that case, the cost to reach subscribers is minimized with satellites in comparison with exclusive terrestrial facilities coverage. Further, satellites can also provide network access through hybrid combinations with terrestrial wireline and wireless systems. Thus, satellites are more cost-

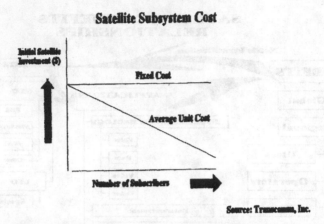

FIG. 9.4.   Satellite subsystem cost.

effective when connected to smaller organizations and/or "thin" routes in rural, distant locations where the higher terrestrial cost of access alone cannot be cost justified. Alternatively, the terrestrial systems are more cost-effective for many applications in urban and suburban locations where the population density is higher with a consequent lower incremental cost. In all these cases, there are profound and diverse socioeconomic benefits and effects on commerce resulting from alternative satellite systems applications.

BENEFIT MEASUREMENT AND CATEGORIES

Many socioeconomic and financial effects of satellite communications can be measured in "self-evident" terms. These are first-order or aggregate benefits for each of the matrix categories defined as those that would not have occurred without satellite communications. They reflect the perceived value by users that flow through the stakeholder hierarchy categories to, for example, operators and carriers in terms of proxy measures. For example, the latter include revenues, employment, and accompanying effects on the supporting industries such as manufacturers, engineering/installation firms, launch vehicle organizations, insurance and financial institutions, and others. Although the specific satellite applications are at the individual user level, the benefits can be perceived as flowing upward through the industry hierarchy on Global–Worldwide and Regional–Local bases.

The specific benefits for each matrix element are derived from the technical characteristics, services, or applications of the respective satellite telecom systems (i.e., the distinguishing technical offerings for the various systems noted in other sections; e.g., voice or video vs. radionavigation, radiodetermination). Clearly the magnitude (measurement) of benefits derived from the alternative systems also varies in relation to the chosen levels of aggregation (e.g., global vs. local). This analysis addresses only selected benefits derived from several GEO satellite systems: Fixed Service Satellites (FSS), Very Small Aperture Satellites (VSAT), and Direct Broadcast and Direct to Home (DBS/DTH). These systems and their respective service/applications are currently the most extensive for which only summary benefits are described. Nevertheless, they demonstrate the pervasiveness of satellite benefits within the defined matrix categories.

IT AND INTERNET-RELATED APPLICATIONS AND BENEFITS

Information technology (IT), including Internet-related services, is a representative application from which telecommunications benefits are derived. This is currently one of the most pervasive satellite applications, available in different forms from numerous satellite systems, and the spectrum of services within this generic category is expected to only increase over time. The G-7 countries, for example, recently spent over 3% of their Gross Domestic Product (GDP) on IT, with the worldwide average being around 2.4%. The United States alone is expected to spend about $1.4 trillion (on related IT services) in 2004, with a compound annual growth rate (CAGR) of 8% since 1999. Moreover, the worldwide IT equipment and services markets, which include satcom, are expected to grow as well through 2005 at almost 9% CAGR (see Fig. 9.5).

The beneficial effects derived from many satellite systems are subsumed within the generalized and comprehensive satellite IT services applications. This broad category encompasses the specific IT services associated with each of the

FIG. 9.5. Worldwide telecommunications market revenues.

different systems. Refining IT one more level, these are generally the voice, data, and video services used mainly (but not exclusively) for Internet access to the World Wide Web (WWW). These services provide the connectivity and access for consumer/business/government transactions from which e-commerce benefits are derived. The generic relationship is shown in Fig. 9.6 and discussed in more detail in the Global–International Benefits section.

It is important to remember that any of the benefits of improved access and network connectivity are related to the "starting point" which is defined by: the state of economic development, the regulatory-political-administrative framework, the existing telecom system infrastructure with associated technical and cost characteristics, and as realized through service offerings available in a specific locality, country, or region. Thus, as noted earlier, the net benefits that accrue will be different in response to the same set of satellite capabilities. In some cases they will be enhanced, whereas in others they will be repressed. These considerations effectively add another dimension to the benefit matrix analysis structure that should be considered in detail for more comprehensive analyses.

Further, care must be taken in defining the appropriate level of aggregation within the hierarchy for which a proxy variable is being used. This is necessary to identify which lower level benefits are subsumed within the category being measured and to recognize or eliminate overlap. For example, revenue at the carrier level is one measure of the subscribers' benefits. At the same time, part of this aggregate benefit measurement subsumes those benefits that can be allocated to other categories such as manufacturers or equipment suppliers.

Within these considerations, the following sections discuss several categories of benefits derived from satellite communications—actual and potential—that are extracted from the stakeholder evaluation matrix. As noted earlier, these benefits are categorized into Global–International and Regional–Local. Clearly these dis-

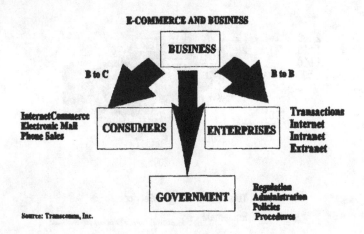

FIG. 9.6.   Commerce and business.

tinctions are not unique. There is and should be extensive overlap between the categories and among the elements of the categories as well. Further, the extent of such benefits is dependent on the unique system–service characteristics available in the marketplace and the environment within which the facilities and services are made available. Equally important to remember is that, although these benefits are traceable in varying degrees to satellite communications and can be measured with proxy variables, the underlying driving force is telecommunications as a whole, which provides the necessary access and connectivity to achieve these benefits.

GLOBAL–INTERNATIONAL BENEFITS

The benefits to users, operators, manufacturers, and others are derived from the distinct characteristics of the various satellite systems (e.g., system application, available services, coverage, cost). Yet many of the benefits that are difficult to measure and attribute are actually self-evident, as mentioned earlier. The fact that these new systems exist at all is *prima facie* evidence of financial and economic benefits to the users, which are passed upward within the stakeholder hierarchy. For example, revenues reflect value received by users/consumers that accrue within the stakeholder hierarchy to facilities and equipment manufacturers worldwide for the manufacture, installation, and testing of satellite systems infrastructures; to domestic and international satellite system launch vehicle organizations; and to satellite system operators and carriers. The latter revenues acquired by satellite system operators reflect the first level of financial gain and are a measure of the economic value or benefits to the users generating that revenue, none of which existed before 1965.

SELF-EVIDENT BENEFITS

The last several years have been some of the most dynamic in the satellite industry since its commercial inception in the late 1960s. New facilities, services, and applications have been instituted in parallel with substantial investments, and, despite some spectacular satellite and launch failures, the total value of facilities and investments has increased—even though the worldwide recession in 2000 and 2001 forced a decline in system-operator revenues and profitability. Consequently, estimates of future satellite communications revenues are problematic but are generally expected to grow through year-end 2007. One source estimates over $100 billion by year-end 2007 (Fig. 9.7).

The largest portion of the current total revenue, about 60%, is generated from entertainment: U.S. and International Direct to Home (DTH) or Direct Broadcast Satellite (DBS) video services. This proportion of the total is expected to decline

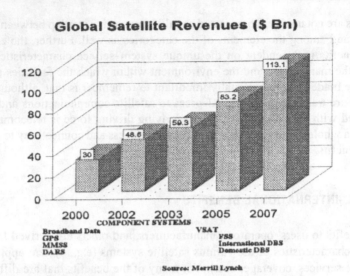

FIG. 9.7.   Diverse forecast for future satellite communications markets.

to less than half by 2007, although in absolute terms DBS services will likely triple in value to some $40 billion from its current level. Moreover, the expectation is that broadband data revenue will grow exponentially during the same period to become the revenue leader, although the unexpected hiatus in broadband growth at this time may make these expectations problematic. Clearly, these revenues, which were nonexistent prior to the late 1960s, are an aggregate measure of the increasing impact on trade, commerce, and reflected value. However, a more specific analysis of the benefits underlying these value measures and their relationship to satellite telecom is appropriate, although much of it may be supported inferentially and qualitatively rather than statistically.

SATELLITES AS A GLOBAL ECONOMIC STIMULATOR

Much of the worldwide, regional, and local economic growth has been attributed to the assistance of parallel—if not leading—telecommunications expansion and growth. There is much specific, if not circumstantial, evidence to support the connection for telecommunications in general and satellites in particular. For example, in the 1950s and early 1960s, the process of internationalization expanded rapidly. Trade and investment between and among countries grew much faster than specific countries' output. This was particularly true in the North Atlantic region, where submarine cables were being installed and expanded and, in the later years, satellite communications systems were instituted. It can be asserted that this was the forerunner of the even more rapid growth of capital flow, exceeding even that of

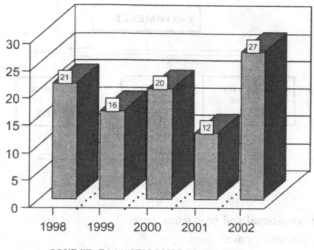

SOURCE: FAA/AST/COMSTAC, May 2002

FIG. 9.8.   Geosynchronous satellite launches.

intercountry trade during the 1970s and 1980s. Further, the increasing satellite capacity in the 1990s clearly provided the access and connectivity necessary for the growth of information transfer among countries—particularly electronic commerce (e-commerce)—solidifying the increasing trend toward globalization.

This is evident from the number of GEO satellites in operation, which increased from 24 in 1979 to 233 in 2000, with the expectation of 186 commercial GEO launches in the next 10 years, according to the most recent Federal Aviation Agency (FAA) report (see Fig. 9.8).

As shown in Fig. 9.9, the e-commerce IT activities can be depicted as a closed loop that perpetuates itself and, at the same time, yields numerous benefits. The e-commerce generates competition by providing revenues and profits that attract new market entry and market expansion while lowering transaction costs. The consequent increasing traffic flows provide incentives for the expansion of national infrastructure and connectivity as well as incentives for the financing necessary to support the changes.

The Business-to-Business (B2B) relationships are increasing in their pervasive use of IT technologies to create increasing value by working with partners, suppliers, and customers virtually in real time. These activities go well beyond the traditional methods of e-mail, fax, and voice mail. High-speed B2B service will improve overall efficiencies by enhancing many of the following:

- expanded and more efficient sales processes
- more comprehensive and faster customer services
- faster market introduction of products and services

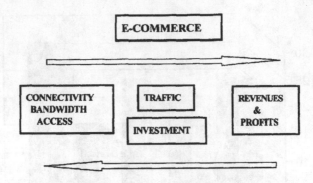

FIG. 9.9.   E-commerce relationships.

- lower operational and production costs
- lower inventory costs

The e-commerce among government–business–residential entities has the additional benefit of fostering the convergence of alternate technologies and worldwide companies (e.g., globalization), the latter mostly through merger and acquisition processes, creating multinational corporations, all of which are designed to increase techno-economic efficiency, which is reflected in market/output growth and profitability. These activities necessarily affect the governmental administrative–regulatory structure and policies in each country, particularly through privatization, liberalization, public/private partnerships, and similar organizational changes. The consequent increased interactions among the countries as a result of these changes reinforce newly enhanced comparative advantages and objectives (e.g., initiating or expanding manufacturing operations or establishing and fostering telecom hub operations). This process leads to increased global economic growth and a convergence of national policies designed to continue such growth through various multinational organizations and agreements (e.g., World Trade Organization [WTO], International Telecommunications Union [ITU], or United Nations Industrial Development Organization [UNIDO]).

INTERNET AND WEB FOCUS

It is the substantial capacity and services available from satellite communications facilities and services that are supporting continued IT expansion worldwide. The substantial growth and consequent benefits rely on the information transfer via the Internet and access through the WWW. The interaction between satellite communications and the Internet is addressed in more detail in a separate chapter. Nevertheless, the extent and nature of this relationship are highlighted here as well.

The distribution of these benefits is proportional to the worldwide supplier–user distribution. The consequent worldwide Internet distribution is not surprising, with the Americas' host density almost six times that of its nearest continental rival, Europe, as shown in Fig. 9.10. Nevertheless, all regions worldwide—as exemplified by the Americas—are growing at a compound average annual growth rate of more than 20%.

While overall user and host densities have lower growth rates (see Fig. 9.11), the absolute number of Internet users follows roughly the same regional pattern.

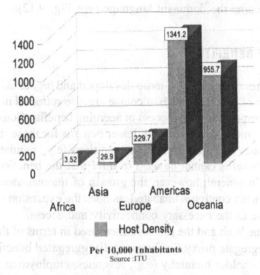

FIG. 9.10.   Regional Internet host density—2002.

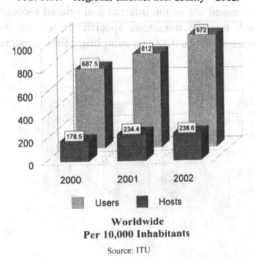

FIG. 9.11.   Internet user and host density.

However, there are several places throughout the world where the subscriber densities are equal to or exceed that of the United States. The Scandinavian countries, Hong Kong, and Singapore are examples, and even Iceland has almost four times the Internet subscriber density of the United States. This clearly demonstrates the perceived value in network access for both developed and developing countries. Not surprisingly, almost two thirds of Internet users rely on the English language, and over 80% of the Web pages are also in English. The nearest competitors are the Spanish and German languages, although some project that Mandarin Chinese may ultimately become the dominant language (see Fig. 9.12).

DISTRIBUTION OF BENEFITS

Clearly, by using these measures, the more developed and higher income countries accrue the larger set of benefits, which of course are also reflected in the respective economic growth, perpetuating the process of accruing benefits. Consequently, major disparities exist among countries, with lesser benefits for those that have lesser economic growth rates and more restrictive conditions (e.g., regulations, pricing). This allocation of benefits cannot be used to infer that the benefits accrue to the most "deserving." In general, however, the growth of Internet users, Web pages, and related e-commerce continues unabated, as does the expansion of satellite services that contribute to the necessary connectivity and access.

The growth of the Web and the Internet, measured in terms of their respective statistics, are an aggregate proxy for the more disaggregated benefits that accrue throughout the stakeholder hierarchy (e.g., revenues, employment) and that can be broken down into more detailed categories of the benefit hierarchy as well.

For example, increased use of the Internet and related e-commerce contribute to economic growth through numerous specific effects on worldwide financial–economic interests. Figure 9.13 is a sampling of the expected growth of the

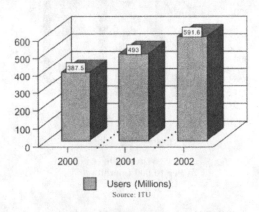

FIG. 9.12.   Worldwide Internet users.

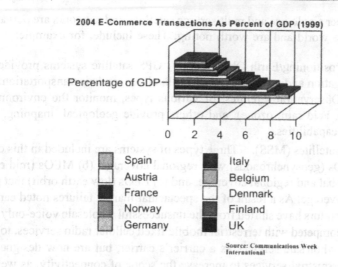

FIG. 9.13. 2004 E-commerce transactions as a percentage of GDP.

overall e-commerce activities, shown in relation to overall economic growth, for several selected western European countries. These effects occur within the following categories:

- trade/trade support
- customs/immigration
- transportation
- banking
- financing
- insurance
- libraries/databases/publishing houses/print media
- telemedicine/health services
- education/training
- security
- cultural interaction and entertainment and broadcasting
- inter-/intragovernmental coordination
- job/skills requirements

These are only a representative sample of the categories affected by satellite communications and for which worldwide benefits are provided. Although their quantification may be approximated at a more aggregate level, it must also be remembered that, unless specifically addressed, it is implicit that associated qualitative benefits are subsumed within the quantification.

Several other single-use and multiservice satellites and networks are pervasive throughout the world and are worth noting. These include, for example:

- Global Positioning/Earth Observation. GPS satellite systems provide invaluable navigation and positioning data for land, sea, and air transportation services; Earth Observation satellites, of various types, monitor the environment (e.g., weather, acid rain, ozone), and others provide geological, mapping, and photographic capabilities.
- Mobile Satellites (MSS). Three types of systems are included in this category: (a) GEOs (geosynchronous) with regional coverage, (b) MEOs (mid earth orbit) with global and regional coverage, and (c) LEOs (low earth orbit) that provide global coverage. As a result of the spectacular market failures noted earlier, the system operators have shifted from the unsuccessful wholesale voice-only services, which competed with terrestrial mobile/fixed cellular radio services, to one that provides wholesale services as a carrier's carrier, but are now designed to complement terrestrial services to increase the scope of connectivity, as well as expanding into broadband data in several cases.

This is not an exhaustive list of satellite systems and their effects. Rather it is representative of the pervasive effects and benefits that are the result of often uniquely increased network access and connectivity provided by these systems. Of specific interest are the more detailed effects on developing countries and on service provision to remote regions.

REGIONAL-LOCAL BENEFITS

The preceding section emphasized worldwide satellite applications and effects. Nevertheless, many of these same satellite systems provide connectivity and capacity for distinctive benefits to local and regional users. Those that are specifically oriented to developing countries, and even moreso to rural areas within these countries, are of particular importance. The benefits accruing in these latter areas provide additional enhancement—from the bottom–up—to increase regional–international–worldwide benefits as well.

The data presented earlier for Internet and Web statistics, application of DBS/DTH, and similar activities, clearly suggest that derived benefits are disproportionately advantageous to developed and economically advanced countries. In fact such advantages available through these services are tending to increase the so-called "digital divide." This refers to the discrepancies in benefits that accrue to the urban–suburban, higher income "haves" which are unavailable to the lower income, nonurban, rural "have-nots." Yet satellite technology advancements are constantly improving the technical, operational, and cost characteristics so as to provide expanded basic telecom services into the latter areas, albeit more slowly than expansion is occurring in the former. Nevertheless, to the extent that financial–economic data alone are used to measure benefits (i.e., without explicit consideration of qualitative rural benefits), the urban areas continue to represent the major benefits. To that extent, the measured benefits tend to understate the totality of benefits derived from satellite telecom.

RURAL TELEPHONY

A simple example of this satellite dichotomy can be shown with telephone density, which is necessary to provide access for basic voice services and connection to the Internet. The relatively high telephone densities necessary for network advantage are taken for granted in the developed countries and even in the urban centers of many developing countries. In contrast, the data for Central and South America show that with average teledensity often below the 10% range for these countries, with few exceptions (Argentina, Brazil, Chile, Uruguay), are well below the 20% density recommended by the ITU as the threshold for economic development.

More important, these density values are aggregate averages for the country; they include the high-density values for urban centers and the much lower rural density values. In Mexico, for example, 56% of the population resides in cities with less than 100,000 inhabitants, whereas 50% of the total access lines are in the cities of Guadalajara and Monterrey. Similarly, 60% of the total lines in Peru are in Lima, covering only 30% of the population. Thus, the averages overstate even the potential for achieving local–rural benefits that can contribute to the country and regional growth.

Telephone density in these rural areas is gradually increasing because the number of technologically advanced satellite systems and their respective footprints is expanding. The increased geographic coverage—in addition to the high urban services revenue coverage—generates a low marginal cost for these areas. This allows carriers to provide satellite services (e.g., voice, fax, low-speed data) at very low prices. As a consequence, the rural telephone density and number of subscribers with access to the local and national network is increasing. Other contributing factors are the declining lower cost of smaller, self-contained satellite earth stations, coordination and integration with terrestrial facilities (e.g., wireless local loop, cellular mobile/fixed), and innovative applications (e.g., Public Calling Offices). The cellular mobile/fixed station access to the main network via satellite is exceptionally important. For example, mostly rural Bolivia has increased its mobile/fixed subscribers at a CAGR of over 98% for 1995 through 2002. This, however, is largely a consequence of starting from such a small base.

MULTIPURPOSE AND SPECIALIZED SATELLITES

There are other examples of satellite communications changes (technology, applications) providing increased local–regional benefits and reducing the digital divide. As distinct from the satellite systems with a more global orientation mentioned earlier (e.g., GPS/Earth Observation), the inherent coverage–cost characteristics noted earlier are the underlying basis for the growth of many fixed station satellite (FSS) systems, which provide specialized networks and rural services. Among these are, for example, Very Small Aperture Terminal (VSAT) systems. These are particularly suited for national/regional business data systems outside of major urban areas, in both highly developed countries (e.g., United States) and lesser developed countries (e.g., Mexico).

The same coverage–cost characteristics underlie the growth in very sparsely populated areas, fostering development of Public Calling Offices and hybrid systems, as noted earlier. The former are small satellite earth stations that use satellite services to provide multiple voice equivalent circuits for direct user access at the satellite earth station location. Alternatively, the earth station is connected to wireless local loop (WLL) systems or other point-to-multipoint systems for geographically distributed, multiple subscriber networks. In both instances, subscribers have access to a terrestrial backbone network and the Public Switched Telephone Network (PSTN). In many cases, these provide the first voice services or "plain old telephone service" (POTS) in these areas. For example, Global Village Telecom (Gilat–GVT) was recently licensed by the Chilean government to serve 1,500 villages, and the company plans similar services and coverage in 18 countries in the near future.

More generally, the VSAT systems are being used for distributed business networks throughout rural areas where terrestrial systems are not effective. Specialized satellite systems such as VSAT are evolving from pure data systems supporting such intracompany operations, particularly for country-wide and nonurban applications (e.g., banking, airline/auto reservations, inventory control), evolving into broadband multiservice data, voice, and video for rural services, particularly in developing countries.

BROADBAND SATELLITES

The most prevalent broadband services are the dedicated Direct Broadcast/Direct to Home (DBS/DTH) video satellites that are used primarily for entertainment (see Fig. 9.14). The major orientation of these systems is toward the urban, high-population density markets, but it covers rural service areas as well.

With the inherent broadband capabilities, these satellites are expected to increase the currently limited two-way Internet access and High Definition TV (HDTV). Hybrid broadband satellites are designed to provide dedicated high-speed data circuits, as well as Internet services, both generally applicable to business use in urban areas. These are in addition to extensive entertainment distribution in the form of separate transponders for DBS/DTH broadcast TV/video services, some of them combined with downlink Internet and high-speed data.

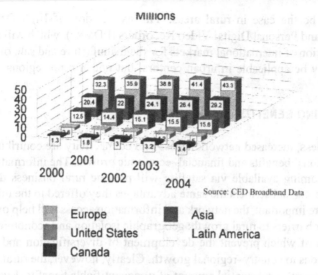

FIG. 9.14.   Direct to Home (DTH) satellites.

Most of the services focus on the high-density and high-revenue urban centers, proportionally increasing manufacturer, advertiser, and operator revenues.

The latter applications have appeared sufficiently attractive to develop specific satellites oriented to expanding coverage and broadband digital applications exclusively for regional and international traffic. However, this has proved to be very fluid in terms of business commitment. The Astrolink project backed by Lockheed Martin, Liberty Media, TRW, Telecom Italia, and Teleglobe Canada canceled their geosynchronous satellite system that would have operated in the Ka band as of late 2001. Subsequently, the other high-capacity, high-performance Ka-band satellite systems—Teledesic (backed by Craig McCaw and Bill Gates), as well as the Skybridge system (backed by Alcatel of France)—also canceled their plans. At this time, only Wild Blue (now backed by Intelsat, Kleiner Perkins, and others) and the Hughes Spaceway system (with backing from AOL Time Warner) are seemingly moving full speed ahead, with dedicated broadband FSS systems that will operate in the Ka-band. Some systems, such as SES Astra and Eutelsat, have much lower capacity hybrid systems that cross-link Ka-band and Ku-Band services, while Hughes Directway and Gilat Starband have systems that operate in the Ku-band.

Broadband satellite services are primarily expected to provide the broadband connectivity on nonurban routes and locations that would not have access to such terrestrial alternatives. Pricing for the latter, of course, will likely be the key. The joint marketing of, for example, the DBS customer facilities with retail distributors, as in the United States, provides distribution coverage, brand recognition, and advertising revenue-benefits in the developed, urban regions. Yet this is unlikely to be the case in rural areas. The expectation of High Definition TV (HDTV) and Personal Digital Video Recorders (PDVRs), which will soon lead to an expansion of international markets for the manufacture and sale of new equipment, may be applicable in urban centers, but not in rural regions.

INCREASING BENEFITS

Nevertheless, increased network access and connectivity are contributing to new rural–regional benefits and financial–economic growth. The information technologies becoming available via satellite will provide rural business development and residential subscribers the same advantages they offered to the urban populations. More important, the network and information access will help overcome the two major barriers to rural growth: geographic isolation and economic specialization, both of which prevent the development of diversification and consequent contributions to country–regional growth. Clearly, however, the rural, "have-not" societies accrue a substantial amount of nonquantifiable benefits. Even to the extent that the quantification uses implicit or proxy variables (e.g., increased wireline and wireless/cellular telephone density), the benefits that are tallied understate the true scope and magnitude of the positive satellite telecom effects.

More specifically, local–rural economic benefits accruing through the satellite-provided network connectivity include:

- substitution of voice/data for travel, providing economic savings and improving productivity;
- improved market access and information for rural businesses and farmers, eliminating distance and cost barriers;
- increased/improved access to administrative and social services (e.g., education, training, health, emergency services); and
- availability of new and expanded jobs, skill upgrades, and reduction of population migration.

In this regard, data strongly suggest that rural economies are responding positively and accruing positive benefits. For example, in some cases such as the United States, agricultural activities have declined from 63% to 2% of Gross National Product (GNP), whereas the growth of industries and services has increased proportionally.

More specific examples abound. Through much of Central and South America, we are seeing increased examples of Voice Over Internet (VOIP) satellite services for long-distance and international calling at *phone cafes*. These are the result of entrepreneurs providing voice services worldwide—where it is legal and sometimes where it is not—at a fraction of the price necessary for traditional telecom networks. Governments are also supporting such rural development where funding is available. The Hungarian government, for example, is implementing 2,500 "Telecottages." These are community centers that provide basic IT infrastructure and related services connected to the public network. Similarly, Bhutan is the first country in a 3-year program with the ITU and Universal Postal Union (UPU) that will use a VSAT network to establish telecom *kiosks* at 38 national postal facilities. These will be combination Public Calling Offices connected to the PSTN, as well as nodal distribution points for e-mail and regular mail in the country. Clearly, these are just several examples of how local–regional growth patterns will continue to be enhanced as the network access expands.

CONCLUSION

Using selected single- and multipurpose satellite systems, services, and applications, this chapter has analyzed some of the socioeconomic and financial benefits attributable to satellite communications, particularly those that can be separated from the more general benefits of telecommunications. Moreover, in establishing specific attributions, a procedural framework for disaggregated analysis was developed, which provides the basis for continuing and expanding the analysis for more detailed benefit investigations. This matrix structure, consisting of stakeholder hierarchy categories, alternative satellite systems and/or services, and geographic areas served, can be refined in subsequent analyses.

This initial analysis has been oriented to the first level of detail: that which addresses the self-evident and/or proxy measures of benefits. To the extent desired, this analysis can be expanded so as to reduce the overlap of benefits between and among the categories used (i.e., by expanding the stakeholder hierarchy) and increasing the quantification of qualitative benefits (i.e., with more specific selection of proxy variables). Nevertheless, even this first-order analysis has shown that distinct satellite benefits exist, and they can be measured in dollar-financial terms or other relevant proxy data. Further the level in the hierarchy for which benefits exist reflects the flow of benefits among the categories, and the benefits vary in accordance with specific satellite system and geographic characteristics.

The analysis also confirms the need for additional work to quantify benefits, especially in regard to the distinctions evident among differing countries and regions, particularly to developing countries. Although benefits are measurable in rural areas, they likely understate the long-term benefits to the society of which it is a part. Further, the regulatory–administrative–policy differentials among countries can create large disparities among similar satellite-benefit categories, which preclude generalizations. These issues can be addressed in later and more in-depth analysis with more precise selection of categories and measurement variables.

Nevertheless, it appears clear that, although benefits are distributed unequally—if not unfairly—they clearly exist and are increasing. Current satellite market and technology trends also suggest that these socioeconomic benefits will continue to increase.

BACKGROUND READINGS

Americas Telecommunication Indicators 2000, Executive Summary, International Telecommunications Union, Geneva, April 2000.

Contribution by the Delegation of Spain, Document 22-E, Plenary, AM-RDC, American Regional Telecommunication Development Conference, ITU, Acapulco, 1992.

Federal Aviation Administration, Associate Administrator for Commercial Space Transportation (AST) and Commercial Space Transportation Advisor Committee (COMSTAC), *Commercial Space Transportation Forecasts—2002*, May 2002.

Fixed Satellite Services, Equity Research-Telecommunications, Bear Stearns & Co., Inc., September 2002.

Global Satellite Marketplace 2001, A Space Odyssey, Pipe Dreams Become Reality, Merrill Lynch Satellite CEO Conference, New York, February 2001.

Hudson, Heather, (1995), *Converging Technologies and Changing Realities: Universal Service in the Information Age*, 7th World Telecommunication Forum, Technology Summit, Vol. 1, Telecom 95, Geneva, Switzerland.

Information Technology, Main Telephone Lines, Cellular Subscribers, 2000–2002, Statistics of the International Telecommunication Union, Geneva, Switzerland, 2003.

ITU: Celebrating 130 Years. 1865–1995, (1995), London, United Kingdom: International Systems and Communications Ltd.

Kaplan, M. H., (2003), "Space Business Facts and Fantasies 2003," Satellite 2003 Conference, February 25, 2003.

Kayani, R., and Dymond, A., (1997), *Options for Rural Telecommunications Development, World Bank Technical Paper No. 359*, Washington, DC: World Bank.

Lerner, Norman, (1991, May), *International Telecommunication Union Journal*, Volume 58, Geneva, Switzerland.

Lerner, Norman, (1998), "Investing in Telecommunications Privatization Around the World," *Proceedings of the Global Telecom Market Forum*, Supercomm, Atlanta, Georgia.

Lerner, Norman, (1998), *Universal Service in a Privatized Environment, Proceedings of the Global Telecom Market Forum*, Supercomm, Atlanta, Georgia.

"Millennium World Telecom," *Proceedings of the World Telecom 99 and Interactive 99*, International Telecommunications Union, Geneva, Switzerland, 1999.

Pelton, Joseph N., (2002), *The New Satellite Industry: Revenue Opportunities and Strategies for Success*, International Engineering Consortium, Chicago, Illinois.

"The Relationship Between Telecommunications and the National Economy: A Pragmatic Approach to Its Quantitative Determination Using a Macroeconomic Cross-Sectional Analysis," *Symposium on Economic & Financial Issues of Telecommunications*, International Telecommunications Union, Geneva, 1987.

Roller, L., and Waverman, L., (2001, September), "Telecommunications Infrastructure and Economic Development: A Simultaneous Approach," *American Economic Review*.

Rostow, Eugene, (1968), *President's Task Force on Communications Policy, Final Report*, Washington, DC: U.S. Government Printing Office.

Saunders, Robert, Warford, J., and Wellenius, Bjorn, (1983), *Telecommunications and Economic Development*, Baltimore, MD: Johns Hopkins University Press.

Technology and Telecommunications, Salomon Smith Barney, Private Client Equity Strategy, New York, February 2001 and 2002.

"Telecommunication Indicators in the Eurostat Area," Working Group, Statistics on Communication and Information Services, Luxembourg, February 1–2, 2001.

CHAPTER TEN

The World of Satellite TV: News, the Olympics, and Global Entertainment

Peter Marshall
Formerly Intelsat and Keystone Communications

Robert N. Wold
Formerly Wold Communications

> *The human mind craves for variety and entertainment . . . soon, communications satellites will be able to fill that need beyond all the dreams of the past.*
> —Arthur C. Clarke (1964, p. 146)

Television news, from the first days of satellite communications, took advantage of this new and highly flexible tool in the journalistic quest to be first and fastest with the news. This quest of being first with the news goes back to at least the Greek wars, when Pheidipides ran (not walked) the 26 miles from Marathon to Athens with the news of victory. The use of the telegraph in the Boer War at the end of the 19th century was another leap forward; so, too, were the arrival of radio, the telephone, the telex, the fax, and now the Internet. Satellite-distributed TV news and the cache updating of Web sites via satellite digital video broadcast technology, in fact, now creates the world's most powerful tandem of news media.

Yet in many ways it is satellites (now coupled with Internet video streaming and conventional TV transmissions) that seem to provide the ultimate technology for news—the combination of ubiquitous coverage, instant communications, and video images.

The speed of its adoption as a medium for daily news in the last decades of the 20th century was limited not by technology, but by global regulation and price. The fact that outside of the United States, the early satellite operators were government-owned national and international monopolies severely restricted the ability of broadcasters and news organizations to realize the full potential of the available capacity in space. By 1975 to 1976, the first private operators—Western Union (Westar), RCA Americom (Satcom), and Comsat General (Comstar)—had started to provide services in the U.S. domestic market, but it took many more years for other countries to follow suit and for international satellite users to reap the benefits of competition. It was global regulatory reform and liberalization of the telecommunications industry that allowed satellites to transcend the last limits of governmental restrictions and censorship. Today, this enables people throughout the world to have access to international TV, plus the benefits that flow from the flexibility and unregulated nature of the Internet. The use of communications satellites as a medium for broadcasting—both TV and radio—had a profound impact on political, cultural, and economic developments throughout the world in the last 30 years. This new capability for events to be seen and heard "live" and simultaneously in every country on earth has changed the way governments behave and the way people live their lives. Although there are numerous examples of memorable and highly significant images of satellite news coverage, some of the most striking involve the role of TV at the end of the cold war, its impact on the course of the final stages of the Vietnam War, and its vivid reports on the conflicts in the Middle East and the Gulf. Today, the world public expects to see its wars "live via satellite," whether from Iraq or the Falkland Islands or Afghanistan.

SATELLITES AND THE COLD WAR

Perhaps the most momentous event of recent years, the end of the cold war and, in particular, the fall of the Berlin Wall in November 1989, owed much to the way in which the people of Eastern Europe, through TV and radio, became aware of the benefits of freedom in the West. A West German commentator put it this way: "Totalitarianism could not survive in the East when the people's antennas were pointing to the West."

During the years of Gorbachev's policy of *glasnost*, the information blockade imposed by the KGB turned out to be "more like a tennis net than an iron curtain," according to author Scott Shane (1994) in *Dismantling Utopia: How Information Ended the Soviet Union*. Even before the wall came down, the communist authorities in East Germany recognized the danger. Walter Ulbricht, the East German leader, said: "The enemy of the people stands on the roof," and ordered brigades of the communist youth brigade to clamber on to housetops and remove antennas. But there were too many antennas. Eventually they gave up the unequal task of trying to stop people from watching West German TV. The desire to placate the population became urgent. The East German regime actually installed cable TV in cities as far to the east as Dresden so that people could receive programs without interference. Yet the people did not want clearer pictures; they wanted a clearer view of reality.

In Moscow, the headquarters of the Central Committee was one of the few locations officially equipped to receive CNN during the 1980s. The officials and staff became so accustomed to watching CNN's coverage of the Moscow Summit in 1988 that they arranged for Gostelradio to transmit it on channel 24. They did not announce the frequency, but it soon became widely known, and what began as a service for the rulers soon expanded to the general public often through the use of illegal and home-made antennas.

It was not only TV that made its impact on the downfall of communism. The role of radio, mainly the Voice of America, Radio Free Europe, and the BBC World Service, is documented in *War of the Black Heavens* by Nelson (1997), and in the foreword, Lech Walesa writes: "If it were not for independent broadcasting, the world would look quite different today. Without Western broadcasting, totalitarian regimes would have survived much longer." These broadcasts came both by satellite and over the air by shortwave transmission, but one way or another they came through.

In Eastern European countries today, one of the most poignant and significant sights is to pass through a small and poor country town where the former propaganda loudspeakers, once the only legal source of information, are now rusting on poles in the streets. Yet on the rooftops there is an abundance of new mini-dishes where the people have chosen to spend their meager incomes to enjoy the pleasures of entertainment and information provided by satellite TV in a free society.

The emergence of the Internet, too, played a part in opening up global information in Eastern Europe. E-mail service began in Russia in 1990, and one of the early services, Glasnet, was able to support 4,000 user accounts and was linked to the global system through a satellite channel donated by George Soros and his International Science Foundation.

TV AND WAR

Public opinion, informed by TV news coverage, first became a major factor in shaping U.S. government policy in the latter years of the Vietnam War. The various reports from the official command of General William Westmoreland were more and more called into question by the reality of the footage that Americans saw in their living rooms each evening. The truth was that the public trusted Walter Cronkite and CBS News. As a result, they were much less willing to trust President Lyndon Johnson and the official reports from the U.S. Military Command. The reports on the My Lai massacre were perhaps the turning point of the Vietnam War. From this point on, the American public became much less willing to accept the war's objectives and meaning. During this period, in the late 1960s and early 1970s, the use of satellite transmission for TV news had not fully evolved in the Pacific Region, but film coverage was flown from Vietnam to Hong Kong, Bangkok, or Tokyo for satellite uplinking.

Nevertheless, if the reporting from Vietnam had been in print only (i.e., in newspapers or magazines), the impact would undoubtedly have been significantly less. There is an immediacy of impact in TV news coverage, which is further enhanced by the use of satellite links. Thus, the public response to the latest news (whether it be coverage of warfare or a terrorist attack) has become an ever more critical consideration every year.

On September 11, 2001, one might wonder whether the reaction by the U.S. government, as well as public reaction around the world, might have been different if the terrorist attacks on the World Trade Center and the Pentagon had not been captured "live" and sent by satellite instantaneously around the world to some 200 countries. What then followed, first in Afghanistan, then in the United Nations, and culminating in the invasion of Iraq in March 2003, were the first examples of both public diplomacy and warfare conducted in the full glare of media coverage—with instant global distribution by satellite (see Fig. 10.1).

In the first Gulf War, which began in 1991, the images of bombs and missiles over Kuwait and Baghdad were seen mainly in Western countries and by the privileged few elsewhere with access to channels such as CNN. By 2003, there were new audiences for satellite TV in many parts of the Middle East. As NBC commentator Tom Brokaw wrote in the *New York Times* before the invasion of Iraq, he had seen for himself that small satellite dishes are now a familiar sight on the rooftops of apartment buildings and other public gathering places, and they are distributing news through four Arabic-language TV channels. The biggest of these, Al Jazeera based in Qatar, now has an estimated 35 million or more viewers. It reflects an Arab point of view, but it is far more independent than the old government-controlled broadcasters that previously dominated the countries of the region. As a result, observed Brokaw, the fundamental structure of Middle East politics has been altered by the widespread dissemination of information.

FIG. 10.1.   These transportable antennas can transmit high-quality TV from remote desert areas for news coverage. Even smaller terminals were used to support "embedded" reporters in the Iraq hostilities during 2003. (Photo courtesy of Gigabit)

Now political pressure develops quickly and independently from the ground up, not just from the top down, which is a dramatic difference from a decade earlier.

At the same time, the Iraq War was seen as a watershed for new media developments. With more than half a billion people worldwide connected to the Internet, the major service providers saw an opportunity to drive usage to new peaks. An executive of AOL Time Warner, CNN's parent company, put it this way in the *Financial Times*: "What the first Gulf War was to CNN, this one will be to CNN.com and its peers."

THE SHTILA REFUGEE CAMP MASSACRE IN 1984

It is hard to say when news coverage of war distributed by satellite actually began. The Vietnam War was actually covered by film crews in the field, but as mentioned earlier the film had to be flown to the U.S. West Coast for land-line transmission to network facilities in New York until Intelsat facilities could transmit the coverage from Hong Kong, Bangkok, or Tokyo.

We do know that one of the most dramatic and memorable examples of the horrific face of war "live via satellite" was the infamous Shtila Refugee Camp massacre in Lebanon in 1984. The cameras of several Western news organizations reached the grisly scene at almost the same time as the main Israeli forces arrived, a day after an advance group had descended on the camp.

While the Israeli government was still meeting and discussing how to explain how their forces had failed to prevent the massacre, and how to minimize the political impact, the first TV pictures were being fed by satellite to the headquarters of the TV news agency, Visnews, in London. From there strong and undeniable images were distributed to broadcasters in Europe and the United States. The facts of the story and the many fatalities were vividly caught on film and in minutes were on TV screens around the world. The worldwide critical response that emerged from other countries around the globe was swift and took the Israeli government by surprise.

Most governments quickly learned the lesson that attacks on civilians led to an instantaneous storm of protest in the court of world opinion. Since the mid-1980s, most governments have reorganized to ensure that their information and diplomatic channels at least keep pace with the new media. Even today, more than one national leader—including a U.S. president—has admitted that CNN or other worldwide satellite TV news media such as the BBC, Reuters, or even the Arabic Al-Jazeera news channels from the United Arab Emirates often serve as the fastest source of information.

The Chinese government, however, through a combination of circumstances, was certainly not prepared for the Tiananmen Square uprising in May 1989. The international cameras and satellite links were already in Beijing for another purpose—a visit by the Soviet leader, Mikhail Gorbachev. Ironically, China was eager to see this event covered worldwide. It was no coincidence that the student demonstrations began and then escalated at this very point in time.

At the time there was little authorities could do to restrict the activities of the news crews for several days. By the time they started to deport news personnel from CBS and other networks, the damage had already been done. The use of tanks against unarmed students and the symbolic picture of a lone student facing down the advancing military units had tremendous impact. This was because these events were caught on videotape and film and then instantly relayed to the world that watched in horror. The rest, as they say, is history.

NEWS EVENTS BY SATELLITE

From the early days of satellites, there were a series of experimental TV transmissions that paved the way for future developments. In the 1962 to 1964 period, two medium-altitude systems—AT&T-built Telstar and RCA-built Relay—and one high-altitude, geostationary system—Hughes-built Syncom—were tested. The winner for the future would turn out to be Syncom, but the highly publicized experiments of Telstar and Relay created widespread public curiosity and enthusiasm.

Telstar 1 was launched in July 1962 and began to fail after 5 months. Relay 1 was launched in December 1962 and remained operational until mid-1965. The medium-altitude orbits of Telstar and Relay were in cigar-shaped patterns that could be accessed for traffic at any single point for only an 18- to 19-minute window every 90 minutes. The TV experiments involved 30-meter antennas at Andover, Maine, and Goonhilly, Cornwall in the UK. There were memorable novelty "firsts" for the BBC and CBS. For example, a BBC special program introduced "live" inserts provided by CBS from Wall Street, Mount Rushmore, and the Mormon Tabernacle choir in Salt Lake City. The black-and-white pictures were hazy and the sound quality was poor, but it was TV by satellite! Telstar 2 was launched in May 1963, and Relay 2 was launched in January 1964.

Syncom 1 was launched in February 1963, but failed to sustain its orbit. Syncom 2 launched in July 1963, and Syncom 3 launched in August 1964. Because of Syncom's 22,300-mile altitude and its geostationary orbit, a fixed position above the equator relative to the earth, the system would require only three Syncom satellites to reach the entire globe except at the north/south polar latitude extremities. In October 1964, Syncom 3 provided continuous transmissions of the 1964 Summer Olympics from Tokyo to the United States. By then Hughes Aircraft Company had been selected by Intelsat to build the world's first fleet of communications satellites.

Among the first uses of Intelsat's fleet were coverages of the various stages of the Apollo space program. The first moon landings were among the events to be seen "live" across the Atlantic and Pacific Oceans (see Fig. 10.2). Then in 1972, three major news events enlarged the essential role of satellites in the news business. The first was President Nixon's landmark visit to China in February, where

FIG. 10.2.   The first moon landing in 1969 was televised live in all regions of the world with an estimated global audience of 500 million. (Photo courtesy of Intelsat)

the U.S. networks used Intelsat to transmit their pictures back to a teleport on the U.S. West Coast for terrestrial relay to their headquarters in New York. *Broadcasting* magazine described this as "a milestone in broadcast history." Then in May, the networks did the same for Nixon's trip to the Soviet Union for a summit meeting in Moscow.

The third 1972 news event came in September, during the Olympic Games in Munich. Satellite links were leased at great expense from Intelsat, through Bundespost, the German national telecommunications administration, by the U.S. network, ABC, which had the rights to show the daily events from the Olympics in America. Yet the Games turned from sport to tragedy when Palestinian terrorists took 11 members of the Israeli team hostage and later killed them. CBS fought to get a share of the only available satellite link, and eventually ABC relented and created a "pool"—and all the broadcasters from the United States and Europe were able to provide "live" coverage of the unfolding drama.

Another major news story came soon afterward in February 1973. The return of American POWs from Vietnam was an event of sufficient importance to justify the costs of satellite transmission to bring the first pictures of the released soldiers at Clark Air Force Base in the Philippines to their families at home.

POLITICAL, REGULATORY, AND TARIFF BARRIERS

It was the high tariffs that slowed the spread of satellite TV and especially global satellite TV news. These were the product of an international government-owned monopoly structure that was originally created to improve global telephony.

There were also those in some governments around the world who saw potential threats in the use of the system for broadcasting and were happy to see this development proceed as slowly as possible.

The tariffs imposed on the broadcasting industry were a combination of the short-term (per minute) rates charged by Intelsat and Eutelsat for the use of the satellite bandwidth, *plus* the 30% or even greater profit margins added by the national PTTs (the telecommunications administrations), *plus* the charges the same government-owned organizations then added for using their terrestrial Earth stations and the essential landlines to the studios of the broadcasters. Therefore, there was ample opportunity for governments to control the use of satellites by price alone.

Through the first decades of international satellite communications, it was the high costs that limited use of the new technology almost entirely to special events. Only in exceptional circumstances could the budgets of TV networks and producers, and the size of their audiences, justify such expenditures on transmission services. Despite persistent lobbying by the broadcasters and their international unions, such as the European Broadcasting Union (EBU) in Europe and the North American Network Broadcasters Association (NANBA) in North America, the decision makers of the international satellite organizations were reluctant to recognize the simple economic fact that lower tariffs would release a pent-up demand for satellite transmission services and would increase utilization, and thus their revenues, dramatically.

CNN CHALLENGES THE SYSTEM

In 1980, the American TV entrepreneur, Ted Turner, launched Cable News Network (CNN), the world's first 24-hours-a-day TV news channel. CNN was transmitted via U.S. domestic satellites to cable-TV systems all around the country, but Turner had wider ambitions to reach the rest of the world. He called in Sidney Pike, an experienced hand from Boston TV, who knew the TV world from some 30 years of experience, and gave him a mandate: get CNN distributed to the world 24 hours a day.

Early on, Pike went to China and made a deal under which he would give free terminals to the Chinese leaders so they could see world news. In return, he would ask only $50,000 for the CNN rights for China for the first year. He figured it would be a loss leader. Then he checked on the satellite tariffs from Intelsat and the Chinese and U.S. signatories who provided the service and found that the TV transmission rates would be well in excess of $20 million a year. This would be a loss leader indeed!

Pike went to Intelsat headquarters in Washington, DC, and visited Santiago Astrain, the director general. He explained that Intelsat's way of doing business was out of date. In the age of 24-hour satellite news, the idea that they would charge for a TV channel on a per-minute basis with a 10-minute minimum made no sense. Further, the existing Intelsat rules charged each receiving country 50% of the uplink charge, which he considered even more ridiculous because it actually cost no more to send a signal to 1 country than to 30 in the same region (except for the receiving earth station costs).

The power inside Intelsat was in its signatory organizations and their representatives on the board of governors. Therefore, the key to any change was to get the signatories to alter their approach to charging for satellite TV. At that time, Randy Payne, of Australia, was chairman of the board. Kerry Packer, a hard-nosed TV broadcaster in Australia, approached Payne to obtain a full-time TV channel. His concepts were very much in parallel with the thoughts of Turner and CNN. His idea was to take his TV programming from Australia and spread it by satellite all over the Pacific region.

After much politicking, the short version is that a few months later a new rate for the lease of international satellite channels on a full-time basis emerged from the Intelsat board of governors. This new TV rate provided the basis on which Kerry Packer could transmit his TV channel by satellite to Australia and the South Pacific, and CNN could start sending its news service to the world. This new Intelsat full-time TV rate was established on a yearly basis and eventually imposed no restriction on the number of down links in the coverage area.

Sidney Pike then persuaded the U.S. Armed Forces Radio and Television unit to sign up for a full-time TV channel and put CNN on it for most of the time. Ted Turner rightly gets much of the credit for innovations in cable TV and satellite distribution, but Pike was one of his secret weapons who managed to get CNN into over 150 countries and even struck up relationships with Saddam Hussein of Iraq and Muammar Gadaffi of Libya along the way.

At the time, the financial people within Intelsat, as well as the signatory organizations from which many of them came, were cautious and concerned that these new full-time leases would cut into the income from short-term occasional use. Even 5 years later, when there were some 15 full-time international transponder leases in place all over the world bringing in tens of millions of dollars, this conservative view still persisted among some of the financial officers at Intelsat and its signatories.

Competition to Intelsat for international satellite routes began in 1983 when U.S.-based Orion Satellite Corporation filed at the Federal Communications Commission (FCC) to operate a private satellite system serving U.S./Europe routes competing with Intelsat's signatories. Then in early 1984, Rene Anselmo's Pan American Satellite Co. (now Panamsat) filed to operate a single private satellite system connecting the United States and Latin America. One other applicant, Cygnus, also filed for United States/Europe. Intelsat's U.S. signatory, Comsat, aggressively fought these applications at the White House and in Congress. In early 1986, the Reagan administration approved the applications with conditions that Intelsat signatories did not suffer economically. Panamsat, having bought out Cygnus in 1987, began operations on PAS-1 in June 1988. The Orion 1 satellite began operations in 1994.

It was only then, in the late 1980s, that international satellite tariffs for TV began to truly open up to the needs of the broadcasting market. However, progress was still slow in many countries because, although the Intelsat rates had been changed, many of the monopoly PTT organizations around the world were still able to maintain prohibitively high end-user rates for satellite TV. Also they were able to use national restrictions to deny "landing rights" to signals from the Panamsat satellite.

DAILY NEWS BY SATELLITE

A victim of this resistance to change was the daily coverage and distribution of TV news. Throughout the 1970s, 1980s, and 1990s, most of the world's TV stations relied almost entirely on the agency services provided by Visnews, based in the UK, UPITN based in London and New York, and the syndication service of America's CBS network. These organizations had their camera crews located in all corners of the globe, and they relied on air freight services to ship their news coverage (originally on film, later on videotape) back to headquarters. There it was edited, scripts were written, and copies were made to be shipped out, again by air courier service, every day, to hundreds of waiting TV stations around the world.

For TV news programs and their viewers, this was the only source of video material from distant locations—not just the top international news, but also the daily sports events, the fashion shows, even the "skateboarding dog" stories so widely used at the end of news bulletins.

The TV news agencies provided slick and efficient operations, but most TV pictures from overseas locations still reached the screens of viewers at home 2 or even 3 days after the event. News commentary writers often evolved creative ways to prepare their scripts to incorporate days-old pictures into a bulletin and do so without loss of immediacy. It was at Visnews in the early 1980s that satellites were recognized as a potential solution to this problem and a whole, new future for TV news syndication. Technologically, it was then possible to collect TV news coverage from anyplace in the world by satellite and then distribute it with a single uplink to reach broadcasting stations in dozens of countries simultaneously. Yet the costs were still impossible for the news budgets of TV organizations to even contemplate.

As with CNN's experience, the breakthrough came in Australia, where the five fiercely competing networks agreed on a joint plan with Visnews by which they would share the costs involved. Protracted negotiations took place in London between Visnews and BT (British Telecom) and in Australia involving Visnews, the broadcasters, and OTC (the national carrier) to persuade them to break out from the institutional straitjacket of regulations and tariffs that did not encompass the needs of a daily video news service. Eventually, a plan was agreed under which a regular 10-minute satellite transmission would take place at a fixed time, 7 days a week, between the UK and Australia.

This pioneering service began in 1982, and TV news in Australia entered into a new era with same-day pictures from Europe for the first time. The transmissions were made over Intelsat's Indian Ocean satellite, and the footprint covered by this bird also reached another important TV market in Japan. So Visnews was soon negotiating with its Japanese broadcaster customers and KDD, the national carrier, for a similar arrangement. As a result, the daily transmission was soon extended to 15 minutes, consisting of 5 minutes of "world news," 5 minutes specially selected for the Australian broadcasters, and a further 5 minutes specifically for the Japanese market. Although the cost of the Intelsat space segment was still high by current standards, it was now made more palatable by being shared among nine different TV stations in Australia and Japan.

The move toward daily news transmissions expanded over the following years in both the Pacific and Atlantic Ocean regions, taking advantage of the new Intelsat video leases resulting from the CNN and Kerry Packer initiatives.

Visnews in London again foresaw the growing demand for daily news transmissions between North America and Europe and the opportunity to develop a form of cost sharing that had been pioneered by Wold Communications and others in the United States for domestic video distribution. The biggest hurdle was that, unlike the services offered by the commercial satellite operators in the United States, the Intelsat leases, as administered by the signatory organizations, did not permit the "third-party" use of the capacity.

By then the early commercial ventures were in the planning stages, and Visnews managers embarked on talks with Orion, Panamsat, Cygnus, Atlantic Satellite, and others that had international ambitions. However, it was only talk at that time; to start providing actual services, Intelsat was "the only game in town," and Visnews set out to find the best way to overcome the obstacle of third-party use of the system. For a transatlantic lease, it was necessary to negotiate a formal agreement with an Intelsat signatory on each side of the ocean, and in this instance it was British Telecommunications (BT) in the UK and Comsat in the United States.

A joint venture was formed between Visnews and Wold Communications to develop a service under which a leased Intelsat transponder would be used to provide transmission services to broadcasters in North America and Europe—in other words, third-party traffic. Although this concept was broadly acceptable in the United States, it took much persuasion on the other side of the Atlantic involving the UK government and the European Commission before BT reluctantly agreed to cooperate. Yet the final condition imposed by BT required another change in plans because they refused to accept the entrepreneurial Wold company in the United States as a party to the service, insisting instead on a "carrier of record." The joint venture was therefore reluctantly dissolved, and Visnews found a new and acceptable partner in Western Union, then active in the satellite business and eager to expand internationally.

In October 1983, the new service was eventually launched as "Brightstar," and it pioneered a further advance into the use of satellites by broadcasters. As markets grew and tariffs reduced, the business expanded. Western Union withdrew from the satellite business in 1987, and Visnews (by then acquired by Reuters) became 100% owners of Brightstar. Ironically, BT's enthusiasm for the TV news market was such that in the late 1990s they bought the Brightstar service from Reuters and continue to operate in the Atlantic region.

The Wold Company manifested itself instead in the Pacific region, where the alliance with Visnews was successful in creating a similar venture for the Japanese broadcasters. They created the Japanese International Satellite Organization (JISO), which entered into an agreement with the Wold–Visnews joint venture to provide daily news transmissions from the United States to all the TV networks in Tokyo. This service began in April 1984 and continued for nearly 10 years until a combination of growth in traffic, availability of more transpacific capacity, and reduction in transponder prices enabled Japanese broadcasters to justify leasing their own, unilateral satellite links.

TV "CONTRIBUTION" AND "DISTRIBUTION"

As noted earlier, it was Channel 9 in Australia that, in 1985, was the first TV broadcaster to lease a unilateral Intelsat transponder for its increasing requirements for news and other program material from the United States. This information was then distributed throughout Australia and the South Pacific. As tariffs were reduced in subsequent years, this became a more widespread practice, and other major networks in the United States, Europe, Australia, and Japan, as well as the news agencies, moved to acquire their own satellite capacity.

In this evolution of the use of satellites by TV broadcasters for special events or news, two separate and distinct applications need to be identified: These are known as contribution and distribution.

Contribution is the use of a satellite link to transmit TV coverage from the location of the event to the broadcasting center or news agency. This is usually a point-to-point transmission, and the technical arrangements are only made with the broadcaster concerned (although such transmissions are often monitored or "pirated" by those with the necessary receiving equipment). The ubiquitous characteristics and widespread coverage of satellite technology are exploited in this application by the ability to uplink the signal from any point on the earth's surface within the footprint or coverage area of a particular satellite. As technology has developed—including the move from analog to digital signals—it has become possible to transmit from increasingly smaller uplink antennae, and operations on location have evolved from the use of trucks equipped with dishes of 3-meter diameter; through the development of "flyaway" units, which could be shipped in boxes as regular air freight to be assembled on site in an hour; to the latest portable one-man shoulder packs. The use of each generation of satellite news gathering (SNG) equipment has been matched by a reduction in the costs of satellite time and provided new mobility for the world's TV news business. Each stage has also brought better technical quality, and at the turn of the 21st century even greater miniaturization and mobility became available through the use of videophone technology for news reports from the front line (albeit at something below usual broadcast quality).

Distribution describes the use of satellites to transmit TV programming or program channels from a broadcasting center to the viewer, either by direct broadcast to the individual rooftop antenna or cable-TV operators for onward relay to local subscribers. Direct broadcasting by satellite (DBS) was developed in the 1980s as a new service utilizing high-powered satellites with a specific global allocation of orbital locations and frequencies to meet national ambitions. However, such systems were high-cost ventures with limited channel capacity, and the early operators in the United States, Europe, and Japan were soon overtaken by new developments. These combined the use of less costly medium-powered satellites, multiple digital channel capacity, and low consumer prices for receivers and dishes.

To the viewer at home, however, these technical distinctions are not significant. It is the range of available programming and the price that have created a new mass market for satellite TV. Furthermore, the two applications are often combined by the broadcaster so that the distinction between *contribution* and *distribution* is invisible to a viewer who may be watching "live" TV from a remote location on the other side of the globe.

SATELLITES AND THE OLYMPIC GAMES

Among the largest global audiences for satellite broadcasting have been those for successive Olympic Games. As long ago as the 1964 Winter Games at Innsbruck in Austria, America's experimental Relay-1 satellite transmitted a small amount of TV programming originated by the European Broadcasting Union's engineers to a handful of international Earth stations. At that time, however, Relay-1 and another U.S. experimental satellite, Telstar 2, in low-altitude orbits were "visible" to the transmitting and receiving Earth stations on each side of the Atlantic Ocean for only 20 minutes at a time.

A few months later, in October at the 1964 Summer Olympics in Tokyo, America's newest experimental satellite, Syncom-3, achieved geostationary orbit over the Pacific Ocean. When the Japanese government learned that it would be in position to transmit live coverage to the United States without interruptions, they lobbied strongly for extensive live broadcasting in the United States. However, NBC Sports had acquired the exclusive rights from NHK, the host broadcaster in Japan, and could not change their program plans as desired. NBC had already sold its commercial time on the assumption that its coverage would be produced on videotapes to be flown to Seattle and then transmitted over land lines to New York for delayed distribution to the network's stations. After considerable pressure from the U.S. government, NBC agreed to a compromise—only the opening ceremonies would be televised live by satellite, and so the black-and-white program, honoring Japan's Emperor Hirohito and the Empress Nagako, began on an October Saturday morning at 1:00 a.m. New York time.

In the *New York Times*, TV critic, Jack Gould, wrote, "Live television coverage of this morning's opening of the Olympic Games in Tokyo was of superlative quality, a triumph of electronic technology that was almost breathtaking in its implications for global communications" (1964, p. B1).

Although NBC's viewers were denied any further *live* coverage, the opportunity was not lost elsewhere because some Olympic events were transmitted from California to broadcast networks in Mexico and Canada, and a handful of satellite feeds were retransmitted from Canada to Europe on the Relay-1 satellite.

Thus it began. For 2 weeks in October 1964, the world had been shrunk; TV, the International Olympics Committee, and the emerging satellite communications industry had collectively taken a huge step forward. Thirty-six years later in September 2000, the Sydney Olympics in full-color lasted 15 days with a mixture of show biz, TV, and Olympics. An invisible fleet of satellites once again delivered the electronic saga everywhere.

THE SOARING VALUE OF TV RIGHTS

In the intervening years, and boosted by the size of the global TV audiences reached by satellite, the International Olympic Committee (IOC) has grown into a $1.1 billion per year enterprise, and today TV rights revenues represent more than 50% of the IOC's total income. Seven of the IOC's 10 leading sponsors are U.S.-based, according to the *Los Angeles Times*, and the Salt Lake City Winter Olympics in 2002 had 64 corporate sponsors and suppliers contributing $859 million in cash, goods, and services. This compares to $480 million raised for the 1996 Atlanta Games and $140 million for the 1984 Summer Games in Los Angeles.

The cost of staging the Salt Lake games was at least $1.9 billion, probably the most expensive Winter Games ever. The U.S. government provided nearly $400 million in funding, with state and local government providing funding at about $225 million. Beijing won the 2008 Games when they pledged to spend $14.3 billion during the next 7 years for sports facilities plus subway and highway construction. The latest budget estimate is $23 billion. This includes spending $1 billion on subway extensions alone, adding at least 60 miles of track and 60 new stations. The growth of the Olympics also has reached a level that it is likely to have a lasting impact on local life for every new venue chosen in the 21st century.

The value of the Olympics to commercial TV networks was well illustrated by the deal struck by General Electric-owned NBC for a multiyear TV rights agreement with the IOC for the U.S. market. The total package was $3.55 billion for the Summer and Winter Games for the period from 2000 through 2008.

Including its basic broadcast network and two cable/satellite networks (MSNBC and CNBC), NBC produced 375½ hours of TV coverage at Salt Lake City. NBC's TV network reached more than 200 million TV households. MSNBC, via cable and satellite, reached 74 million homes, and CNBC reached 83 million. The number of additional people reached via video streaming on the Internet was undoubtedly in the tens of millions.

These huge audience figures led to another leap in the value of TV rights when the IOC held a three-way auction during 2003 for the 2010 Winter Olympics. Although the venues had not yet been decided, NBC won the U.S. media rights with a bid of $2 billion against competition from the News Corporation's Fox network and ABC/ESPN, owned by the Walt Disney Company. This amounts to a staggering 33% increase on the $1.5 billion that NBC had previously agreed for the Winter Olympics in Torino, Italy in 2006, and the Summer Olympics in Beijing in 2008.

The $2 billion from NBC is further augmented by a deal worth nearly $200 million, in which General Electric (NBC's parent company) becomes a worldwide sponsor, which means that the IOC can expect to receive 46% more for the Games in 2010 and 2012 than for those in 2006 and 2008.

SECURITY CONCERNS

One area of increasing concern (and cost) for the IOC and national organizers is security. The memories of the Munich Olympics incident on September 5, 1972, certainly live on. As described earlier, on this occasion, 11 Israeli athletes and officials, after a period of standoff, were massacred by a group of eight Palestinians. Because of the facilities and transmissions networks already in place, the Munich disaster also became a worldwide live drama via satellite and TV. Then there was the shooting death and exploded bomb at the 1996 Atlanta Olympics. In short, the Olympics have become a global electronic bullhorn that captures and amplifies every story or event as it occurs. Any happening suddenly becomes somehow more newsworthy just because it took place during the Olympics. If terrorists want to make an impact by planning an attack or staging an event, they know that at the Olympics they have a global stage and every news occurrence will be instantly linked across the planet by satellite news reporting.

SATELLITES DELIVER GLOBAL TV AUDIENCES

There have been many claims for the biggest global TV audience—not only from the IOC—and in truth actual figures are difficult to ascertain. When astronauts Neil Armstrong and Buzz Aldrin were NASA's first people to walk on the moon in July 1969, NASA's live TV pictures sent down to Earth for worldwide dissemination were estimated by Intelsat to have "reached" 500 million people. By the time of the 1976 Montreal Olympics, the number of estimated viewers had jumped to an estimated 1 billion. Now Intelsat and others estimate the numbers who watch some part of the Summer Olympics to be close to 3 billion. Of course this is not one TV show, but over 100 different shows in many dozens of different languages (where the home team always seems to win).

In 1969, the world had just above 4 billion people, and Intelsat had but 69 signatory countries. By 2000, the world population had expanded to over 6 billion, and Intelsat had grown to 143 signatories (with total connectivity to over 200 countries, protectorates, and entities). In 2000, the World Cup and Olympic Games were held; again, according to Intelsat, both these events "reached" global audiences of 3 billion people.

The FIFA (Association Football) World Cup Tournament (known as soccer in the United States) is another sporting event that attracts huge global audiences every 4 years. Each year of the event, claims are made for even greater cumulative audiences, and these audience figures lead to a similar escalation in the TV rights for the event, which are reported to be 1.3 billion Swiss francs for the 2002 World Cup tournament in Japan and Korea. What is not in dispute is that this series of football (soccer) matches, involving national teams from every part of the world, attract huge audiences for the networks screening them "live" at whatever time of the day or night it happens to be. In the absence of definitive and validated audience research, the global viewer counts for these, and other major events often seem to be an aggregation of those with access to TV sets in the countries reached.

THE NEW MILLENNIUM ARRIVES

However, one special event that may well have broken all the records was the Millennium weekend, which inspired a myriad of TV programming combining culture, entertainment, and news gathering. Months, even years, of planning brought together a seemingly endless stream of events from every corner of the globe, with cameras, reporters, satellite uplinks, and transponders taxed to the limit.

An imaginative compilation called "2000 Today" reached more than 1 billion people around the world, according to the BBC. The 25-hour production epic was hosted and choreographed in London by the BBC, working closely with WGBH Boston. It was beamed live to the United States, where it was morphed into "PBS Millennium 2000" for the Public Broadcasting System's 348 member stations. In the final consortium count, "2000 Today" had 81 broadcaster members in 80 countries. Fifty-seven broadcasters contributed content, and the remaining 24 were "receive-only." Twenty-one satellites were used for contribution backhauls from the 57 locations, but only 3 satellites—and some transatlantic fiber paths—were needed for relays from the hubs to London.

One remote feed came from Antarctica, relayed via the TDRS 6 satellite to a NASA downlink facility at White Sands, New Mexico, from where it was second-hopped via GE-2 to New York for a third hop to London, all in real time. An especially busy period in London was at 2300 GMT (11:00 p.m.), when European countries celebrated midnight. Sixteen satellite feeds from most of the capitals of Europe had to be received at the BBC simultaneously. For the outbound global distribution of "2000 Today," six satellites were used to distribute the program to the other 79 member countries. They were PanAmSat 3, PanAmSat 5, Eutelsat W3, NSS 703, AsiaSat 2, and Intelsat 701. The program had a modular format that enabled seamless cutaways for local hosts and commercials. Other elements crucial to the international success of "2000 Today" were a live English-language "guide" audio channel from each remote site. Productions originated in more than 30 different languages, and an Internet Web site called LiveScript enabled the family of broadcasters to coordinate in real time with the BBC's program production rundown and timings.

A BBC executive we talked to after the telecast enthused that not one satellite feed was lost. This was truly amazing given the huge number of live feeds that came in from all over the world. Indeed it was probably the most complex program the BBC has ever mounted. What made it different from programs such as the Olympic Games was that it had contributions coming in from many diverse locations, in addition to and simultaneous with the outbound distribution.

Another measurement of the Millennium weekend's satellite business was a count of occasional traffic over and above existing full-time leases. Among international carriers, Intelsat's 18 international satellites carried 30 supplementary channels representing over 1,650 hours. Panamsat announced an extra 1,300 hours of event coverage and approximately 300 transmissions among 40 customers using its 12 U.S. domestic and international satellites. Amsterdam-based New Skies Satellite, with five birds previously owned by Intelsat, generated 687.5 hours of special traffic. Not included here are the millennium traffic counts for other major satellite carriers such as AsiaSat, Eutelsat, Teleglobe, Loral Skynet, GE Americom, and others that likely added several thousand more hours of programming.

CNN described its "Millennium 2000" project as "a most extraordinary programming initiative . . . the biggest journalistic undertaking in peacetime." For 100 hours over the entire 4-day weekend, CNN delivered "comprehensive, global news coverage of millennium-related events and turn-of-the-century issues." Nine international and five U.S. domestic satellites were broadcasting these events based on over 200 different feeds from 73 locations. Of course when we celebrate the next millennium in 3001, the feeds will come not only from all over the world, but perhaps from different sites in the solar system.

WHAT WILL COME NEXT?

Communications satellites made a remarkable contribution to the growth of global broadcasting in the last 40 years of the 20th century. In turn global broadcasting has impacted politics, information, entertainment, and the daily lives of people in every corner of the world.

Now in the new millennium, there are the early glimpses of an even more exciting and challenging future. The convergence of digital TV and computer technologies, which has been predicted and discussed for many years, is now beginning to emerge with the development of broadband services to the business sector and the home. Services are now becoming available in the United States and elsewhere, whereby one small elliptical-shaped satellite dish with multiple feeds can receive TV, the Internet, and actually support a range of new interactive digital broadband services.

The rate of market growth of such interactive broadband satellite services, which combine entertainment and information services at this relatively early stage of development, is influenced more by economic factors than by the technology. In the new competitive digital environment, satellite-based services have to compete with the terrestrial broadband services provided by cable-TV operators and telecommunications carriers installing DSL and fiber networks. In fact different types of satellite systems designed for so-called broadcast, fixed, and mobile satellite services are starting to compete with each other.

The advantages of satellite applications are to be found in multicasting services, where a single up-linked broadband signal can be received by large numbers of receivers without additional transmission costs. There are also large sections of the potential market in most countries of the world—possibly as great as 30%—where the costs of deploying terrestrial systems are not economically viable when compared with the installation of a small satellite antenna. This may be because of mountains, deserts, swamps, sparse population, or even urban areas where utility conduits are not practical to install.

Whichever system is in use, whether satellite or terrestrial networks, however, it is ultimately a combination of the program and information content available to users and the price of the hardware and access that will determine the rate of market growth. The rate of growth will, in turn, influence the ability of the service providers to achieve the critical mass necessary to reduce prices at the consumer level. The potential of Internet- and IP-related services to ride on the back of popular entertainment services to form integrated digital satellite offerings is one of the truly interesting developments of the new millennium.

Technological development is making possible new features that will help to grow the market—applications such as movies on demand, interactive services, multiple split screens of information, digital TV, high-definition TV, and so on. In time we may even see multiple-rastered or three-dimensional video with incredibly realistic surround sound.

Meanwhile, journalists in the field, using laptop equipment and small transmitters, can now transmit text, pictures, audio, and even video material by satellite directly to their newspapers and radio and TV stations. During the military incursions into Afghanistan and Iraq, for example, the use of videophone technology enabled reporters to report graphic accounts "live" by satellite from the battle zones in city and mountain locations. We also saw infrared images of events in the dark of night.

Predictions of specific new technologies and applications are dangerous in that some new products and offerings might even become available—or even be overtaken by newer developments—before the proofs of this book are read. What is clear is that broadband video services, higher definition images, and combined entertainment and information services are all part of the future and satellite systems will be a key part of making it so.

REFERENCES

Clarke, Arthur C., (1964), *Everybody in Instant Touch: The Coming of the Space Age*, New York: Meredith Press.

Gould, Jack, (1964, October 5), *New York Times*, p. B-1.

Nelson, Michael, (1997), *War of the Black Heavens: The Battles of Western Broadcasting in the Cold War*, New York: Oxford Press.

Shane, Scott, (1994), *Dismantling Utopia: How Information Ended the Soviet Union*, Chicago: Ivan R. Dee.

BACKGROUND READINGS

Butrica, Andrew J., (Editor), (1997), *Beyond the Ionosphere: Fifty Years of Satellite Communications*, Washington, DC: NASA.

Hudson, Heather, (1992), *Communications Satellites: Their Development and Impact*, New York: The Free Press.

Martin, Donald H., (1996), *Communications Satellites 1958–1995*, Washington, DC: The Aerospace Corporation.

Via Satellite, (2000, November and 2001, April), Potomac, MD: Phillips Publishing.

IMPACT ON SOCIETY

CHAPTER ELEVEN

New Opportunities and Threats for 21st-Century Life

Joseph N. Pelton
The George Washington University

> *The future of humanity lies far beyond Earth.*
>
> —Carl Sagan
>
> *Men have become the Tools of their Tools.*
>
> —Henry David Thoreau

These two dramatically different quotes from Henry David Thoreau and Carl Sagan represent the dialectical extremes of philosophical thought and about where advanced technology is taking modern society. There can be no doubt that satellites and other advanced electronic technologies bring us both new opportunity and new types of concerns about humanity's future.

Brockman's (2000) book about the greatest inventions over the past two millennia provides some useful insights. It reveals, in a series of short vignettes, what progress we have or have not made in the last 2,000 years. Brockman managed to con some of the world's greatest scientists, engineers, and scholars into contributing their thoughts on this subject. The results are actually quite a mixed bag. Some took the assignment quite seriously and others much less so. One rather dry response was: "Nothing much."

The most remarkable thing about the majority of these "great inventions" as inventoried by Brockman is how ordinary or routine most of them seem to us today. Most are virtually "invisible" in our everyday lives. Electricity, batteries, telephones, TVs, radios, computers, running water, modern medicines, anesthesia, vaccines, sterilization, birth control, books, printing, automobiles, trains, jet airplanes, modern cities, specialization, fiber optics, satellites, automation, manufacturing, democracy, and, heaven help us, even marketing made the list of nominations. All-time favorites such as astrophysics, spandex, and chewing gum, however, did not make the list.

Only a few suggested great inventions such as rockets, atomic weapons, lasers, robots, or artificial intelligence (AI) currently seem to hold any special or ominous fascination—or rather more accurately "fear factor"—for modern-day citizens. The fact is that in most cases, the greater the invention, the more completely it is enmeshed in our lives and now taken for granted.

In the early days of satellites, the public was fascinated to see TV broadcasts of news, sports, or coronations from around the world emblazoned with the title "live via satellite." In those days, an evening newscast that did not have one such report was certainly considered a dud.

Today, however, we see events such as the Gulf War, the war against terrorism in Afghanistan and Iraq, or even the Olympics on our TV screens without a thought as to how it got there. If there is any recognition of satellite transmissions in the process, it is almost always likely to be in the context of a momentary service interruption. We are perhaps peevishly annoyed when a football game or soccer match is momentarily interrupted due to "satellite transmission problems." (In fact the problem is typically the result of a technician who has not flipped a switch or the local coax feed has been disconnected. Satellites, as an exotic technological device, however, represent a convenient and easily believable culprit, although they are actually 99.99% reliable in their service record over the past decade.)

So what credentials might qualify satellites as a great invention in the overall course of history? Actually a good case might be made on several bases:

• *Universality of Connectivity.* Satellites are the first and only technology that allows the possibility of connection to anyone, anytime, and anywhere and for an amazing array of existing as well as yet-to-be-created services.

• *Flexible Broadband Links.* Only satellites allow easy, affordable, and flexible broadband links around the globe. Satellites can provide broadband services to smaller and smaller terminals whether the demand is for TV, radio, telephone, or Internet connectivity. Because satellites are not restricted to physical media like wire or fiber optics, they can provide connections with total flexibility and ease. They can provide extremely wide coverage to the most densely packed city or thinly populated areas, on land, sea, or ice cap, for any type of service. Some 10,000 satellite TV channels are now in operation globally, and the existence of Internet links to the majority of some 200 or more countries is directly due to satellite. Fiber only connects the big guys.

• *A True Breakthrough Technology.* The breakthrough nature of satellites is due to their ability to solve a problem in a totally new way. This is the concept that Buckminster Fuller (1971) considered the essence of human intelligence or "ephemeralization." Modern long-distance communications were first limited to wire connections (expensive and cumbersome) and then terrestrial wireless (constrained to the curvature of the earth and variations in its atmosphere and ionosphere). Today, satellite technology allows any connection to be made virtually without constraint to an infinite number of potential locations. Only if we can find a way to effortlessly communicate through the earth and across space without delay or bandwidth limitation (sort of a super ESP) would there seem to be a more powerful communications technology.

It is the ability of satellites to use much less resources and to do so much more in terms of connecting everyone that makes these devices truly a "breakthrough technology." For instance, satellites weighing only a few thousand pounds (or a few metric tons) can do the work that heavy cables spanning an ocean and weighing hundreds of thousands of tons are required to do. Even so the cable is capable of only linking two specific points—not 40% of the planet and billions of people at once.

It is not difficult to set forth the many strengths and new capabilities that satellites bring to human society. The list of telepower capabilities represented by telecommunications satellites is thus long and impressive. One can certainly argue the case as to how satellites significantly assist consumers, businesspeople, schools, or doctors accomplish their work or lead more convenient or enjoyable lives. Satellites enable global news. Satellites bring connectivity especially to developing countries and emerging markets. Satellites help global trade, airline safety, educational exchange, health and medical services, electronic fund transfer, military defense systems, scientific inquiry, and navigational systems.

Satellites are now an essential part of modern telecommunications systems in developed and developing countries, and they are especially effective in providing global mobile, navigational, and broadcast communications systems. Internet would be much more exclusively a system for wealthy countries and urban networking without satellite connections. The case for the benefits of satellite telepower is not difficult to make.

Yet if one looks more deeply, there are clearly satellite-driven problems and issues as well. It can certainly be argued that satellite systems can be used to extend the sway and influence of multinational corporations, foster so-called *neo-colonialist* influence over less powerful nations, and bring unwanted cultural and religious messages to societies that wish to be left alone. If one looks for isolated examples of "satellite teleshock," they are not hard to find. Residents of barrios in Sao Paolo, Rio de Janeiro, and even more isolated cities in Brazil—many with dirt floors—have responded to satellite-distributed TV commercials to buy commercial brand floor wax as a status symbol.

Those who picket the World Economic Forum, the World Bank, and the World Trade Organization with signs and giant puppets, protesting the negative aspects of globalization, could very well turn against satellite technology as one of the ultimate instruments of global trade promotion and global economic integration. Norm Lerner's chapter 9 (this volume) about satellites, global trade, and economic impacts helps define the benefits that such systems bring to the modern world. It can reasonably be said that satellites are almost always instruments of change with both positive and negative impacts. Often these impacts are quite different for different groups of players.

There is no doubt that satellites and their kin are creating new capabilities for people, business, news organizations, broadcasters, and governments that might be called *global telepower*. Likewise they are creating new problems that might be called *global teleshock*. Table 11.1 attempts to sort out what types of changes that satellite communications and related IT technologies are making to our world, and it examines the pluses and minuses that frequently occur.

As one analyzes Table 11.1, it becomes clear that many dozens of other impacts of satellites and IT networks could be identified beyond those listed. The importance of the table is in helping identify clear examples of major forces or trends of change. These key elements of change are diverse and often quite powerful.

SATELLITE TECHNOLOGY

Satellite technology tends to bring speed, rapid updating, broadband access, instant reaction, global coverage, global marketing and markets, and systemization and consolidation (or what might be called "electronic centralization" despite "geographic decentralization"). This technology often tends to favor centralization over decentralization, large scale over small scale, globalism over localism, and "efficiency" over traditional cultural or individual artistic practices. Perhaps more profoundly the rapid generation, transmission, reporting and use of information, tends to depreciate the value of all information. When people, businesses, governments or organizations are constantly barraged by torrents of information that is streaming around the world 24 hours a day, 7 days a week, important data or trends can be overlooked or discarded before it is seriously evaluated.

ECONOMIC CHANGE

Then there are the economic forces that global information technology, rapidly updated intelligence reports, and soon 24/7 stock and commodity markets tend to generate. These economic forces, as driven by global news broadcasts, Bloomberg reports, stock market trends, earnings reports, government indexes, and so on, include transnationalism—the ability to quickly target particular or specialized markets, especially with computer program-driven trading and so on. The global trend to place more and more value on trade in services and give lower value to commodity means that instantaneous global information systems have ever greater economic impact. Those with the greatest information to key and proprietary information can often avoid laws, regulations, and taxes; can quickly sys-

TABLE 11.1

Satellites as Global Change Agents in Society, Culture, and Economic Systems

Satellite-Enabled Capability	Telepower Gain	Teleshock Concern
Global connectivity via the Internet to over 200 countries	Increased global trade, lessening of the digital divide	Vulnerability of global databases, adverse cultural impacts, poorer countries have much less access
Global tele-education via Internet and satellite telecasts	Increased access to education in even rural areas including over 7 million in China and some 1.5 million in India	Local input into educational content can be limited
Global satellite telemedicine and diagnostic help	New outreach to even the most remote areas in developing countries and in scientific outposts	Opportunities for misapplication, inadequate training of local staff
Global disaster warning and emergency recovery systems	The ability to save lives and direct rescue teams and aids to most adversely affected areas	Lack of local training and most affluent and urban areas are most able to use assistance
Ability to centralize and consolidate databases and consolidate global commercial enterprises	Increased cost-efficiencies and ability to use international assets and human resources more effectively	Increased vulnerability of international headquarters and facilities to attack, technoterrorism, poor wage laborers at multinational plants
Ability to provide broadband and multimedia services to global community cost-effectively	Increased productivity, increased competitiveness for multinational enterprises	Information overload, takeover and consolidation of local enterprises
Ability to provide just-in-time facilities and training to global network of employees	Increased productivity and rapid roll-out of new products and services	Tendency toward mega-training rather than education of workforce
Ability to create VSAT networks and DVB-RCS anywhere, anytime	Ability to operate global enterprises on a 24/7 or 168-hour work week	Erosion of local economic and political ties, stress of nonstop business
Ability to provide latest economic, news, sports, and political information on a nonstop basis	Ability of governments and businesses to act with great agility and respond rapidly to both opportunity and risk	Danger of "Internet addiction" and mental health breakdowns (the state of California actually now has Internet addiction centers)
Ability to share scientific, research, technical, and medical information at nearly instantaneous speeds	Rapid spread of new knowledge and international collaboration	Espionage, opportunity for terrorists and ruthless political and business leaders to exploit information for venal or criminal purposes, identity theft
Ability to fight flexible, quick-strike wars and use reconnaissance information to pinpoint enemies	High-efficiency and low-casualty rates for advanced nations with modem satcoms and Nintendo warfare tools	Digital divide concepts in warfare, tendency of less advanced countries to use terrorism and civilian hostages
Global access to sophisticated databases, low-cost networking, and open trade under WTO guidelines	Ability of global corporation to use electronic immigrants and favorable open trade regulations to cut cost and expand international reach	Inability of local governments and business to compete on a level playing field due to lack of economy of scale and scope, cultural imperialism

tematize and enhance competitive processes; can recruit from a large (and more cost-effective) labor pool; and can exploit wage differentials from region to region.

However, global satellite technologies can, at least in the short term, raise liabilities. Those international service organizations such as airlines, banks, insurance companies, retailers, and so on can find themselves exposed to sudden political risks, arbitrary national laws, or quirky regulations, tariffs, and trade barriers. Developing countries can now start to exploit information technology (IT) to create new lines of communications that evade neocolonialist systems and negotiate improved prices for goods and services. Large international corporations with the latest in management, billing and marketing systems, and communications technology, plus financial resources, can, at least in the longer term, be more efficient and establish dominance over local businesses.

E-commerce—operating through satellite, wireless, and cable technology—has served to redefine business practices in the area of marketing, advertising, billing, inventory control and warehousing, maintenance, and customer services. Satellite-based e-commerce has extended the reach to nearly 100 additional countries, sometimes with unfortunate results. Hughes decided to withdraw from the direct broadcast satellite TV market in many parts of Asia simply because it could not develop a reliable and consistent way to bill for its entertainment services and respond to various national regulations that governed their services. Thomas Friedman, the *New York Times* reporter, noted that when he goes to a country, he starts by asking what is the strength of local telecommunications networks and how many international satellite circuits are in operation.

SATELLITES AND GLOBAL POLITICAL AND CULTURAL IMPACT

Satellite services and their impact on political and cultural systems are perhaps equally strong as in the business and commercial sectors. Several political observers as diverse as Ted Turner to Robert Kaplan have suggested that many of the changes that led to the breakup of the Soviet Union stemmed from the rapid and uncontrolled spread of information, news, sports, and entertainment across political borders. The Soviet Union, within the Committee on the Peaceful Uses of Outer Space and other parts of the United Nations, spent two decades in vain fighting to control broadcasting into their country and place limits on countries and satellite networks that would seek to beam programming across international boundaries. The power of satellite broadcasting is indeed enormous.

For instance, citizens of small South American countries can write more often to the president of the United States than their own head of state or send communications to the Pope rather than their local church officials. Thus, satellite systems can and often do open new horizons, sometimes with good effect and other times with unfortunate results. Usually satellites serve to weaken local political, social, or cultural influences.

Ways in which satellites tend to create supranational political, social, and cultural influences and weaken local allegiances include:

- Creation of jobs, income, and economic opportunities across political boundaries.
- Allow workers to escape from local adverse working conditions or onerous social, cultural, or political controls.
- Bring a new awareness of conditions and opportunities in other communities or countries as well as closer ties to relatives who live abroad in other countries.
- Bring education, training, skills, literacy, and health care to areas without such opportunities.
- Bring new sociocultural awareness with regard to birth control, AIDS prevention, and religious and ethnic hatred or tolerance.
- Create higher levels of resistance to social intolerance as a result of global news about the activities and speeches of political, social, religious, or cultural leaders around the world. (Satellites and other free-flowing electronic technologies can also serve the cause of revolution or terrorism by allowing dissidents to talk freely and avoid censorship.)

In general, complex, geographically extensive, and technologically sophisticated electronic systems aid the cause of economic growth, efficiency, and political consolidation and patriotism. This is particularly true when the local political and economic conditions are perceived as good and beneficial to the local population. When the conditions are bad and there is economic stagnation, poor education and health care systems, and racial, religious, social, or cultural misunderstandings or distrust, things change. In these cases, the existence of efficient and low-cost communications can be a source of revolution and terrorism. In such circumstances, the political leadership will attempt to control, and control absolutely, all forms of communications.

These extremes are actually not hard to see and understand. The more subtle issues raised by satellite communications is that a new communications capability often seems to improve social, economic, and political conditions, but it can also lead to new difficulties and new teleshock conditions that are hard for local governments to control.

History has shown us time and time again that technology tends to create totally new and unanticipated concerns. One needs only to cite the case of the London City Council at the turn of the 19th to the 20th century. At the time, the Council was about to go bankrupt due to its responsibilities for removing literally tens of thousands of tons of horse manure from the city streets each month.

At the time, a presentation was made about a new technology that would solve the major environmental disaster that was threatening the economic livelihood of the city. The technology was, of course, the automobile, and its cure of the horse manure problem inevitably led to the current problem of carbon dioxide, carbon monoxide, and nitrous oxides that threaten the clean air of our cities all over the world today.

Satellite TV, at least to some extent, gives us an informed populace, but satellite-distributed X- and R-rated movies are also thought by some to lead to problems of increased crime, drug addiction, violence, sex attacks, rapes, and child pornography. Satellite-based Internet gives us truly democratic access to new knowledge and information, but it also enables the distribution of the most vile types of ethnic, racial, and religious propaganda that incites people to avarice and criminal behavior. Table 11.2 explores the dual character of national and international satellite systems.

The Janus-like nature of modern electronic technology (with a smiling face on one side and a menacing face on the other) is, of course, far from new. It is the speed of innovation and change, however, which suggests that we are becoming less and less adept in dealing with the torrent of change. Not only are there new types of problems, but the speed and magnitude of the impact has also changed. New technologies can create problems and issues, and at speeds that are too fast for human response in a timely manner. Mere mortals cannot turn off a switch or reroute traffic in a few nanoseconds. Just the issue of trying to deal with satellite and fiber-delivered "spam" is enormous. The latest estimates from the U.S. Federal Trade Commission (FTC) have suggested that by 2004, 40% to 50% of all traffic on the Internet will be unsolicited electronic ads for everything from hair restoration, to penis enlargement, gambling sites, to porn cams in coed dorms. With the introduction of the "Do not call" telephone call registry in the United States to establish telemarketing control, the thrust to increase the level of e-marketing and spamming on the Internet will only increase. Most filtering systems are ineffective against technically sophisticated spammers.

There are those who have seriously suggested that weapons of mass destruction should be automated to allow a "machine response" to attacks because human operators could not respond quickly enough to the new "smart wars," with tens of thousands of weapons interactions per second that are now projected to occur in future decades. In this new environment, potentially 1 million weapons interactions could occur in a matter of minutes. This new environment, in which we are forced to respond to an attack literally before it occurs, as reflected in the George W. Bush new preemptive engagement policies in Iraq, represents a 21st century filled with uncertainty and angst. This new satellite-linked world of instant connectivity, smart bombs, and GPS-guided weapons for instant interdiction is presumably to be filled with fear and anxiety. The 21st century seems to have brought us not only instant connectivity, but constant worry of orange- and red-level alerts against out-of-the-blue terrorist attacks that can happen anytime and

TABLE 11.2
Dual-Edge Nature of Modern Satellite Technology—Coping in Real Time

New Capability	Benefit	Ensuing Problem
Instantaneous sharing of music worldwide	Rapid sharing of high-quality music at low cost	Evading payment of reasonable copyright fees
Ability to view virtually all movies ever made	Ability to see quality movies at low cost and on demand	Easy access by children to pornography; spread of hate, genocide, and other messages
Mobile satellite communications systems	Ability to talk and communicate from anywhere at any time	Increased ability to eavesdrop; variety of criminal and terrorist uses
Ability to move money via EFT worldwide—up to $400 trillion per year—perhaps $100 trillion via comsats and wireless systems	Ability to create worldwide corporations and 24/7 commercial operations	Computer crime, money laundering, tax evasion
Ability to communicate via telephone, Internet, or other electronic messaging systems	Rapid dissemination of news, business information, scientific knowledge, and ease of connection between friends and family	Ability of terrorists to plan attacks and facilitation of criminals' and drug dealers' operations
Distribution of movies, educational materials, sports, news, and video games to millions of subscribers	Low-cost entertainment and ready access to global news and educational materials	Increased addiction to TV and video games; information overload; deskilling; potential increase in violence
Electronic monitoring of remote sites	Ability to use CCTV and sensors for security, pipeline, elevator, and other equipment monitoring from remote locations	Extension of "Big Brother" surveillance systems throughout society
Satellite navigation, Web mapping, and integrated digital services	Ability to carry out a range of services via integrated communications and digital devices	Unauthorized surveillance and tracking capabilities
Ability to combine video, data, and information in multimedia format and provide interactive satellite services to users	Home banking, voting, governmental services, and educational and health services	Possibility of increased gambling addiction, fraudulent offers, and other criminal applications

anywhere. We seem to be moving toward a "Hobson's Choice" of having to make an "inhumane choice" to invade rogue countries or assassinate terrorist leaders in the name of trying to save peace and humanity. Recent movement toward creative, active offensive and defensive weapons in outerspace would suggest a further escalation of long-term worries about weapons systems that could impact large numbers of people before the end of the 21st century.

For years we have looked to the future and anticipated the problems that advanced technology, automation, and too much communications might bring. Thus, we are aware of such perils as technological unemployment or underem-

ployment, nuclear attack, and the abuse of drugs to escape the pressures and difficulties of our "advanced and high-pressure culture." It is the element of constant
insecurity from technoterrorists that has been the most unsuspected new development of the past 5 years. It is the antiterrorist concern that will drive the implementation of new satellite and IT systems over the next 5 years, with results still
to be clearly understood.

The last few years has brought us the Internet, continuous satellite and cable
TV entertainment, and, now, always-on automated security systems. The new
antiterrorist environment certainly adds unwelcome elements to the developed
world, and few have clear proposals as to what the alternatives are. These
teleshock effects include always-on "Big Brother-like" surveillance cameras
and computer systems, enhanced security concerns, and inspections when
boarding an airplane, going to a football game, or even entering a shopping
mall. We seem to have entered into a new type of *risk environment* as of 2002,
where each technological advance in communications, command, control, surveillance, intelligence, and counterintelligence seems to escalate to a higher
form of insecurity.

SATELLITE-BASED OPPORTUNITIES

In an earlier chapter, important economic benefits that satellite communications
and associated information networks bring to the world were addressed in some
detail. In this chapter, however, we are looking beyond just the strict economic
profile to examine the broader sociocultural context. The list provided in Table
11.3 identifies sectors that are being aided by satellite communications and related information networks and the perceived benefit.

The interesting thing about these variously perceived areas of economic, social, or cultural benefits is that virtually every innovation gives rise to new concerns and social problems. In a social environment, technology is almost always
a two-edged sword. Various satellite and fiber optic networks can become much
more efficient in the delivery of up-to-date and accurate information. Yet this
process can also lead to the reduction of jobs and indifference among workers
who are facelessly servicing customers who are hundreds or even thousands of
miles away.

In the *Theory of the Business Enterprise*, Veblen (1954) discussed how the
increasing separation of production and consumption in modern business and
government can create so-called *economic efficiencies*, but also can give rise to
boredom, indifference, or even worse. There seems to be a point of diminishing
returns in highly automated and distributed networks. There are thus a number
of cases where modern satellite and electronic networks created economic efficiencies, economies of scale, faster response time, broader economic, and social
reach. Yet these so-called *gains* also give rise to new types of concerns and
problems.

TABLE 11.3
Sectors of the Global Social and Economic Systems
That Are Aided by Satellite Communications

Key Sector	Type of Perceived Benefit
Banking	Reduction of "transactional float of money." Increased speed, homogeneity, and accuracy of national and international transactions.
Cultural interaction	Increased opportunity to exchange cultural programming. Improved cultural interaction and global educational programs.
Customs/immigrations	Greater efficiency of processing international visitors and detecting illegal aliens.
Disaster warning and recovery	Greater ability to anticipate, predict, prevent, and/or respond to disasters more effectively.
Education/training	Opportunity to provide education and training to a much wider audience and at much lower cost with more up-to-date materials.
Financing	Ability to communicate more quickly and provide assessment and evaluation information rapidly to a wider range of institutions.
Insurance and risk management	Rapid response to damage and risk assessment. Opportunity to serve a broader clientele in developed and developing countries.
Inter- and intragovernmental services and coordination	Ability to provide a growing range of governmental services online and allow governments to more rapidly exchange information with regard to crises and respond to misunderstandings more quickly.
Job skill improvements and automation	Provision of on-demand remote training as well as ability to serve teleworkers more rapidly with lower cost connection to corporate headquarters.
Libraries/information databases/archives	Ability to share books, journals, and other materials more efficiently and provide online services to business, government, and schools.
News services	Global and instantaneous coverage of political, economic, and social news from around the world and convey it directly to end users.
Scientific investigation and large-scale experimentation	Ability to interconnect observatories, supercomputers, and high-tech laboratories to carry out real-time experimentation, observation, and other international scientific research on a collaborative basis.
Security and public safety	Improved security systems for airlines, public buildings, transportation systems, as well as failure reduction for automated systems.
Telehealth/telemedicine	Opportunity to share educational, health, medical, and other facilities and software on a local, national, regional, or global basis.
Trade/trade support	Faster movement of traded goods and services and more streamlined handling of trade regulations and tariffs.
Transportation and transportation planning	Faster and safer transportation systems in terms of fault detection, aircraft, ship, railroad, and automobile routing. Ability to monitor flows of traffic and provide alternative routing around areas of congestion.

Satellite-Based Concerns

Some of these elements of concern and opportunity that satellites and other related electronic information networks have particularly fostered include the following.

Information Overload. The problem of information overload was identified as a potential concern prior to the age of the communications satellite. Paul Lazarfeld and Robert Merton, two American sociologists, began to seriously study the phenomenon in the 1930s. They first started with overloading the sensory perception of white laboratory rats and found that this gave rise to serious mental and physiological problems in the experimental animals. They found that over time the information overload led to lethargy, loss of appetite, inability to sleep, nervous disorders, loss of sex drive, and ultimately death. A reviewer of this scholarly research once observed, with presumed wit in mind, that this is disturbing news, but at least the sequence seems right if this is to be our fate.

Today there are many studies that show the overloading of information channels gives rise to such issues as political and civic apathy, a desire to tune out social problems (i.e., the "couch potato" syndrome), and inattentiveness to key issues (such as the inability to distinguish priority matters or differentiate major from minor problems). It turns out that over the last 30 years the percentage of information consumed (i.e., carefully read) by a newspaper subscriber has decreased by an order of magnitude—from nearly a third to less than 3%. Satellites, by being able to provide consumers with hundreds of TV and radio channels, access to billions of Web pages on the Internet, and so on, has only served to escalate this problem. Some individuals have even become addicted to the constant link to real-time information to the extent that they cannot sleep, eat, or take care of essential bodily functions. They cannot disconnect from the Info Niagara. In California, there are counseling and rehabilitation centers that only address Internet addiction problems.

De-Skilling and Mega-Training. The advantages of modern electronic tele-education and distance-based training are clear cut. In countries such as China and India, millions of students are able to obtain an education that they would not otherwise be able to receive. There are, however, increasingly sophisticated techniques to provide just-in-time training in such a way that workers have no way to judge the educational or social context of automated task instructions. In short, these employees are provided with no background or context as to why, how, or the wherefore of what they do. This process can, over the longer term, be destructive because it does not provide a social, ethical, or educational context for what they are being trained to do in a rote instructional way—to perform a mind-numbing task.

Super-Speed Living.   It is possible to transfer broadband information from corporate facilities in Singapore to Los Angeles and back at unprecedented speeds. In his *TeleGeography* report published each year, Staples (2002) showed how the modern world of financial, business, and other international services is defined not by physical distance, but the speed of network interconnectivity. We now have the nonstop corporation that can address problems 24/7 and employ electronic teams that can be simultaneously drawn from scores of companies. This intensity of action is leading us to the continuously operational stock exchange, travel agency, bank, funds transfer, and trading organizations. Although satellites and computers now perform in gigabits/second or gigaflop speeds and without any pause, people do not operate without sleep and respite. There are sociologists and psychologists who believe that the rise in divorce, increase in child and spousal abuse, higher incidence of road rage, increase in drug addiction and violence, decrease in student test scores, and increased rates of functional literacy are all tied (directly or indirectly) to the pace and stress of modern technological life.

Various studies at the Annenberg School of Communications at the University of Pennsylvania over the past decade have correlated increased levels of TV watching, intensity of computer gaming, and other new patterns of super-speed living to harmful patterns of social behavior and higher levels of stress. The Massachusetts General Hospital Stress Center claims that substituting meditation for normal patterns of high-stress, high-technology lifestyles can have a remarkably positive impact on both mental and physical health.

Although such studies are still sketchy and episodic, there is mounting evidence that people are getting less sleep, working longer hours, and interacting with machines and TV for longer periods of time, thereby generally increasing their stress levels.

One index of the toil that super-speed living is placing on modern living is the number of energy-boosting drinks now on the market. According to the University of Pittsburgh Medical Center, these offerings include Red Bull, Jolt, Lizard Fuel, Mad River, Energy Hammer, Red Devil Energy Drink, and Sobe Adrenaline Rush. Doctors at the sports center at the Pittsburgh Medical Center have described the continuous effort to boost awareness, energy, and performance as just craziness.

At some point, the futility of humans trying to match the 168-hour work week that our machines can now manage will need to be faced with more sophistication than ever higher caffeine levels in our fast-energy drinks. A recent national study of sleeping habits in the United States concluded that Americans are sleeping less and thus are now more on edge and angrier. This conclusion seems to confuse the fact that more stress produces a "lack of sleep" rather than the reverse—namely, that the stress of modern super-speed living causes people to become angrier and more stressed, and this in turn impacts on people's ability to sleep. In short, it is not the loss of sleep alone that is causing our problems. Loss of sleep is just one of many symptoms.

Globalism.   Closely tied to super-speed living is the rise in globalism—not only in terms of trade in goods and products, but in the past decade especially in terms of increased international trade in services. The rise in global trade is undeniable. This applies to trade in goods and products, trade in services, and entirely new phenomena such as international telework and the increase in so-called *electronic immigrants*. This move to integrate the global economy has been long predicted by the trilateralists and others who have seen how computers, fiber optics, the global media, and satellites have increasingly webbed our world together. Some have feared the rise of what some have called "electronic neocolonialism."

Technological Unemployment, Underemployment, and De-Skilling.   The initial impact of modern automation and advanced information systems was to create a decline in manufacturing jobs. Actually this process began a half century ago, and the decline in industrial manufacturing jobs has been evident in economically advanced countries of the OECD for some 40 years. The advanced telecommunications, IT, and AI systems today are increasingly geared toward replacing service jobs. Occupations such as appraisers, accountants, pharmacists, and bankers, among others, are either being reduced in number or "de-skilled" so that lesser trained people augmented with "smart software" can perform many of these functions. The ability of satellites to move skilled positions around the world at virtually zero cost (in comparison with labor costs) tends to globalize this trend ever more quickly.

Global Culture and the "E-Sphere."   The ability of satellites and fiber optics to transfer information, movies, video games, and money around the world at spectacular speeds has created a world that is dramatically different than just a few years ago. Within the next two decades, we will see a new world of cyberspace-based knowledge, computing power, and information that is astonishing in its scope. In another 20 years, a new rendition of the Internet, based on current trends, might well contain some 10,000 times more content than now. Further, the speed of access could be much, much quicker than today's dial-up machines or even DSL or cable modem systems. This new environment will change almost everything—the nature of jobs and employment, cultural and entertainment systems, and even political, legal, and judicial systems. These advanced digital systems will create not only sociocultural unrest, but, as noted by Professor Mowlana (chap. 13, this volume), it could lead to social and political unrest of much larger proportions and impact major geographic areas of the world.

For 300 years, the political systems of the West have been built on the fundamental rules shaped by the Treaty of Westphalia and the primacy of the nation state. Today advanced digital technology and systems of private enterprise and capitalism, on the one hand, and fundamental religious and culture beliefs, on the other hand, are challenging these "rules of the road" as the 21st century begins. The "Coalition of the Willing"—countries that joined with the United States, the

United Kingdom, and Spain to invade Iraq as a preemptive strike against a future use by Saddam Hussein of weapons of mass destruction—defines a new day in global politics. In many ways, it represents a redefinition of *national sovereignty* for the 21st century. Clearly many weapons systems and technologies have worked to create this new environment. Nevertheless, global communications and navigational satellite networks have certainly served to make such "national interdiction" policies possible and aided the speed with which it occurred.

The rule of world society is being challenged in new and fundamental ways, and modern, broadband satellite systems are now a part of that struggle. Satellites now help to provide surveillance, help to target weapons, provide command and control for weapons systems, allow electronic accounts and networks to be monitored, track someone's viewing and spending habits, and much more.

Less dramatic, but still important, satellite networks not only make our lives more enjoyable, informed, efficient, and convenient, but also contribute to information overload. They heighten our levels of stress, increase our exposure to pornography and violence, facilitate an increase in globalism, heighten opportunities for technological unemployment and electronic immigrants, and potentially increase levels of risk for an advanced technological society. Without satellites and related technology, our global economy would literally shut down, starting with over $1 trillion in electronic funds transfers each day. Ours is a complicated world, and satellites contribute to that complexity, not to mention serving to increase both our prosperity and our angst.

REFERENCES

Brockman, John, (Editor), (2000), *The Greatest Inventions of the Past 2,000 Years*, New York: Simon & Schuster.

Fuller, Buckminster, (1971), *Utopia or Oblivion: The Prospects for Humanity*, New York: Bantam Books.

Staples, Gregory, (2002), *Teleography 2002*, Washington, DC: Teleography Inc.

Veblen, Thorstein, (1954), *The Theory of the Business Enterprise*, New York: Harpers.

BACKGROUND READINGS

Bernstein, Peter, (1998), *Against the Gods: The Remarkable Story of Risk*, New York: John Wiley and Sons.

Broad, William J., (2000), "Snooping's Not Just for Spies Any More," *New York Times*, April 23, 2000, WE 6.

Brown, Harrison, (1965), *The Challenge of Man's Future*, New York: Viking Press.

Friedman, Thomas L., (2000), *The Lexus and the Olive Tree*, New York: Anchor Books.

Gould, Stephen J., (2003), *The Hedgehog, The Fox and the Magister's Pox*, New York: Harmony Books.

McLuhan, Marshall, (1964), *Understanding Media: The Extensions of Man*, New York: Signet Books.

Naisbett, James, and Aburdene, Patricia, (1990), *Megatrends 2000*, New York: William Morrow and Co.

Pelton, Joseph N., (1992), *Future View: Communications, Technology and Society in the 21st Century*, Boulder, CO: Baylin Publications.

Pelton, Joseph N., (2000), *E-Sphere: The Rise of the World Wide Mind*, Westport, CT: Quorum Books.

Roszak, Theodore, (1980), *The Cult of Information*, New York: Pantheon Books.

Toffler, Alvin, (1980), *The Third Wave*, New York: William Morrow and Co.

CHAPTER TWELVE

Satellites and the Promise of Teleeducation and Telehealth

Joseph N. Pelton
The George Washington University

> *The next ten years are a critical window of opportunity for satellites with respect to the countries with developed economies and perhaps the next twenty years for developing countries with less developed terrestrial telecommunications systems.*
>
> —Joseph N. Pelton and Alfred MacRae, et al. (1998, p. 17)

There are somewhere between 250 and 300 geosynchronous satellites in orbit today. A typical satellite uses digital TV transmission technology; if used exclusively for TV, the latest systems launched can each send some 500 plus channels to the earth. A large, late-generation satellite such as an Intelsat-9 could conceivably transmit some 1,500 separate TV channels at once (or about 10 billion bits of information or the equivalent of 5,000,000 pages of a very, very long book—all in 1 second). If only 5% of the world's in-orbit satellite capacity were to be reserved for educational and health care purposes, this would create a downpouring of information unprecedented in the history of humankind. In a single day, using only 1/20th of the world's available satellite capacity 3,300 terabits of information would have streamed to earth (the equivalent of 400 Libraries of Congress)—all in just 24 hours. In a year's time, this would represent a tidal wave of information (or perhaps one might call it a *title wave* of knowledge equivalent to 150,000 Libraries of Congress). In short, the key issue involving satellite-based education and health care today is not whether sophisticated networks to support improved global education and health care systems could be devised, but how it might be done. How could world governments, the United Nations, or even private enterprise organize a new and open economic and social system that would allow expanded global education and health care, live via satellite, to occur? To succeed must national governments or even localities be in charge?

The problem is clearly not delivery capability from the satellites, but creating the educational and medical content to flow through the systems that already exist and making affordable, convenient, and user-friendly user terminals available to teachers, students, doctors, and health care providers alike. Believe me there are huge numbers of cultural, social, political, and economic problems to be overcome before low-cost and easily accessible information could flow to an eager world community.

Education, training, and health care delivery systems are a multitrillion dollar business around the world, and this alone means huge economic barriers to providing such services directly to world consumers at very low cost and with highly productive modern systems. The cultural and political barriers to global satellite education are even higher.

One might ask whether there is actually a real need to augment conventional educational and health care systems with satellite or other electronic media to meet currently unmet needs? The answer is most definitely yes. A recent UNESCO study concluded that there are more people to be educated and receive health care in the next 40 years than all of those people who have lived since the time of the Ancient Greeks. These same studies concluded that over 2 billion people today are living with little or no educational systems or health care. It was also found that, without significant changes in methods of delivery education and health care, the gap would likely increase rather than decrease in size. In short, the digital divide will only widen unless exceptional methods are taken.

The most common concept of teleeducation or telehealth envisions a group of young students gathered around a TV set or people lined up to talk to a doctor via a TV monitor. This is essentially a wrong-minded view of the process. The most common use of satellite systems, based on experience to date, is to use space communications networks to train teachers, doctors, as well as educational aides and paramedics to do their jobs.

The great ambition of the Chinese National TV University (that I helped to begin under Intelsat's Project SHARE that ran from 1984–1986) is, in a single day, able to reach most of its 10 million teachers via a simultaneous video broadcast as well as to reach millions of students. Certainly experiments have shown the power of satellite to perform in the health and education area. Almost two decades ago in 1985, again as part of Project SHARE, it was shown how to assemble electronically over 50,000 doctors and medical aides on four continents to provide them simultaneous information on the latest information with regard to AIDS testing and diagnosis, treatment, and ongoing research. The combined electronic audience in North, Central, and South America; Europe; and Africa was able to learn of the latest research and medical findings about AIDS, in the most powerful and intense way possible—namely, by using simultaneous satellite TV, telephone, and computer hookups. Despite this remarkable achievement in the mid-1980s, little has been done to exploit this type of capability.

The key element to bear in mind is that satellite TV, multimedia, voice and data channels, radio, computer networking, and more can be used to meet many needs in terms of training of teachers and doctors, updating teaching materials and medical information, as well as assisting in the direct training of students of all ages. The biggest challenge today is not in developing the transmission networks, but in developing the right educational, health, and medical information and programming, and in the right language and, most important, the right cultural context. The challenge of meeting diverse needs in a diversity of countries and cultures via satellite, the Internet, and other media has never been greater, more urgent, or more difficult.

Satellites and other advanced digital information systems are certainly not a panacea, and their use must be altered for each practical application. For instance, there is a large difference between educational and health applications. Telehealth applications are more difficult, require higher resolution and expensive equipment, are less developed, involve greater legal liability, can invoke life-and-death decisions, and often involve more cultural and religious concerns. Today telehealth applications are 100 times less prevalent and extensive around the world than teleeducation.

Yet if we do figure out how to use the Internet, fiber optic networks, and satellites to cope with educational and health needs and applications, this may ultimately be the key to human survival.

Marshall McLuhan and others such as Sir Arthur Clarke suggested for decades that the ability to achieve global electronic connectedness and instantaneous information exchange calls for a new paradigm for living, working, and learning in the 21st century. Few political leaders have recognized the importance of developing such a new vision. Most economic, political, cultural, and educational systems are a century behind our advanced information technology. Humanity will never reach its potential until it finds ways to use its advanced technology in creative and emphatic ways to address sociocultural ways.

Socrates once said: "There is only one evil—ignorance. And there is only one good—knowledge." In the golden age of satellite communications, when system capacity is plentiful and the low cost and convenience of ground user terminals make such systems widely available to a global audience, we have the potential to spread good to the world in unprecedented ways, first to teachers and doctors, and ultimately to the entire global population. The barriers to doing so, however, are large and resilient.

GLOBAL EDUCATION AND HEALTH CARE— KEY TO HUMAN SURVIVAL?

The survival of our species on planet earth for the long haul is actually now in question. This is due to such challenges as global warming; the changing albedo of the polar ice caps; the rising threat of radiation damage due to the depletion of the ozone layer; the misuse of weapons of mass destruction; biological or chemical warfare; the loss of rain forests, wetlands, or other crucial flora and fauna critical to sustaining life; or other misuse of our technology and economic systems. In 2002, the Global Resource Institute released its report on the world's forests. This report suggests that if the deforestation process does not change from its current pace, we may reach a point of permanent and unrecoverable damage in only a few decades. In Indonesia, for instance, over half of its primal forest has been lost in the last 40 years, and forests the size of Massachusetts are being lost each year. There is irony in the current situation, in that we seem to be caught in a trap of too little and too much use of technology at the same time.

Thus, we seem to be held in the pincer of too little education and health care in the least developed portions of the world as well as in urban and rural slums. In this case, well over 2 billion out of some 6 billion humans have little or no basic services in terms of both health care and education. However, we find that the misuse or destructive application of our most advanced technologies constitutes grave danger as well. The connecting link is the earth's biosphere. We have not yet learned how to view our world as an integrated system. A lack of health care, a lack of educational systems, and a lack of knowledge about how our technology affects our world's environment and survivability as a species—these are all threats to our long-term success as a species.

This dilemma is the special challenge of the 21st century. There is at least reason to believe that satellite communications technology, when used for educational, health, and environmental purposes, can ennoble our race, improve our quality of life, refocus our values (from materials to survival and intellectual growth), and extend the chances of human survival as a species at least another millennium into the future.

Because of the great reach and low cost of satellite coverage, it offers special opportunities to address global needs in both education and health care. Already thousands of teachers and doctors, as well as tens of millions of people, are receiving rural and remote education via satellites, and soon large numbers will be able to receive remote health and medical care as well. The Malaysia government, for instance, just committed itself to provide satellite instruction and interconnection to every village and rural settlement within the next decade. Indonesia, India, China, and other countries have already made similar commitments and are today already providing enhanced instruction to nearly 10 million students.

Some of the world's greatest problems stem from lack of education and health care in many areas of the world. Yet in other regions of the world, problems of political peace and 21st-century survival stem from the misapplication of our exploding technology. We see a pattern of too many material demands being placed on our biosphere, which is stimulated in part by too much satellite broadcasting of radio and TV and too many commercials that urge more and more consumption. We live within a Niagara of specialized and unconnected data that are spread unevenly across the world. The availability of well over 10,000 video channels (most for commercial and entertainment use) often compounds rather than lessens the problem. The United States has the most commercial broadcasting and TV viewing of any country in the world. The "average number of hours that a television is on in a home in America" is over 7 hours a day. It seems no accident that the United States—with 6% of the world's population—consumes half of the energy produced and generates half of its garbage and refuse.

The ability of *Homo sapiens* to survive thus involves several challenges. One challenge is to use satellite and other electronic media to share education and health care more thoroughly and equitably around the world. The other challenge is to create not only an effective electronic global village, but to create humanized smart cities that allow progress to mean more than just increased technical information, but a more just, free, energy efficient, and environmentally aware world. To achieve such objectives will require better education, better health care, better telecommunications, more accessible information systems, improved energy systems, and ultimately more survival-oriented values than exist today. The digital divide is a problem that extends across both parts of the world's economic spectrum. Better educational systems are needed in both the north and south parts of our complex world.

The truth is that the 21st century can be one of two things. Either it can be a vital connecting link to the future, in which we use our most advanced satellite and information systems to deploy improved health and education networks that in turn create a viable economic, industrial, political, and ecological system for human survival, or we do not. The danger is that if this transformation does not occur in the next century, there is serious risk that *Homo sapiens* will not make it as a species. In the next single century, we have the potential to use our information and telecommunications technology and systems well or much too unwisely.

Simply put, we can indeed fail as a species. We can exhaust our forests—the source of our supply of oxygen. We can pollute our atmosphere with noxious gases. We can deplete our ozone layer, which protects us from radiation illness. We can overheat our planet and melt our ice caps. We learned from the Biosphere 2 experiment in the deserts of Arizona that we are not yet able to design and sustain a self-contained, livable environment. To avoid these unwelcome future possibilities, there is much we can do. One key option available to us is that we can deploy satellite-based health care systems and new educational systems via satellite to address our global societies' many needs.

The challenge of the 21st century is not whether we will live better or worse than our forebears in a material sense, but whether we will use our smart telecommunications, information, energy, and transportation systems to create a viable econiche for *Homo sapiens*. We need to learn how to devise and operate satellite and other information and knowledge networks that can create a world that is better educated, that has reliable and effective health services, and one that can nurture sustainable development.

How can we fail in the crucial 21st century? There are many ways open to us. One of the clear ways is simply by not developing more effective global education, not delivering improved medical and biological knowledge, and continuing to promote economic growth rather developing survival-oriented information systems and building a sustainable world.

Here are some of the rather unhappy choices now open to us:

- We can continue an increasing pattern of oil spills that ultimately change the albedo of our polar caps. The end result of these oil spills and pollution can be the flooding of our towns and cities. (Once melted the seawater will be enormously difficult to freeze again.)

- We can create enormous holes in the ozone layer and mutate ourselves and other flora and fauna, like today's frogs already seem to be doing—ultimately risking our very existence.

- We can develop a steady-state global population, but not adapt our economic systems to operate effectively within such limited growth systems. The problems of the Japanese economy may well be sending us such an early warning today. One recent estimate is that Japan will lose one third of its population by 2050, and its economy will take a tremendous economic hit if this indeed occurs. The educational needs to cope with environmental pollution, radiation dangers, global population shifts, and cultural and terrorist issues as well as those issues associated with lack of education in developing countries are enormous. The fact that these could be two parts of the same communications problems is not widely understood in political or educational circles, and the satellite and telecommunications world certainly does not recognize this as "its problem" at all.

- We can fail to use our best teleeducation and telehealth systems to provide for global equity. We can fail to teach doctors and educators and their support staff how to address the world's health care and nutritional needs. We can also fail to create curricula that address the needs of an increasingly interdependent world with a human population that likely does not stabilize until it reaches 12 billion people around 2075—yet is in danger of shrinking in most developed countries starting in the next decade. Satellites are helping us to create an increasingly seamless worldwide economy, but these systems are not yet being effectively used to develop in parallel global health care and educational systems that can expand fast enough to deal with the billions that are now underserved and that will swell into many billions tomorrow.

- We can allow the process of desertification and the destruction of our wetlands and rain forests to continue unabated. This will reduce our food supply while also raising the levels of carbon dioxide in our atmosphere to dangerous levels. (To cope, our remote sensing and information-processing systems need major changes to become much smarter and user friendly. These also need to be integrated with our global satellite educational and health care systems.)

- We can continue levels of energy consumption of hydrocarbon-based fuels that mount to higher and higher levels until the cost of fuels and pollution levels reach disastrous levels. The day we can achieve a hydrogen-based economy and transportation system cannot come too soon. We can design homes and buildings that consume a tenth of what conventional structures do today, but our research priorities are driven by economic growth and expansion and not survival.

• We can fail to use wide-ranging, up-to-date, and cost-effective global satellite networks to reach the 2 billion underserved populations to provide adequate health care and education services to rural and remote population and thus further extend the digital divide.

In short, we have a lot of choices to make. Implementing the right technology at the right time and at the right cost is becoming increasingly difficult.

In rising to meet this challenge, we must recognize that space-related technologies and applications, information and communications technologies, and effective use of artificial intelligence and expert systems could well be the keys to success—or failure. Only these smart systems can allow us to think and act synoptically on a planetary scale and create the tools to establish global cooperative behavior. The primacy of space in coping effectively and comprehensively with planetary problems such as pollution, education, health care, communications, smart transportation, and navigation is often overlooked. Key space applications are not a luxury, but in many cases may be basic to human survival—as important as national defense. In this case, it is planetary defense. It is because satellites can be global in scope that they are critical to making planetary changes not only in our education and health care systems, but our changing needs for global survival.

SPACE-BASED TELEEDUCATION AND TELEHEALTH AS A CRITICAL STEP

At the opening session of the International Programme for the Development of Communications, organized by UNESCO in the 1970s, Sir Arthur C. Clarke explained that our emerging computer and communications technologies could eventually produce something he called the *electronic tutor*. It would be portable, cheap, and contain a vast encyclopedia of information that could be electronically updated or interconnected to via satellite to respond to questions. It could be programmed in multiple languages. When asked whether this electronic tutor was going to replace teachers, he sensibly replied: "No, of course, not. We must use the technology to train teachers first." He went on to explain that we will always need teachers. Come to think about it, he said that any teacher who can be replaced by a machine probably should be.

Today this technology that Sir Arthur Clarke described by the phrase "electronic tutor" is likely less than a decade away. There are already "anytime, anywhere" satellite hand-held transceivers to be bought at under $1,000, and soon these will be broadband and support multimedia services with terminals that cost under $200. There will soon be 50 "smart" virtual reality arcades from Disney strewn about our globe. It would be quite possible to use parallel technology to create fun but effective training centers for teachers and doctors. We can also buy Dragon software today (that allows spoken word to be transferred to written text). Such technology could allow us to accelerate the development of needed new educational and health care programming in many languages.

New satellite and user microterminal technology now on the horizon can revolutionize our global educational and health care systems by making these universally available and at a price that is an order of magnitude less than we pay today. In fact much has already been done with today's satellite technology when it comes to education.

Experiments carried out under Intelsat's Project Satellites for Health and Rural Education (SHARE) in 1985 and 1987 led the way almost two decades ago. This activity, which involved over 40 tests and demonstrations and over 100 countries, showed the potential of space-based teleeducation and telehealth. The Chinese National Television University, for instance, was started under the Project SHARE tests and demonstrations, and it now has over 5 million students and over 90,000 operational terminals. This system transmits many hours of educational and health programming each week. This programming is all locally produced by either Central China Television or the Ministry of Education. It carries programming that not only furthers basic and advanced education, but also provides basic health care information on nutrition, the setting of broken bones, and other fundamentals of medical care. The cost of this system is tallied in only millions of dollars and, as such, is hugely cost-effective. This system reaches rural populations that would otherwise receive no educational services or health care information.

The so-called SITE experiments, which started with the ATS-6 satellite in India some 30 years ago, have likewise stimulated the operational teleeducation and telehealth programs now carried on the Insat satellites. Insat programming now reaches over 1 million students each day, and soon this number will be 2 million.

Project SHARE showed that space-based teleeducation and telehealth succeeded not on the basis of sophisticated technology, but on the basis of local-initiated programming that responded to locally defined needs. As director of Project Share, I personally insisted on locally produced educational and health-based programs whenever possible. In fact none of the internationally imported educational programs under Project Share continued after the tests were over.

MISSION TO PLANET EARTH—SATELLITE TOOLS FOR SENSING AND COMMUNICATIONS

If we are to begin to save the earth's biosphere, we need more effective urban and transportation planning, better use of our land and resources, increased energy conservation, and pollutant containment. Here remote sensing and communications satellite networks working in tandem can also be of major assistance—again in developed and developing nations alike. This is really not as difficult as it sounds. Already one can see how this can be done, and the evidence is actually now viewable from outer space. The Indian province of Uttar Pradesh has, through the long-term application of space-based remote sensing, been transformed from a semi-arid desert to a "green and agriculturally productive area." Amazingly the processes of replanting forests, rechanneling streams and rivers, and redeveloping arable lands have also served to create new industries and jobs, empower women cooperatives, and improve education and health care systems.

There is no reason that the lessons learned in Uttar Pradesh cannot be repeated again and again in Sri Lanka, Burma, Bangladesh, Pakistan, China, Thailand, Indonesia, Vietnam, the Philippines, Malaysia, Ecuador, Bolivia, Nigeria, Ghana, Uganda, and so on throughout the so-called *developing world*. These tools can and indeed are being used in Japan, Canada, the United States, and Europe as well. Some of these systems are being used today in the rebuilding of Iraq and Afghanistan. The powerful new information tools that emerge from the use of Global Information System (GIS) together with remote sensing and communications satellites are only beginning to be understood.

Today there are plans for integrated development and modernization projects that combine solar electrification, information and telecommunications systems, teleeducation and telehealth, microbanking, microenterprises, and so on. These are being coordinated through the World Bank, Intelsat, the Solar Electric Light Foundation, and the new Clarke Institute for Telecommunications and Information. The point of such projects is to create a model that generates sufficient revenues to sustain the capital investment and is also environmentally sound as well.

It is likewise important to develop new economic and technological models for developed countries to sustain their long-term steady-state technological, economic, and environmental development. Satellite technology and applications will provide critical solutions in the areas of education, telemedicine, reduced air pollution, sustainable and productive fishing, recycling of arable land, reversal of desertification, conservation of wetlands and mountainous regions, and coping with ozone depletion. In the third millennium, we can use space, telecommunications, and information systems to deploy smart transportation systems, provide interdisciplinary urban planning, promote effective use of telecommuting, and create new "clean" jobs in industry and service corporations. Extensions of these projects could seek to utilize GISs in new and innovative ways, create effective disaster warning and recovery systems, provide effective search and rescue, connect to electronic libraries and museums, as well as provide job retraining, professional recertification, and so on. It is a mistake to draw strict boundaries around satellite networks for education and health care and another set of walls around social, economic, scientific, and planning needs.

The key is as simple as seeking a win–win approach that advances knowledge, improves the environment, and creates an economic system that contributes to a "spiral of improvement" by integrating satellite and other information networks with other planning and economic systems.

These are not changes that can be made in 1 year or even a decade. As the third millennium begins, there are over 20 active armed conflicts occurring around the world. These conflicts exist in parallel with the Internet, fiber networks, supercomputer centers, and sophisticated satellite networks that are collectively shaping a global economic network. We humans are a complex group of beings at once magnificent and munificent, but also at times monstrous and malignant. Building a global village and ultimately a planetary electronic culture seems to be our destiny. We are tool makers, and each year our tools become more and more sophisticated.

In short, if our descendants evolve continuously on this planet and go on to settle other planets and star systems over the next billion years, this future will hinge on the steps we take in this new century of a new millennium to open a new path to global knowledge, education, health services, and satellite-delivered social services. Although fiber optic networks, computer networking, the Internet, and other technologies will help, the global reach and penetration needed to make these changes in a timely matter will only come via satellite.

Everyone will have to become proactively involved in developing better educational and training programs for the 21st century if true reform is to be achieved. This is because contemporary challenges to developed and developing cultures are both extremely great. If we were to believe outspoken social and education critics such as Lewis Perlman (1998), author of *School's Out*, we would have to consider his conclusion that today we are running on a reserve tank of intellectual capital. Currently 17% of the U.S. citizenry is functionally illiterate, and the numbers are headed up, not down. Even the numbers for Japan and Korea are headed in the wrong direction.

Certainly a few people and some institutions and companies have caught on to and implemented some key new ideas. These organizations are developing improved ways of learning as well as providing health care more effectively and at a lower cost. Unfortunately, these are the exceptions to the rule.

The noble traditions of higher education have been etched in stone for over 1,000 years, but unfortunately the university of that tradition is better suited to the second and first millennia than the third. Everywhere across this six-sextillion-ton planet of rock, water, and vegetation that we call planet earth, there are new educational and training possibilities open to us. These opportunities for satellite and electronic teleeducation and new forms of "interactive" and "experiental learning," as spelled out at the end of this chapter, are largely being squandered. If one looks at productivity figures for education and health care, in comparison to other U.S. industries, for instance, we find that they are at the bottom. Industry and even government have applied new processes plus computer, IT, and telecommunications technology and have shown gains in terms of work done per dollar spent. Yet productivity in education and health care is much less than it was two decades ago. Doctors and educators resist innovations that have been widely implemented elsewhere in service industries.

We are increasingly acquiring the right to plug into a global knowledge base. In many cases, this vast reservoir of information is based on a combination of humanistic, scientific, and increasingly ecological values. Yet education and learning are in general not improving. Why not? Why is there so much resistance to change and innovation?

The creative and productive learning experience for the new millennia needs to be life-long, self-directed, interactive, multidisciplinary, and self-paced—yet with team-based opportunities in mind. They could even be enjoyable.

Modern education today can represent the best of times and the worst of times. The linkage of satellites, fiber networks, the Internet, and other health and education networks creates enormous potential, but there are also more opportunities for things to go wrong. We have to understand and exploit the enormous educational opportunities of the Internet, online education, and media-based health and teaching systems. These are not a panacea, but only tools—at ones of great potential. Yet with the wrong approach, these tools can be misused and can lead to dangerous abuse.

The key to this new environment is what may be called *network learning*. This new approach to education depends more on fresh educational concepts than on Internet or new digital technology tools or software. It depends on the ability to achieve global sharing, to engage in both experiential and life-long learning and 10 other basic reforms set forth next.

The key is to exploit and harness the power of online and interactive learning, of multidisciplinary team synergies, and to create new educational systems for a new millennium. The single most important point is to create a clear and challenging framework within which learning can occur. Some reality checks need to be built into current educational practices to determine what is and is not working.

Our educational institutions, our methods of teaching and learning, and especially many of our universities in the Western world today appear ill-suited to their assigned tasks. A few years back, my colleague, Professor Eli Noam of Columbia University, made a series of presentations about what he saw as the dim future of universities in terms of high cost, low productivity increases, and failure to innovate.

BASIC CONCEPTS OF LEARNING IN THE AGE OF GLOBAL SATELLITE NETWORKING

It is now nearly 2,500 years since the Age of Socrates, Plato, and Aristotle. Since that time, we have created rocket ships, biotechnology, genetic engineering, lasers, radio astronomy, nonlinear math, chaos theory, satellites, super computers, bungee jumping, spandex, talk-show TV, and artificial intelligence. Most have been for the good, and the majority of these innovations we would view as progress. Yet what about our educational process? Two-and-a-half millennia later, we are putting students in a classroom with an authority figure, of presumed knowledge and intelligence, in front to teach. This person's defined job is to lecture or maybe occasionally pose questions to the students for prescribed periods of time. This we call education.

Education in the United States and many other countries has been reduced to well-known but essentially effete formulas. We typically equate education in colleges and universities to completion of a set "jail sentence." A three-credit-hour course is equivalent to "doing time" for some 40 to 45 contact hours mandated by accrediting organizations. The teacher serves as the warden, while education is the punishment that is quantitatively measured by the time spent in the classroom or "cell." Seldom in the current formula for education do we find the idea of measuring education on the basis of retained knowledge and actually acquired skills. Why is it so hard to accept that cyberlearning is best when a person is actually acquiring and applying new knowledge rather than being a passive listener?

In this educational concept or model, the swift and slow learners seemingly acquire the same amount of learning or knowledge by exposure to a more or less equal amount or quanta of information. Now we do have laboratory and design courses, and there are indeed some research activities associated with theses and dissertations. Except for a few modest testing mechanisms called Graduate Record Exams, LSATs, or GMATs, we end up having little to go on about acquired knowledge. It is actually difficult to find out exactly what a college graduate or high school student has really learned from these standardized tests. Amazingly, when we subject U.S. high school teachers to basic competency tests, a third or more of them fail. The standards of some education degrees in some teachers' colleges are shockingly low.

Most college students with 4 years of residency within the halls of ivy are simply "more ripened" rather than necessarily "more learned." The "educated student" presumably emerges with adept reasoning abilities and "a great deal of acquired knowledge." Yet little in the current educational process guarantees that this will actually happen.

Perlman (1998) and others would say that the modern, prepackaged, and standardized schooling process of the late 20th century, which we call *formal* (as in straitjacketed and overly prescribed) education, is about 90% wrongminded. Perlman suggested that the predominant educational paradigms in the United States and many other Western countries are intellectually bankrupt and highly cost-inefficient to boot.

Furthermore, these same critics would argue that the rest of the developed countries are really not much better off when it comes to effective educational systems. The lack of sufficient skills to read, perform simple math, use a map, or follow the plot of a graph or even a soap opera is now estimated to represent some 4% to 5% of the population in Japan and South Korea and, as previously noted, perhaps over 17% (and rising) in the United States.

The key to our educational problem seems to be that the teacher is provided a monopoly on the right knowledge, and the tutelage of the student is thereby strictly controlled and constrained. The student had to learn the correct or ideal knowledge according to the teacher, which in some cases may be information that is over a decade old with respect to the latest scientific research.

The student must not be allowed to think or learn "outside of the box" thoughts with respect to the accepted intellectual orthodoxy of a textbook or formatted lesson. The basic idea in education for the last 2,000 years has been that knowledge is defined from above. In this model, wisdom must be stored in vats or bins and poured out in measured quantities. This elitist model of education, with the teacher as the dispenser of knowledge, has survived in a resilient way for a long time.

The current course of events in U.S. education is not likely to be effectively reformed in the near term or probably even the longer term. Nevertheless, the idea that the Internet can largely substitute for a complete educational system is no more plausible than to say that libraries or museums can substitute for a complete learning system. There are no simple answers here.

THE SCOPE AND NATURE OF PROBLEM IN GLOBAL EDUCATIONAL REFORM

The challenge of reforming, improving, and extending quality education and training to a global population is huge. Billions and billions of people must be educated in coming decades, and old methods (in terms of process or distribution) are actually poorly suited to the task.

The main issue to attack is not the obvious target. Clearly we need better and more up-to-date educational materials and more and better trained teachers. It is, however, the institutional barriers and the traditional images of education that represent the roadblocks to reform. Using new networking technology and software is another key component. In essence, it is the current structure, paradigm, and objectives of schools and universities that are the biggest obstacles to progress. Here are 10 reforms for serious consideration. Although these reforms are aimed at education and educators, in many ways these changes could and should be used with regard to medical education and in concepts related to the delivery of remote health services or training doctors, nurses, and paramedics at a distance.

Global reform of education to overcome the constraints of "the medieval educational model" and to extend adaptive learning systems throughout the planet is a huge task. It is comparable in scope to such human undertakings as the Apollo Moon Project, the Great Wall of China, or the Great Pyramid of Egypt. In this case, the materials we must work with are neither stone, steel, nor construction plans.

Table 12.1 indicates what are the biggest problems to attack. Although these focus primarily on educational problems, they also relate to telehealth issues and health education programs as well.

TABLE 12.1
Reinventing Satellite- and Internet-Driven Systems for 21st-Century Education

Some of the Major Problems in Education Today

- Lack of self-paced learning systems and databases that can augment "fixed-course" lectures and text materials and be easily updated on demand.
- Lack of opportunity for independent learning and challenging self-paced instruction.
- Absence of experiential or hands-on design exercises and problem-solving assignments to develop critical thinking, enhance writing skills, and work interactively within design teams.
- Discouragement rather than encouragement of interdisciplinary and multidisciplinary learning exercises.
- Absence of process to learn new skills for information acquisition and "screening" and other ways of knowing when information may be incorrect, out of date, and so on.
- Few mechanisms to distribute high-quality information and educational programs through global satellite, fiber, Internet, or other electronic networks that are culturally and linguistically "transparent" to students.
- Most "high-tech" distribution mechanisms now available for teleeducation typically are based on content developed externally rather than by local educators. Further the assets of satellite delivery systems are seldom considered in the development of teaching systems, rapid updates, multiple languages, and text materials.
- Few systematic requirements for education and skill sets to be periodically updated and modernized. More doctors, nurses, and engineers need to be "recertified" on a scheduled basis, and technical innovations to assist with this task are seldom considered.
- Little recognition of the need for life-long learning and new skill acquisition.
- Creation of a system to measure actual learning that is better than hours of incarceration. The current award of grades, degrees, and active disincentives for team learning all need to be rethought.
- The need to develop, in a period of increasing global prosperity, new methods of assessing the cost of providing education versus "not providing education" to everyone on our planet (e.g., population growth, spread of disease, etc.).
- Consider the dangers of "mega-training" that can derive from "automated learning" or abuses of new teleeducation techniques.

The 10 steps in Table 12.2 could help transform and revitalize global education and especially higher education and research in the 21st century. The advantage of these proposals is that they would allow us to educate more people more effectively by releasing the learning capacity of every individual. These steps could help exploit the power of team learning and interactive critical thinking, and they could also aid in health care. Amazingly, it could possibly all be done with a potential saving of money, particularly if long-term social and resource costs are considered. However, to accomplish this we need to reinvent our schools and universities, our medical colleges, as well as institute life-long adaptive learning; we need to recognize that the status quo is the enemy of effective 21st-century education.

The power and economics of global networking and the potential impact that worldwide education reform could make is hard to deny. Nevertheless, the issues of language, culture, and societal norms will pose major challenges that have yet to be seriously addressed. Just because new technologies and applications will

TABLE 12.2
Programs That Would Parallel New Global
Educational and Health Care Reform

1. At the institutional level, deregulate education and stimulate competitive systems
2. Instill concepts of teamwork, critical thinking, and continual learning
3. Experiment with new global educational systems such as interdisciplinary and multidisciplinary centers of excellence
4. Seek new ways of keeping score in education: Credit hours and degrees are increasingly passé. Recertification of knowledge is key
5. Reinvent academic research, and especially reform the "publish or perish" paradigm
6. Emphasize experiential learning, and rethink the classroom lecture as a measure of acquired knowledge
7. Exploit the best of new educational technologies and applications especially from the satellite and computer world
8. Beware the danger of mega-training (i.e., training to task rather than educating)
9. Higher education offerings need to be relevant to current social, economic, and cultural needs
10. Adapt to globalization of education and interactive networking

make global education more plausible and cost-effective, one should not assume that the problems of content, software, and educational process are issues that can be easily solved.

Although the problems of creating global distribution networks and accessible user terminals around the world will remain great, the problems of software and the "right" content of learning systems will be much greater to agree. Even so, the potential use of Internet, satellite, and wireless communications technology to reach the 2 billion people now dramatically underserved by effective educational systems remains brighter than ever. Potential, however, is almost always well removed from reality.

THE CYBERSPACE UNIVERSITY AND MEDICAL SCHOOL

For a long time, schools and universities have been defined as buildings with teachers, classrooms, and students. In the coming age of cyberspace education, interactive learning, and globalization of the learning process, the rules will change. Bricks and mortar cost money and require the consumption of energy and complicated transport systems to shuttle students. The idea of schools without walls, universities and laboratories webbed together by electronics, and learning systems based on new academic paradigms will not be easily denied in the 21st century.

Cracks in the barriers to teleeducation are now occurring. Stanford University, for instance, has embraced distance learning and is offering its courses via satellite across the United States. The Massachusetts Institute of Technology (MIT) has put all of its courseware on the Internet. Today's prototypes such as the

Knowledge Network of the West (KNOW), the International Space University, or the University of the West Indies will become tomorrow's mainstream of educational procedures. There are even lessons to be learned from such innovative yet unconventional institutions as the University of Phoenix, which has broken ranks from conventional universities in many ways.

Generations from now, a new "elite" and a new university establishment will arise at the top of cyberspace educational institutions. By that time, Internet will have greater controls, greater security, and massive information storage, processing, sorting, and distribution capabilities. If the so-called *information highway* ever comes to mean something in a societal sense, it will mean the increasing triumph of cyberspace learning systems over traditional schools and universities and new access to health care systems. The time to begin prototyping and perfecting new health and education systems for both the developed and developing world is certainly now.

The importance of these changes should not be neglected nor downplayed. These reforms are needed to create a global educational system of information exchange and more democratic access to health care. Table 12.3 seeks to summarize again the key issues that need to be considered and some of the opportunities for reform that our new technology can offer us.

SATELLITES AND GLOBAL HEALTH CARE

The first stage in using advanced information systems to improve our 21st-century world will come in educational programs. Satellite health and medical care applications will follow. This we know because the trend is already firmly established. There are millions of students already using satellite and Internet systems at every educational level from prekindergarten to advanced college degrees. These systems are easier to establish, involve less socioeconomic costs, require much less sophisticated equipment and digital throughput rate, and involve much less risk in terms of legal liabilities, potential loss of life, and so on. A small difference in coloration in an instructional material has little potential for adverse affect in a math or physics lesson, but in terms of medical diagnosis it can be a life or death matter.

The most important use of satellite technology in the health and medical sectors in the near term will be in broadcasting information to the broad public on such matters as how to combat the spread of infections, nutrition, ways to cope with epidemics, finding access to medications or vaccinations, and so on. Today these applications have already saved millions of lives in Asia, Africa, South and Central America, and other parts of the world. In West Africa, the loss of life among newborn infants due to dysentery has decreased in the hundreds of thousands largely as a result of satellite and ground-based radio broadcasts.

TABLE 12.3
Challenges and Opportunities of New Technology

*Current Difficulties and Challenges in Modern
Education and Health Care in the Information Age*

- Lack of basic health care and education for 2 out of 6 billion people on the planet (these people also often lack electricity, potable water, and other basic services)
- Rising functional illiteracy in most Western nations as well as increased obesity, inadequate exercise, and rising cost of health care
- Deployment of satellite and electronic networks to serve business and commercial needs with little or no integration with social services or health needs
- Lack of interdisciplinary, integrated, or systems-oriented educational programs and lack of coordination between educational and health care programs
- Lack of self-directed, online, experiential, and interactive learning or health care programs
- The needs of most rural and remote areas of the globe in health care and education are today still largely ignored

New Opportunities Stemming From Global Satellite and Networking Systems

- Encourage interdisciplinary teams and renaissance thinking especially as it relates to education, life-long learning, health care, and so on
- Create systems analysis capabilities that can capture the power of worldwide networking and the latest in medical and education knowledge
- Creation of new programs in life-long learning, health care, nutritional development, and self-improvement
- Development of new flexible adult retraining courses and creation of self-directed health regimes
- Implementation of online, on-demand interactive learning programs in all disciplines including health care and education
- Initiation of interactive Web site learning labs that promote "experiential learning"
- Creation of more effective uses of global satellite networks to provide worldwide, regional, and national teleeducation and telehealth offerings
- Recertification of professionals at specified intervals
- Ability to begin to totally rethink academic and medical degrees

The second wave of satellite-based telehealth applications will come as a result of satellite- and Internet-based training for medical practitioners and researchers in remote locations. These medical personnel can use these systems to learn the latest treatments, order needed vaccines, train their nurses and medical aides, and otherwise support their operations even in the most isolated locations. For instance, for several years the Satellife operation has allowed remote clinics and hospitals to order medicine or seek information from the latest medical journals in a matter of hours. Satellite networks now allow remote clinics—whether in a jungle or an off-shore oil rig—to train their nurses, request diagnoses based on high-quality video scans, and otherwise assist their operations.

Eventually satellite networks will be able to provide a variety of medical services to patients in remote locations as a matter of routine. Techniques such as those developed by NASA for the remote treatment of astronauts, including surgery by remote robots, may one day enter the realm of routine practice.

CONCLUSIONS

The broad application of teleeducation systems, both in terms of satellite terminal equipment and software, will lead the way to telemedicine applications in the decades ahead. The bottom line is that this technology is too powerful, too versatile, and ultimately too cost-effective to not be used to provide humanity with improved education and health care in the decades ahead. The most important thing to bear in mind is the transmission technology is the least of the challenges. It is the applications that need to drive these programs forward with a true sensitivity to the social, cultural, and moral implications of how these technologies can serve humankind.

REFERENCE

Perlman, Lewis, (1998), *School's Out*, New York: New American Library.

BACKGROUND READINGS

The Global Access to Tele-Health and Education System (GATES) Study, The International Space University, Research Report, Barcelona, Spain, 1997.

Kress, Gunther, (1999, Summer), "Educating for an Unknowable Future," *TeleMedium*, Volume 45, Number 1, pp. 7–10.

Pelton, Joseph N., (1996, November–December), "Cyberlearning vs. the University," *The Futurist*, Volume 30, No. 6, pp. 17–20.

Pelton, Joseph N., (2003), "The Future of Broadband Satellite Communications," in T. Varis, T. Utsumi, and W. Klemm (Editors), *Global Peace Through the Global University System*, Hämeenlinna, Finland: University of Tampere Press.

Pelton, Joseph, MacRae, Alfred, et al., (1998, December), *Global Satellite Communications Technology and Systems*, Baltimore, MD: ITRI.

Pelton, Joseph, MacRae, Alfred, et al., (1999), *Global Satellite Communications Technology and Systems*, Baltimore, MD: ITRI.

Speight, K., (1999, August), "Gaps in the Worldwide Information Explosion: How the Internet is Affecting the Worldwide Knowledge Gap," *Telematics and Informatics*, Volume 16, No. 3, pp. 135–150.

Varis, T., Utsumi, T., and Klemm, W., (Editors), (2003), *Global Peace Through the Global University System*, Hämeenlinna, Finland: University of Tampere Press.

CHAPTER THIRTEEN

Satellites and Global Diversity

Hamid Mowlana
American University

> *Things are in the saddle and ride mankind.*
> —Ralph Waldo Emerson (n.d., p. 99)

The last decades of the 20th century had the vibrations of an earthquake about them. The world shuddered and is shuttering still in the wake of 9/11. The new Millennium has not brought us the "end of ideology" or the predicted muted convergence between capitalism and socialism as sociologist Daniel Bell had predicted. Nor has it resulted in "the end of history"—the unabashed victory of economic and political liberalism that political analyst of the American conservative wing, Francis Fukuyama, had anticipated.

No, quite simply, history has continued to crack open. The quest for new ideologies continues. New leaders and followers have come darting out—dark surprises for some, yet lights of hope for others. The forces of change are everywhere: American-fueled capitalism is riding on the back of the Internet, satellites, and fiber optic networks! Yet Islamic culture is finding new vigor from Morocco to Indonesia! European socialism is struggling for relevance! Western technology everywhere is reshaping entertainment, patterns of living and working, political systems, and religious beliefs all over the world. Yet its impacts on the East and the South seem the most abrupt and harsh in the wake of free trade agreements and the expanding reach of the World Trade Organization (WTO). This is to cite only a few of the many tides of change. The forces of change have ventured into new and experimental regions: uplands of new enlightenments or mining new valleys of despair. The implications of the U.S.-led coalition and its forcible regime change in Iraq are still too early to read, let alone interpret.

The last 40 years—the age of satellites—have been pivotal. It has produced vivid theaters of operations: Vietnam, Iran, Lebanon, Afghanistan, Nicaragua, Panama, China, Iraq, and, of course, Eastern Europe and the Soviet Union, not to mention the developments from Northern Ireland to South Africa and from Bosnia to Kosovo. Finally, the new reality of the attack on the World Trade Center Towers in New York that has created a crescendo of tensions. A spirit of change seemed to pervade the political, economic, and cultural climates of the world.

It is precisely in this environment that cultural, economic, and technological forces have come into play globally. Satellite networks that link us all together via telephone, fax, and now especially the Internet can most certainly be given both credit and blame for these occurrences. The question is whether Western technology and economic systems represent for world society a panacea, a pandemic, or both. The tools of automation, electronic information networks, global electronic funds transfers, and much, much more are powerful forces of change that are cracking open global society. These electronic tools, just as Thorstein Veblen (1963) predicted in the *Theory of Business Enterprise* a half century ago, are separating people from their local environment and culture and depersonalizing our lives.

A SATELLITE- AND FIBER-DRIVEN TRANSITION

Although one can point to the advantages of widespread education and health care and rising economic wealth—at least in Western and some non-Western societies—the negative impacts must not go unnoticed. The Western-fueled system of "communications, capitalism, consumerism, and continuous change" contains seeds of a new form of conquest. This now surging e-Sphere of information, communications, and capitalism seems to be seeking to conquer the culture and diverse human capacities of the world.

This may seemingly be a quiet and subtle conquest, but the power of this technological ocean of electronic change must not be underestimated, nor the increasing speed of its accelerating current. As noted in the first chapter, the speed of this satellite- and fiber-driven transition has reached historic highs and is far from subsiding.

As Western writers such as Jacques Ellul, Victor Ferkiss, Lewis Mumford, and Thorstein Veblen have noted, the unchecked spread of technology can change everything—both the good and bad aspects of our lives. Even Walter Buckingham, an advocate for automation and technical innovation, noted that all technological improvement is not a net gain.

Our electronic information systems and Western methods of innovation and production separate and divide producer from consumer, depersonalize our economic systems, reduce our diversity, and even destroy so-called *primitive* or *non-Western* languages and cultures. At the core, we find Western and non-Western values in conflict. Especially at the start of the 21st century, we sense Christian and Arabic/Islamic cultures in conflict—a real consequence of the 9/11 attacks on the World Trade Center, but also a cultural, religious, and value rift that loomed for many years before this catalyzing event.

As international relations expand into a multitude of diverse interests and structures, ranging from military to political, from economic to cultural spheres, we find today a different world. We are finding new global values in conflict and new questions of what might be called *communication ecology*. The growth in recent years of both forums and literature in the area of ethical and moral dimensions of international relations illustrates the centrality of conflicting East versus West value systems. This serves to increase the attention given to the symbolic environment that is created by information technology and global satellite networks. The ideological, religious, and spiritual struggles of the last 10 years, and punctuated by 9/11, highlight both the urgency and depth of cultural clashes in international relations.

Long before Canadian Marshall McLuhan's (1964) phrase "the global village" became popular, American Wendell Willkie, with an enthusiasm touched off by modern air transport, popularized the phrase "one world." The complex feelings and ideas now generally associated with these phrases, however, were not originally Willkie's or McLuhan's discourse. These phrases have a long history. The fact that the world is one in an astronomical or a technological sense—the fact that the world is a single planet located in the gravitational field of a definite star—is not of political, economic, and cultural importance. Historical geography depends on an Einsteinian rather than a Newtonian function. Despite technological and scientific development, including the tremendous growth of communication and information hardware/software over the last several decades, the large majority of residents of this "global village" live in undignified conditions of illiteracy, disease, hunger, unemployment, and malnutrition and are still deprived of the basic tools of modern communication information and knowledge.

THE NOTION OF DIVERSITY

The notion of diversity has been used frequently in our contemporary discourse. The word *diversity* appears in book titles ranging in subject from the diversity of moral thinking to the diversity of high-tech capitalism. Diversity has been written about in relation to religion and conflict, career choice and feminist theory, animal diversity and biodiversity. Some recent titles include: *Celebrating Diversity: A Multicultural Resource* (Clegg, Miller, & Vanderhoof, 1995), *A Diversity of Dragons* (McCaffrey, Woods, & Howe, 1997), *Evolution and Diversity of Sex Ratio* (Wrensch & Ebbert, 1993), and volumes such as *Contemporary Poets Cele-*

brate the Diversity of Their Art (Finch & Varnes, 2002). Diversity is talked about in the form of human relationships and the politics of intolerance; it has been referred to when talking about the order of the state as well as when discussing holiness. The list of resources about diversity as well as the contexts in which the term has been applied now seems endless. The notion of diversity addressed in this presentation focuses on the diversity and interchange of symbols and ideas. It evolves around the subject of communication, culture, and social realities.

It is said that we live in a time of revolution, a historic juncture that goes under various names: a communications revolution, postmodernism, or a new world information-communication order. We are led to believe that this is a time in which the nation-state recedes into the forces of globalization, and with the withering of the nation-state, we are promised either a world of prosperity or impoverishment, a utopic or an apocalyptic end of history. The phenomenon of international communication was born of revolution. International communication is brought forth with the revolutionary political rhetoric signified by the end of the nation-state. The rhetoric of international communications is revolutionary, celebrating or condemning a technological mutation that is altering the spatiotemporal configuration of reality and the very nature of nature. In contrast to the age of Enlightenment, and its scientific, technological, and political revolutions (that which is conspicuously absent from our revolutionary rhetoric), is a sense of philosophical reflection. That is, we are lacking a sense of reflection concerning the conception of revolution. Are we to believe that our revolution brings an end to "philosophy" as did the "end of ideology" debates of the 1950s? Or instead should we entertain the end of diversity? Are we to continue working as if science, technology, and political economy have severed their philosophical roots? Has epistemology given way to methodology? Has philosophy given way to the instrumental framework of political science? Instead of asking fundamental questions concerning the political and its relation to the technological or concerning revolution and the time of history, we continue to assume the enlightened foundations of the political, placing our faith in the dialectics of critical reflection/practice, if not the specter of revolution. Yet, insofar as we are unwilling to bring a sense of philosophical rigor to the discourse concerning global communication, we are forever threatened by the disappearance of the space and time within which to reflect and act. We may get that for which we are asking. The last spaces left for philosophical reflection (academia and intellectual associations) may indeed disappear, becoming nothing more than another medium of communication integrated into the feedback mechanisms of a cybernetic society.

AFTER HEIDEGGER

In his last will and testament, Heidegger left us with a final thought regarding the question concerning technology: "Only a God can save us" (1976, pp. 193–221). Here we pick up from where Heidegger left off. From within this historic juncture, we are concerned with a *philosophical revolution* in the deepest and most profound sense of the term. This philosophical revolution evades the rhetoric of our communications revolution. In contrast to the rhetoric of our communications revolution and its prophecy of a new world information-communication order, we recognize, with the end of the cold war, the return of the same old hegemony, although with the geopolitical axes shifting from east–west to north–south. With this shifting of hegemonic axes, there is a limit that is disclosed from within the Western philosophical tradition. With the geopolitical and geophilosophical tensions between liberalism and Marxism, the limits and tensions articulated were reconfigured from within the Western philosophical tradition and its dialectics of enlightenment. The geopolitical and geophilosophical struggles between North and South lie along two axes. One is the Western philosophical tradition, and the other is religion.

Let us be more precise. The epistemological battlefield with which we are concerned is not simply between philosophy and religion, but instead between differing conceptions of the relation among philosophy, politics, and religion. Regardless of their political and ideological oppositions, both liberalism and Marxism inherited the philosophical tradition handed down by Plato, wherein the political emerges only with the retreat of the religious. With the dominant liberal and Marxist readings of the Western philosophical tradition, philosophy is bordered, on the one hand, by religion, and on the other hand, by politics. With this reading, religion is a remnant of metaphysics, whereas politics is of the factual world of contestation and struggle. Yet, this distinction between the religious and political is foreign to those philosophical traditions that lie outside the West. For example, within the Islamic tradition, this philosophical opposition between the political and religious does not exist. The same is true in the teachings of Buddha and others.

With the coming of the so-called *information explosion* and *information society*, crucial questions face Islamic societies. These are the ultimate control of information processing and technology in the contemporary electronic age and the gradual disappearance of oral or traditional culture that has been a major resistance force in the face of cultural domination. The concept of secular society was introduced into the complex life of Islamic lands at the time when the forces of resistance were at a minimum. In many other so-called *ascriptive cultures* of the Southern Hemisphere and of the East, the same wholeness of view that mixes politics, religion, tradition, and culture can be found.

A MORAL AND ETHICAL COMMUNITY?

With the new awareness and the degree of mobilization and cultural resurgence that we see in non-Western communities around the world, the introduction of the information revolution and the entry into the information society seem to land on rocky soil. The crucial question for non-Western societies is whether the emerging global information-communication community is a moral and ethical community or just another stage in the unfolding pictures of the transformation in which the West is the center and the others are the periphery. Throughout history, especially in the early centuries, information was not a commodity, but a moral and ethical imperative in many cultures and especially so in Islamic culture.

Is information society a kind of network community in which a new rationalism is likely to impose a policy of radical instrumentation under which social problems will be treated as technical problems and the common citizen is to be replaced by outside experts? Will the new technologies of information encourage the centralization of decision making and the fragmentation of society, leading to the replacement of forms of community life with an exasperated individualism? Is information society in a position to produce qualitative changes in traditional forms of communication and eventually to transform social structures? Will such new structures require new ethics? Thus, it seems that the discourse and concepts of global order now at the center of world politics both celebrate the arrival of a new communication ecology and also hold the key to greater information control.

The notion of diversity that we are addressing here is also related to the process of so-called *globalization*. Dash (1998) noted that *globalization* has become a buzzword that is overused and inappropriately applied, such that it is often interpreted as synonymous with global capitalist expansion. In this light, he offered a list of provocative questions that he sees as arising from this newly framed notion of globalization. He asked, what is the nature of the contemporary state–capital relationship? Is class formation changing? Does room remain for national "developmentalistic" projects? If so, what and where is it? What notions exist for countering national hegemony and global capitalism? What roles will be played by whom and within what alliances or coalitions? Can these coalitions work in the international arena? Dash's queries prompt us to look beyond the limited notion of globalization as worldwide consumerism and interrogate the deeper issues of identity and consciousness that lie at its heart.

In the same vein, Van Dusen Wishard's (1999) notion of a successful globalization is one that would result not only from technological achievements, but also from the "harmonization of human attitudes and perspectives." To achieve this, close attention would have to be paid to the unique experiences of individual nations as they struggle to become part of the integrated world. Love's (2000) description of globalization encompasses blurring of experience and shared anticipation, but not common thought or value(s). She claimed that our similarities reside not in the content of our thoughts, but in our preparedness to open our wings and fly, knowing that the wind that supports us is the same wind that disperses dust around the globe.

Rieff (1993/1994) described the creation of a global culture stemming from the "astonishing homogeneous" U.S. American society that is exported globally by the consumer market. It is not surprising that the tastes and artifacts of a superpower would infiltrate other nations. Historically, this phenomenon has occurred repeatedly. What is noteworthy, however, is that even at a time when U.S. influence has waned in other domains, American mass culture proliferates and is "everywhere triumphant."

According to Rieff, the irresistible seductiveness of U.S. American culture has led to the "victory of culture that makes money over all other forms." The culture of traditional societies is succumbing to bottom-line thinking and, in many cases, is on the brink of extinction. "American consumer culture is corrosive of all traditions and established truths" (p. 77), Rieff claimed. Although reference is made mostly to popular culture, the implication is that, with popular culture, thoughts and traditions are exported. Plainly, Rieff stated, "Those who yearn for authenticity, for the preservation or restoration of the traditional, will not prevail" (p. 78).

A NEW ERA OF GLOBAL DISORDER

Anthropologists, too, have entered the discourse on the emergence of global culture. Rather than speak solely in terms of diversity and/or convergence, however, they have made note of the cyclic pattern of empires that rise to power and in turn strive to fashion cultural hegemony before inevitably collapsing and losing the control necessary to propagate their singular cultural model. From this standpoint, the West is portrayed as presently on the downward slide—in the process of decline. For some anthropologists—namely, Richard Wilk of Indiana University—this shift is seen not in terms of the death of the West, but rather as the resurgence of local identity and ethnicity. Wilk's (1998) view seems to speak to the revival of diversity rather than any convergence or loss of uniqueness in the global village. Not all of his colleagues see this as reassuring, however. Friedman (1999) depicted aspirations for local culture, which still cling to bits and pieces of Western cultural forms, as symptoms of degeneration and disorder. The picture painted is one of various groups struggling for identity and authenticity in the face of downward mobility, fragmentation, and a new era of global disorder.

With the overpowering force of technology incised into the North–South axis of our new world information-communication order, now increasingly amplified by satellites and fiber optic networks, we encounter a philosophical revolution. This revolution, as Heidegger would have it, turns around the unthought essence of the Western philosophical tradition. In an attempt to open up a sense of communication that does not remain deadlocked by the geopolitical and geophilosophical hegemony inherited by the Western traditions, we need another way to think about the relationship among philosophy, politics, and religion.

Those philosophical frameworks that set religion in opposition to the political would hardly be hospitable to the work and demands of international communication. The demand and work of philosophical reflection in the field of international communication needs to move beyond the conceptual oppositions, which, for example, inform the politics of our communication revolution as an overpowering force. It uproots people from their national-ethnic roots, cultural identities, and religion as a violent reaction to this uprooting. This conceptual opposition perpetuates the double violence inherent to our dialectics of enlightenment (the violence of our uprooting and the violent reaction to this uprooting).

With Heidegger's reading of the history of metaphysics, the transcendental and theological return to haunt our enlightenment with the technological creation of that which Freud would call the "prosthetic God." As developed in France (through the thoughts of Emmauel Levinas, Maurice Blanchot, Jacques Derrida, and Jean-Luc Nancy), not only has the Western philosophic tradition never been capable of cleansing itself of the political and theological impulses, but this attempt to cleanse itself spawns the double violence of separating the political and economic systems from unyielding value systems or vice versa.

The fundamental question thus becomes: "Does the global information community now emerging facilitate or impede the social use of information and help to meet the genuine social aims and needs of non-Western society?" The answer lies in examining the elements of the so-called *information society*, which is central to the dominant model of economic, political, social, and cultural activities of the United States and a number of other countries. It also calls for an examination of the broader conception of social life that underlines the non-Western model of community and the state.

TWO VISIONS OF SOCIETY

At the center of the controversy are two visions of society: "The Information Society Paradigm" and the "Non-Western Community Paradigm." On the intellectual and philosophical levels, the philosophy and theory of information and communication have replaced transcendental discourse as the prime concern of philosophical reflection in the West. On the practical and policy levels, the Information Society Paradigm in the West has come to portray the ideology of neomodernism, postmodernism, or postindustrialism without abandoning the capitalist economic and social systems that continue to characterize its core. Thus, the Information Society Paradigm is presented as "the realization of society that brings about a general flourishing state of human intellectual creativity, instead of affluent material consumption." According to this assertion, the relationship among the state, society, and individual will be determined by the production of information value and not material value. Thus, information society, as it is argued, will bring about the transformation of society into a completely new type of human society. Some of the characteristics of the Information Society Paradigm, according to its proponents, are "spirit of globalism," "the satisfaction of achieved goals," "participatory democracy," "realization of time value," "voluntary community," and a "synergetic economy."

Furthermore, we are told that such an information society is based on services; therefore, it is a game between persons. What counts is information. The central person is the professional who is educated and trained to provide the kinds of skills that the information society requires. The information society is also supposed to be a super-secular knowledge society based on the nation-state system.

In a number of fundamental ways, the notion of the Information Society Paradigm and the emerging global information community runs counter to the basic conception of the non-Western community, particularly with regard to a number of principal tenets of Islam. The contemporary information revolution that underlines the Information Society Paradigm should not be portrayed as a unique phenomenon in human civilization nor should it be treated as a separate phenomenon from the Non-Western Community Paradigm. In all three stages of technological and societal development—agricultural, industrial, and now the postindustrial—information and, in a broader sense, technology have been the central, most pervasive, and common elements in their development processes. Information in the form of skill and knowledge preceded capital formation and, in many ways, characterized all three stages. If we accept this assumption, it simply means that information and knowledge are not the exclusive property of industrialized societies unless *information* and *knowledge* are defined in terms of Western epistemological and technological content. An example is the amount of scientific information and knowledge produced in the non-Western world in such fields as medicine, mathematics, geography, history, astronomy, philosophy, literature, architecture, and the arts, not to mention the development of communication, transportation, navigation, paper industry, printing, and book making.

The Information Society Paradigm is based on secularism, whereas the Non-Western Community Paradigm is founded on a religiopolitical, socioeconomic, and cultural system based on an elaborate legal code and jurisprudence. Thus, secularism becomes alien to non-Western social and political thought when it attempts to separate religion from politics, ideas from matter, and rationality from cosmic vision. Many in non-Western cultures, as well as numerous previously mentioned critics of currently prevalent Western paradigms, view all these issues and concerns as an integrative whole. Although such views were at one time central to the Renaissance, these ideas have been lost in Western political and educational systems.

This global intercultural communication today stands at a crossroad. This crossroad—intersecting the past, present, and future—points in different directions depending on how one maps our present juncture. Do we conceive of the intersection among past, present, and future as a grand or master narrative, sorting out the archive and our sense of collective memory, in an attempt to establish and defend our identity? Or do we evoke the gaps in memory, giving voice to its silences, re-reading the texts excluded from our archives, responding to the call and demands of those who remain outside the traditions constituted by our institutional identity, in an attempt to open our horizons, which have yet to be opened and journeys that have yet to be adventured? Do we map our historic juncture as an ongoing continuity predetermined by the ongoing debates between dominant and critical paradigms, or not? Do we instead map this current historic juncture as a discontinuity, demanding a re-reading of the narratives through which to link the past, present, and future?

As a discontinuity, our present juncture is not located historically, but instead opens the horizon of what is and can be thought to the traces of a past that has been erased from memory, returning to haunt the present with its absence, opening the hermeneutic and ideological horizon of our discursive and interpretive community to the "unthought," if not unthinkable, after effect of a future yet to come. Our historic juncture discloses the gaps, fissures, and silences inherent to any and all master narratives, demanding a recasting of its past and a re-envisioning of a future yet to come. International communication solidifies itself, establishing its identity and destination in advance, charting and determining its course along the well-tread paths of dominant paradigms.

THE END OF THE NATION STATE?

The appeal of international communication (both in a social and technical sense today) is the hope that a confluence of technological, economic, political, cultural, and social formations is possible. There can at least be the aspiration to create a framework through which to recast and rethink the problems and questions of our age, opening the juncture of our present to unthought border-crossings between East and West and between North and South. We can seek to reconceptualize the problems and questions that haunt our present and past Western traditions of knowledge and civilization and broaden our horizons. As Benjamin (1969) would have it, it is to demand of our present juncture to "seize hold of a memory as it flashes up at a moment of danger" (p. 255). The dangers and anxieties that haunt global communication are many. One of the dangers unearthed by an analysis of the technological, economic, political, cultural, and social configurations of our current world is the prospective end of the so-called *nation state*. With a de-constructive re-reading, the institutionalization of international communication is predicated on a proactive process, whereby the intellectual community constitutes and legitimizes its authority via an exclusionary gesture. The imaginary identifi-cations or, as Foucault (1972, 1980) would have it, the sociodiscursive networks of knowledge/power are intertwined with the conceptual territories of the nation state. Within this context, the memory that "flashes up at a moment of danger" concerns the trauma, anxiety, and melancholy that implicate our sense of identity within the so-called end of the nation state and the images that the global village, worldwide mind, or e-Sphere call forth. All of these images are essentially based on the centrality of Western thought and capitalistic paradigms. In this sense, the communication satellite systems of the world and the related information technology become the tools of Western capitalism. It implies, at the very least, exploitation and cultural domination in what Thomas McPhail (1987) and others have called *electronic colonialism*. From the non-Western perspective, this becomes teleshock indeed.

From within the framework of political economy, the electronic and cybernetic vectors of global information-communication economies erode the political sovereignty of the nation state, opening the cultural and linguistic borders of different people and mentalities to either. Depending on how one looks at it, this is an enlightened and progressive shift from national chauvinism to global village or a decline of political and cultural values, ranging from the leftist critique of commoditization and homogenization to the rightist critique of a sociocultural anarchy uprooted from any sense of value and tradition. Stated in this way, the problem is informed by perspective, staging the battleground mapped between (a) the technocratic apologists who, propagating the freedom of information, would conceive of our discipline as a medium of communication, an information clearinghouse, and policy research institute; (b) those theorists who conceive of information as a commodity that masks its class-structural relations of production and consumption would view our discipline as a ritual of resistance; and (c) a conservative morality, which, both celebrating the wealth of information economies and admonishing its uprooting of traditional sociocultural values, would erect a barrier between the technological and economic progress of information economies and a return to the values of a classical humanist tradition.

Each of these positions refuses to address the complexity of *international communications* as broadly defined. Today this is a field configured by the cleavage and tensions that yoke together enlightenment and its ruins, progress and its uprooting, expanded horizons and colonization, representing instead a special-interest predefined in reference to a set of political, economic, cultural, social, and philosophical values.

If we are to address the problems and questions that haunt the field of international communication, we must develop a framework that moves beyond the hegemony of philosophical enlightenment and Marxist critique, deconstructing the unity in division of right–left polarities. In an attempt to develop the framework through which to recast and rethink the problems and questions brought forth, we must deconstruct the philosophic oppositions and the historic master narratives that inform our worldview. The dangers and anxieties that haunt international communication, as unearthed with the ruins of the nation state, demands of our present juncture an archaeological excavation of the fractures, cracks, and crevices in our archives, opening the philosophical traditions of Western civilization to its forgotten pasts—a selective process of forgetting that is forgotten. It is from within the gaps and silences in our historical archives, and the fractures and fissures in our archaeology of knowledge, that the nondeterminance and discontinuity of our present juncture is unearthed. The incision of our wounded present is sutured into the progress of scientific enlightenment. The innovative destruction of technological and economic development forms a critical turn in philosophical enlightenment, the abyssal ground uprooted with the enlightenment of the modern subject, and the political-moral reaction of classical humanism.

THE CONCEPT OF *UMMAH*

The Islamic notion of community or *ummah* has no equivalent in either Western thought or historical experience. The concept of *ummah* is conceived in a universal context and is not subject to territorial, linguistic, racial, and nationalistic limitations. Thus, the non-Western Community Paradigm (or in this case the Islamic Community Paradigm) is not based solely on either politicoeconomic or communication-information foundations. The community is a universal society and polity whose membership includes the widest possible variety of nationalities and ethnic groups, but whose commitment to Islam as a faith and ideology binds them to a specific social order. Sovereignty belongs to God and not to the state, ruler, or people; therefore, the concept of *ummah* is not synonymous with "the people," "the nation," or "the state," which are the vocabulary of modern international relations and are determined by communication grids, information flows, geography, language, and history, or a combination of these.

It is precisely here that the Information Society Paradigm and the non-Western Community Paradigm have philosophic and strategic conflict, and this conflict has a cultural base and information-cultural consequences. In summary, the nation-state system has divided the *ummah* into smaller parts—a process of disintegration and political frontierism that emerged in the early part of the 20th century. The failure of the ruling elites of non-Western countries to challenge the dominant political and economic paradigms that have laid the foundation of the existing world order has undermined the historic grassroots nature of their culture, and their traditional lives have brought us to today's juncture.

For centuries, the non-Western culture maintained a fine balance between oral and print cultures and between interpersonal and intermedia communication. The print and electronic culture, primarily in the West, helped concentrate power in the hands of a few and contribute to the centralization of the state apparatus and corporate monopoly. On the contrary, the oral mode of communication in traditional societies helped decentralize and diffuse the power of the state and economic interests and establish a counterbalance of authority in the hands of those who were grounded in oral tradition. The individuals maintained their ability to communicate within their own community and beyond despite the influence of state propaganda and modern institutions.

The political and revolutionary movements within the non-Western countries in general have been led by traditional authorities and institutions such as the *ullama*. This is but one example of the potential use of oral culture and its confrontation with modernism inherited in mediated electronic cultures. Today, the information needs of many countries are manipulated primarily by market interests and the media, and by the dominant global powers, and not necessarily by the articulation of the genuine needs of individuals, groups, and societies.

The scientific foundation of international communication serves to legitimize the meaning of the international approach as a model of oppositional interests. This, in turn, reduces the meaning of communications to a mechanical and cybernetic model of information flow subsumed under the expansionist interests of technological and economic development. Although the standard historiography reads the political significance of the international as derived from technologies of communications, the non-Western perspective is different. The meaning of *international* is not mapped in reference to the expansion of global communications media. Instead it is the ideology of expansionism, deterrence, and containment that informs our mechanistic conception of communications. Stated simply, communications media did not create the international system, but instead the technological framework and political-economic interests of the international system created the image we have of communications. This image has, for the most part, been developed by the self-serving interests of the information elite.

How do we conceive of the *inter* in international, intercultural, and interpersonal communication if not as the irreducible and irreconcilable difference broached by the space of the in between? It is from within the flux, intensity, and agitation of this interstitial space that we seek the ferment in the field of international communication. If we conceive of the international as a framework through which to address the social, political, cultural, and economic problems of our day, then it is our contention that we must reflect on and question the design, structure, and border of this framework. As denoted by the *inter* in interpersonal, intercultural, and international, the space of communications is of the interstices in between singular terms (be they persons, cultures, or nations). Not only does the interstitial space, as opened with the cracks and crevices among different people, languages, and cultures, deconstruct the unity assumed by the global field theories of our new world information-communication order, but it signifies a flux, tension, and movement that both precedes and exceeds the sovereign geometries of the nation state. Hence, the space opened with the international is of a space that is neither global nor national. The international in international communication should not be conceived of as a mediating term between the national and global, but instead as the spatiotemporal cracks and crevices that resist and subvert the totalities, identities, and oppositions assumed by this mediation. It is from within the interstices of a field of irreducible and irreconcilable differences that the field of international communication opens the horizons between and bridges across the great divide among people, languages, and cultures.

Does not the significance of the *inter* in international communication demand a rethinking of the relation between language and communication, and, consequentially, a re-reading of the *inter* in intercultural communication? With the borders crossed in between different people and cultures, language can no longer be considered a transparent medium of communication. The semiotics and poetics of language maps the imaginary and symbolic space of a desire—a metaphysical desire—that opens thought to communication and identity to the other. Language opens communication to the cracks and crevices bordering heterogeneous civilizations. With the interstitial space opened by international communication, language is neither a transparent medium of communication nor a tower of Babel, but instead serves to unearth the differences of a relational context.

Language can be a powerful tool within the field of international communications to open and explore the cracks and crevices that exist between cultures—what Foucault (1972, 1980), would call the *archaeology of knowledge*.

COMMUNICATION AND CULTURE AT THE CROSSROADS

There are many key questions that can guide our re-reading of the history of communication and culture. The crossroad of our present juncture demands that we carefully examine the border crossing between uniformity and diversity. As mapped by the often-used term of *North-South axis* of global communication, the philosophical and epistemological foundations of Western civilization must open the horizons of its thought and unthought to other cultures. To conceive of diversity as the *other* and specter of Western philosophy as the *valid* is where the deconstruction must begin. The irrationality seen in one culture (i.e., non-Western) and the rationality observed in Western philosophy and science must be abandoned before meaningful analysis can begin.

In truth, the history of Western philosophy is an open book that was never capable of foreclosing its cultural dimension. From Kant, Marx, and Freud to many other Western philosophers, the discourse has been more cultural and religious than ever truly noted. At this critical juncture, it is perhaps most productive if we shift our attention away from such notions as "the end of ideology" and "the end of history" to address the "end of diversity?"—a discourse that is highlighted with a question mark.

It is well and good to examine the historical, economic, educational, developmental, political, and informational impacts of satellite communications as previous chapters have addressed in some detail. Yet fundamentally the cultural impact of satellites on diversity, world peace, and human survival is the core issue. Are satellites instilling just one value system? Are they destroying or enhancing global diversity or perhaps both? Satellite systems can be a powerful tool to bridge the gap between the people of the world and to increase knowledge, but are they truly making the world a better place for all humans in all lands? The jury on this question is still out. International communications must be viewed in a broader sense than just technology and gigabits/second. Questions as to the ability of satellites to engender understanding and facilitate and define intercultural diversity for all humans are the bigger issues. These will likely remain the more important concerns to explore a decade or even a century from now.

REFERENCES

Benjamin, Walter, (1969), "Theses on the Philosophy of History," in *Illuminations*, New York: Schocken Books.

Clegg, Luther B., Miller, Etta, and Vanderhoof, William, (1995), *Celebrating Diversity: A Multicultural Resource*, New York: Delmar.

Dash, Robert C., (1998, November), "Globalization: For whom and for what?," *Latin American Perspectives*, Vol. 25, No. 6, 52–54.

Emerson, Ralph Waldo (n.d.), *Poems*, New York: Hurst and Company Publishers.

Finch, A., and Varnes, K., (2002), *An Evaluation of Forms: Contemporary Poets Celebrate the Diversity of Their Art*, Ann Arbor: University of Michigan Press.

Foucault, Michel, (1972), *The Archaeology of Knowledge*, New York: Pantheon Books.

Foucault, Michel, (1980), *Power/Knowledge*, New York: Pantheon Books.

Friedman, Thomas, L., (1999), *The Lexus and the Olive Tree*, New York: Farrar, Straus, Giroux.

Love, Kaleen E., (2000, January 27), "Millennial Visions: Metamorphosis," *Far Eastern Economic Review*, Vol. 163, No. 4, 60.

McCaffrey, Anne, Woods, Richard, and Howe, John, (1997), *A Diversity of Dragons*, New York: HarperPrism.

McLuhan, Marshall, (1964), *Understanding Media: The Extensions of Man*, New York: McGraw-Hill.

McPhail, Thomas, (1987), *Electronic Colonialism: The Future of International Broadcasting and Communication*, Newbury Park, CA: Sage.

Rieff, David, (1993/1994), "A Global Culture?," *World Policy Journal*, Vol. 10, No. 4, 73–81.

Veblen, Thorstein, (1963), *The Theory of Business Enterprise*, New York: New American Library.

Wilk, Richard, (1998, April), "A Global Anthropology?," *Current Anthropology*, Vol. 39, No. 2, 287–288.

Wishard, Van Dusen, (1999, October), "Globalization: Humanity's Great Experiment," *The Futurist*, Vol. 33, No. 8, 60–61.

Wrensch, Dana, L., Ebbert, and Mercedes, A., (1973), *Evolution and Diversity of Sex Ratio in Insects and Mites*, New York: Chapman & Hall.

PART SIX

FUTURE TRENDS

CHAPTER FOURTEEN

The Future of Satellite Communications Systems

Joseph N. Pelton
The George Washington University

Takashi Iida
Naoto Kadowaki
Communications Research Lab of Japan

Any sufficiently advanced technology is indistinguishable from magic.
—Sir Arthur Clarke's Third Law (McAleer, 1992, p. 169)

In the field of satellite communications, just as in many other fields, one can speak of the problem of which came first, the chicken or the egg? In the satellite context, the issue is whether important new applications drive the satellite industry forward or whether key new technologies unlock new, and heretofore undiscovered, opportunities that drive new services. The answer over the past 40 years seems to have been both. There is some wisdom in the seemingly nonsense advice of the famous New York Yankee baseball player, Yogi Berra, who once said: "If you come to a fork in the road, take it." In this context, planners for future communications satellite systems need to pay close attention to applications and advanced technology.

We know that many important new satellite applications and innovative service demands have emerged from the 9/11 terrorist attacks in the United States, as well as from military missions, natural disasters, and educational and health service requirements. Terrorist attacks and military operations, for instance, have given impetus to telecommuting, electronic decentralization, advanced methods of monitoring and analyzing flight information for aircraft, and expanded use of remote sensing and GPS space navigation information to protect critical infrastructure, both in the United States and abroad. The great Kobe earthquake in Kobe led to the creation of a national VSAT network in Japan to ensure communications even after natural disasters occur.

Within a decade, there will likely be satellite-based systems that will use radar, optical imaging, and Micro Electro Mechanical Systems (MEMS) to monitor everything from water supplies to energy plants as a result of antiterrorist concerns. Satellite and sensor technology will play new and creative roles. These applications will include use of advanced satellites to monitor and even remotely control commercial aircraft, warn of impending aircraft equipment failure and other on-board problems, and provide instant and interactive communications to "smart automobiles and smart highways." These satellite systems will make transportation and energy systems on the ground and aircraft in the skies safer and more efficient. The same path to new innovations has come in response to natural disasters and has also helped provide education and health services in remote locations and respond to other demands for services in isolated or dangerous areas.

The other side of the coin, however, is also true. New multiplexing systems such as code division multiplexing (also known as spread spectrum), turbocoding, new satellite antenna and user terminal designs, new versions of the Internet Protocol, and new types of batteries and solar cells—among other technical innovations—have allowed an amazing range of new services. The Internet alone has led to e-commerce, online educational and health services, and literally thousands of new applications. So how does one predict the future of communications satellites? Is it through new social, political, or economic applications, or new technology? Or is it, more logically, through an interactive consideration of all these factors and more?

It has been suggested that the safest path to predicting the future, whether it be satellites or some other technology, is to offer long-term forecasts so that no one will remember to check on them 50 years or further into the future. We have perhaps unwisely ignored this advice and seriously sought to examine where satellite technologies and applications might evolve to over the next one to two decades and identify some of the critical success factors that will help us achieve that future more quickly and efficiently. What we do know is that satellite technology has many new avenues open to it today and a host of new applications on the horizon. This is key because 21st-century satellite systems will need to continue to evolve quickly to remain competitive with (or at least complementary to) fiber optic networks. Thus, the analysis in this chapter seeks to explore what technologies can and will evolve over the next 10 to 20 years and what the key market forces are that will stimulate that development.

ALTERNATIVE PATHWAYS TO THE FUTURE

In concept, the satellite communications technology of the 21st century could evolve in many ways. These include the extrapolation of current satellite technology so that it simply gets bigger and better. However, this seems unlikely because the demands of the marketplace suggest that new technologies and new efficiencies are needed, both in space systems and in terms of new and much more cost-efficient user terminals. The rapid advance of fiber optic technology and broadband terrestrial wireless systems suggests that satellite technology needs to pursue some new breakthrough technologies.

There are many possibilities to consider. There are the prospects of new types of orbital systems, but recent experiences in the market place suggest that most low-earth orbit and medium-earth orbit systems, in particular, will be postponed due to technical challenges associated with broadband switching and uncertain business models and system economies.

Another option is the possible evolution of high-altitude platform systems (HAPS) that might be based on a variety of technologies such as dirigibles, much more efficient jet and conventional aircraft engines adapted to high altitudes, and solar-powered aircraft such as those demonstrated in the form of the Centurion stratospheric platform. These new HAPS could allow for a new type of telecommunications and IT infrastructure to be created in what has been called *protospace* or *subspace* so that many forms of mobile and broadcast service needs could be met without overextending the technical capabilities of either terrestrial or satellite systems. This new tier of communications and remote sensing capability actually has much to recommend it, and the most recent World Radio Conference of the International Telecommunication (ITU) allocated new frequencies in the EHF band for this new type of communications service.

One could also envision a high-performance optical ring of satellite relays that allows broadband relay of data via a global beltway in the sky. Because there is no atmosphere to interfere with optical signals in outer space, this type of network could move data across the sky at a cost comparable to fiber optic cables laid on the floor of oceans.

In short, there are many new types of satellite architecture and antenna systems that could produce major payoffs in terms of new cost-efficiencies and new and more effective use of spectrum. For satellites to remain market competitive, it seems both necessary and likely that some new breakthrough technologies will need to be developed and deployed within the decade.

If one projects current satellite technology forward in terms of mass, spectrum, and performance, the likely design parameters would be expected to be as reflected in Fig. 14.1. If it were possible to develop new technology and systems that reflect a new type of architecture that outperform this extrapolation, then good things may happen for the satellite industry. The reverse, however, is also true. If there are no breakthroughs for the satellite industry in terms of the design and conceptualization of satellites and user terminals, then the potential of satellites in the 21st century must be downgraded significantly.

Fortunately, there are lots of new options. There could be optical rings to create a global information beltway. There could be new "smart beam" and turbo-coded IT networks in the sky that are supplemented by HAPS to create new frequency and cost-efficiencies. These would enable the deployment of high band-

Development

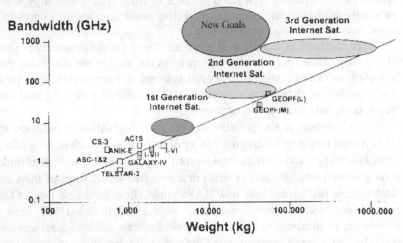

FIG. 14.1.   New goals for future satellite communications development.

width interactive user terminals that can support telecommunications and entertainment applications that are also small, affordable, and potentially user-friendly and even wearable.

Another option, which could combine with several of these possibilities, would be large aperture but low-mass phased-array antenna systems (or low-mass parabolic antennas with phased-array feed systems). Such new architectures, combined with frequency efficient turbo-coding techniques, would allow new, higher performance breakthrough satellites to become plausible. There is not time here to explore all of these exciting new possibilities. The object of this chapter is not to provide a description of all the new technical and service possibilities, but rather to show that a range of exciting new satellite and user terminal technologies is now on the horizon.

THE HIGH-PERFORMANCE, LOW-MASS SATELLITE OF THE FUTURE

Large-scale, high-performance types of satellite systems of the nearer term future (i.e., the next 20 years) would likely be deployed in geosynchronous orbit (GEO). This is because one cannot likely achieve the needed scale of efficiencies in either medium-earth orbit (MEO) or low-earth orbit (LEO) constellations. Further, there are problems with high-speed, broadband switching (especially in LEOs).

The research plan to create new, high-performance, low-mass, and truly innovative satellites, which is implicit in the new goals for space systems identified in Fig. 14.1, would likely have a number of specific objectives. These new and quite demanding space telecommunications development goals might include the following:

1. Develop 10 to 100 times more usable radio frequency spectrum for global satellite telecommunications. This could, in part, be derived from a geoplatform that would dramatically reduce the cost of all types of space telecommunications and information services—private and public—and in part from "smart beam" technology, which allows all spot beams in a satellite network to re-use the transmit and receive frequencies. (This process would be strongly supplemented by new turbocoding techniques that are spectrum-efficient and might achieve 4 bit/ Hz throughput within 5 years and improved performance thereafter.)

2. Allow the introduction of new forms of user microterminals. These would not only be extremely low cost and low powered, but also sufficiently small so that they could be extremely portable or even eventually become wearable devices. (The advent of low-power user transceivers will also have long-term health as well as an economic advantages.)

3. Create system economies that would radically reduce the cost of global telecommunications networks, broadcast and mobile services, and international teleeducation, telehealth, telesafety, and disaster warning systems.

In short, the new and highly advanced satellite systems would likely be based on large-scale space telecommunications systems that might called *geoplatforms* because of the new scale and performance economies. However, because they would be low in mass and density, these would not resemble a space station or earlier concepts of geoplatforms that would weigh 10 to 100 metric tons in orbit. These would be much more like gossamer structures—almost like a spider web in the sky. (Alternatively, clusters of smaller versions of these extreme aperture systems might also be deployed to create a virtual geoplatform.)

These new gossamer, low-mass geoplatforms could stimulate new public- as well as private-sector applications and provide economic breakthroughs in the cost of access to orbital spectrum. It could also aid in the creation of new jobs and information services. Discussion of possible new applications that these new space systems might make possible are addressed later in this chapter.

Such advanced geoplatform designs could, in theory, involve the deployment of less mass and less complex systems in space than today's most advanced space communications antenna systems. More specifically, these advanced systems might actually weigh about the same as today's largest satellites (i.e., about 5,000–7,000 kilograms), yet they could also achieve performance that is significantly better (i.e., 10 to 20 times) than today's most advanced satellite designs. This assumption is based on the idea that there can indeed be revolutionary design capabilities in an advanced geoplatform vis-à-vis conventional satellite design. Also built into this view of the future is that governmentally based research programs by national space agencies and national communications research centers would in some form or another assist in the research and development (R&D) needed to deploy such high-performance systems.

For such advanced design concepts to be successful, they must be more than just competitive with today's communications systems, but rather they must seek to be competitive with fiber optic and laser telecommunications systems some 15 years hence. Further, they must deliver system capacities and quality of service that would be compatible with system user needs that might be anticipated in the 2010 to 2020 time period.

This is obviously a difficult objective in light of the continued expansion of system users and exponential growth of broadband user needs as driven by high data rate Internet applications, growing numbers of terabyte databases as being deployed via Internet 2, exploding use of multimedia applications, and explosive global growth of e-commerce and worldwide radio and TV broadcast systems.

Therefore, it is assumed that the prime driver of advanced geoplatform design will be the ability to deliver huge amounts of spectrum from orbit (orders of magnitude greater than today's satellites) and at commercially competitive prices. The key goal of this advanced system design study of an advanced geoplatform would appear (without trying to be too glib) to be threefold: spectrum, spectrum, and more spectrum. Furthermore, to achieve these demanding goals, these systems may need to be a combination of new technologies (i.e., large-scale, high-efficiency, new lightweight antenna systems, "smart beam" antenna systems, and perhaps "super satellites" that combine with optical, high-performance optical rings, high-altitude platform systems, and new patch antenna designs deployed with new types of "wearable user terminals").

If all of the technologies could be developed so that satellite cost-efficiencies could be greatly advanced, new demand for satellite services can be envisioned that are 10 to 50 times greater than today's service requirements in such areas as broadcasting, multicasting, aeronautical safety, navigation and mobile communications, and especially mobile applications involving multimedia over IP, and new broadband business, teleeducation, telehealth, and other social services.

Figure 14.2 summarizes some of the primary goals that future satellite systems would need to meet.

EXAMINING NEW ARCHITECTURAL DESIGNS FOR ADVANCED GEOPLATFORMS

To illustrate how such new satellite architectures could be developed and deployed, three basic designs for advanced geoplatforms are presented next. These are presented not as engineering models, but simply as concepts to outline how totally new goals for cost-efficient geoplatforms might be achieved.

A brief summary description of these advanced geoplatform design concepts are described in the following list. These are progressively more ambitious to

- Much smaller beam footprint on Earth
- Increased spacecraft transmitter power requirements and much lower user terminal power requirements
- Implementation of many simultaneous independent beams with "smart beam coding" to allow much more intensive frequency reuse
- Less interference between antenna elements within a GEO "slot"
- Significantly more frequency reuse within a GEO "slot"
- Significantly more capability, but with lower total weight and cost per data, video, multimedia, or voice channel
- Significantly greater on-board processing capability to utilize such systems more effectively in meeting consumer and business demand
- Ability to service, repair, or update elements incrementally
- Greatly enhanced spectrum efficiency

FIG. 14.2. Goals for future satellite systems.

achieve, and each one presents increasing performance capabilities and progressively more advanced technology. These ideas are addressed in greater depth in Iida, Pelton, and Ashford (2003). These three advanced systems are:

- **Geoplatform Option 1:** Implement a constellation with a number of conventional antennas placed far apart in space and fed by a central unit. The location of the antennas in this constellation would be maintained principally by passive means most likely achieved by a rotating tethered system. Alternatively, there could be a free-flying cluster of these satellite elements.
- **Geoplatform Option 2:** Implement one (or at the most a few) very large aperture antennas using advanced technologies with little or no structure. These new materials might allow one to even two orders of magnitude reduction in weight and cost reduction as compared with conventional antennas. The advanced antennas involve phased-array systems or multiple illuminators to create many narrow beams.
- **Geoplatform Option 3:** This involves the implementation of an extremely large sparse-array antenna with a large swarm of tiny elements acting in concert to form a coherent phased array without any structure. The dimensions of the antenna would be maintained principally by passive means that derive from the earth's gravitational field. This would take the idea of satellite clusters to the ultimate extreme.

These three designs can be evaluated against such factors as cost, reliability, technical feasibility, complexity of deployment and operation, need for new material development and new R&D, space environment/space debris concerns, human health and safety issues, and legal or regulatory constraints. Clearly such developments are expensive and will take a substantial amount of time and research effort. Fortunately, there is a good deal of potential synergy between research for large-scale geoplatform systems and future space solar power systems. This is because many of the critical technologies involving piezo-electric structural materials, polyimide materials, tethers, and phased-array antenna systems are common to solar power systems and advanced geoplatforms for telecommunications services. Thus, there is an opportunity for shared research costs in properly planned programs. This is now a growing awareness among space researchers, but not yet among policymakers and high-level government officials.

The idea of a large-scale geoplatform that could provide intensive frequency reuse and high throughput capabilities is only one possible pathway to the future. Such concepts will be tested against other concepts, and the ultimate solution may indeed be a combination of design and systems concepts.

The questions to be addressed in carrying out such comparative analysis would definitely include the following: efficiency of frequency reuse, degree of latency especially for highly interactive service, cost-efficiency, quality of service, ability to provide seamless interconnection, throughput capability, system availability and size, cost, and user-friendliness of user terminals. Of course one must ultimately consider the technical, economic, regulatory, and business feasibility of all these designs.

CRITERIA FOR EXAMINING THE PERFORMANCE OF ADVANCED DESIGNS

To access conceptual designs against objective standards, one must have some criteria against which to judge proposed designs. These might include the following:

• Assess the proposed designs against projected user requirements and applications for broadband space communications appropriate to the 2015 to 2025 time frame (in terms of throughput/second, access to common spectrum, and ability to respond to a wide range of broadband user applications and services envisioned for the future without requiring a "wide range of ground based user terminals that vary significantly in aperture size, cost and complexity" that would obviously retard the use of such systems and escalate the cost of user interconnectivity);

• Determine whether the proposed advanced space communications systems could possibly be financially viable in comparison to alternative communications systems. (Note: There are many requirements for future geoplatform-based services [i.e., in the mobile communications, broadcast, and multicasting services area] that cannot be provided via fiber-based systems. Thus, such financial benchmarks only indicate what the relative economics of fiber and geoplatform systems might be rather than seeking direct economic competitiveness.)

• Ask whether the designs are compatible with the relevant social, legal, liability, health, and international issues and concerns that might accompany the implementation of such a new type of system. In short, it is not sufficient to design possible future space antenna systems and deem them to be a *successful design* if they can be conceived only in terms of adequate technology and economic performance. Consideration must be given to whether it gives rise to other problems such as health standards, environmental or regulatory problems, or lack of appropriate frequency allocations. If there remain such regulatory or policy issues, there would still be major problems with the implementation of an advanced geoplatform system.

• Extend the broadband offerings and throughput capabilities provided to small, compact, and low-cost user terminals by an order of magnitude beyond today's communications satellite capabilities or more.

• Offer significant new consumer advantages in terms of lower power and healthier user terminals, longer battery life for portable terminals, wider range of multimedia services, and newly integrated telecommunications, computing, and navigational services within a single portable or compact user terminal.

• Utilize totally new types of space technologies and architecture that "embrace the concepts unique to the space environment," rather than building "earth and gravity-centric structures in space." (This last point is particularly important in that this design goal seeks to exploit the advantages that a low-gravity and high-vacuum environment might afford to new systems design concepts.)

CRITICAL TECHNOLOGIES ESSENTIAL TO FUTURE GEOPLATFORM SYSTEM

Some of the technologies that may well be necessary to create the new satellite architecture include the following considerations:

• Deployment and interconnection of various types of large antennas with "high-tensile strength and low-mass" tether structures.

• Preliminary look at the solar, gravitational, micrometeorite, and cosmic/radiation environmental effects on advanced geoplatform antenna design. Also look at requirements that are unique with regard to heat exchangers, system statics and dynamics, and power distribution systems.

• Design concepts for the large-scale space power systems needed to sustain advanced geoplatform operations.

• Advanced components and piezo-electronic semiconductor modules that may be imbedded in the polyimide materials to support needed phased-array antenna shape and long-term operations.

• Preliminary strategies for interconnectivity with regard to ground user equipment (i.e., microterminals) and efficient terrestrial interface architecture.

• Consideration of the best deployment strategy for space user equipment and antennae systems so they would be spaced far enough apart to reduce interference and intermodulation products to required levels.

Beyond these broad technical goals, however, some specific technologies may well be needed. Some of the technologies currently thought to be the most important of these are set forth in Fig. 14.3.

Although Fig. 14.3 highlights specific technologies that would likely be crucial to the realization of future advanced satellite systems, it is not totally comprehensive. Other technologies specific to particular new designs will also be necessary, and some of these are provided by way of example in Fig. 14.4.

The variously proposed advanced systems briefly discussed next would appear to offer major economic benefits. Preliminary engineering and economic studies indicate that the potential financial benefits that could be derived from such systems could be substantial. These systems, at least in concept, could provide business and scientific dividends valued at many billions of dollars (just in frequency reuse and spectrum conservation techniques alone). More refined calculations are to be developed in further studies, but it is reasonable to estimate that the value of spectrum today is at least 10% to 15% of the revenues gener-

---

Technologies for Advanced Satellite Systems—2010 to 2020

- Advanced-phased array antenna concepts
- Polyimide materials (less than one kg/square meters, including electronic components)
- Advanced materials for advanced tether structures for antenna deployment (e.g., fullerene-based or integrated carbon materials commonly known as "Bucky-balls")
- Tether structural design for forming large, passive constellations in space
- Adaptive electron beam-shaped piezo-electric membranes
- Liquid metal solid state Field Effect Electric Propulsion units
- Broadband time-delay microcircuits capable of modulation
- Large-phase shift capability microcircuits that can be modulated
- 10- to 12-bit digital samplers at gigahertz frequencies
- Integrated, self-contained silicon or gallium arsenide-based picosatellites that can be deployed in swarms
- Highly efficient photovoltaic thin films (i.e., rainbow solar cells)
- Piezo-electric bimorph membranes resistant to the space environment
- High-temperature, superconducting long wires
- Precision optical sensing of surface figures
- Integrated multifunction components
- Precision metrology for position control in formation flying constellations
- Advanced turbo-coding, multiplexing systems
- "Smart Beam" technology to allow intensive frequency reuse capabilities

---

Authors' Note: A number of technologies are applicable to each of the geoplatform concepts. Although it would be desirable to describe these technologies in greater detail and then discuss their applications, this is not possible within this short chapter. For further background, please see: Takashi Iida, Joseph N. Pelton, and Edward Ashford, (ed.), (2003), *Satellite Communications in the 21st Century*, Washington, DC: AIAA. Finally, it should be noted that Mr. Ivan Bekey, president of Bekey Designs, has been a primary contributor to the study of this subject. He identified a number of these technologies for research and is the "father" of some of these key concepts and architectures.

FIG. 14.3.   Critical technology goals.

OTHER KEY TECHNOLOGIES ON POTENTIAL INTEREST

- Advanced constellation management for tightly configured, free-flying geoplatform components
- Nano-power technologies and improved power systems technologies
- Advanced modems and super processors (especially modems that can perform in the 3 to 8 bits/Hz spectrum efficiency throughput levels) (see turbo-coding)
- Multi-Gbps optical interplatform links and optical space-to-earth links
- Stratospheric or High-Altitude Platform Systems (HAPS) for high-capacity RF communications systems with optical links to geo satellites

---

FIG. 14.4. Supporting technical goals and recommended research areas.

ated by today's satellite systems, which currently stand well in excess of $40 billion per year in global revenues.

This would suggest that the global satellite-related spectrum is worth at least $4.0 billion to $4.5 billion per year as of the 2003. If one could find a way to expand the spectrum available from space by one to two orders of magnitude, it is clear that such a development has huge economic value—perhaps in the range of $50 billion to $100 billion when viewed over a span of a decade. This value would be in addition to the economic value of being able to produce space communications throughput at costs significantly below those now required by conventional contemporary satellite communications systems.

It is thus probably not realistic to seek to make satellite systems directly cost comparative with fiber optic cable systems, but rather to design future systems that are still viable as complementary networks, especially for broadcast, mobile, remote, or very large network applications. Such new and highly cost-effective systems might be able to attain the following objectives for the future as shown in Fig. 14.5.

Future Cost Performance Objectives for Advanced Satellite
Systems in the 2015–2020 Time Frame

- Achieve by 2015 basic throughput of 2.5 Gigabits/seconds for rates of $3600/hour or $1.5 M per year (expressed in 2003 dollars).
- Achieve a 1 to 2 Terabit/second system in geosynchronous orbit at a capital cost of under $2 billion and with operating cost of under $200 M per year (expressed in 2003 dollars).
- Develop design concepts so that the power and antenna systems are sufficiently low in mass and volume and technically viable within 15 years so that a single country or corporate entity might be able to launch and operate the system individually or as joint projects.

Authors' Note: These "rough order of magnitude" objectives largely derived from the analysis of contemporary fiber and satellite cost trends. Further, the above indicate "milestone costs" only relate to basic economic costing and do not relate to the actual pricing of services offerings to businesses or consumer and certainly do not relate to the pricing of value-added offerings that would likely be at much higher levels.

FIG. 14.5. Measuring financial performance.

A number of technologies are applicable to each of the concepts discussed later. Although it would be desirable to describe all of these technologies fully in this chapter and then discuss their potential applications, this goes well beyond the scope of this chapter and the overview nature of this book. Those who wish to read further about these topics are referred to the Reference section at the end of this chapter.

PRINCIPAL ARCHITECTURE OPTIONS FOR ADVANCED SATELLITE SYSTEMS 2010–2020

The earlier technologies can be combined in various ways to yield entirely new and useful architectures for implementing advanced systems for the 2010 to 2020 time period. The principal objectives of these new space communications architectures would be:

- Large-scale frequency reuse
- Maximizing capability within an orbital slot
- Generation of large numbers of simultaneous spot beams
- Generation of very small footprint spot beams that can be steered
- Minimization of the weight and cost of large spacecraft

These objectives result in a need to implement large antennas (or clusters of antenna systems) with many spot beams whose frequencies do not interfere with each other or other spacecraft or ground systems.

There are many architectural (or rather conceptual) options that could implement one or more of these objectives. Only three conceptual architecture concepts are included in this chapter to illustrate the possibilities of which there could be many others. These basic designs, as mentioned earlier, are briefly summarized and illustrated next. These advanced systems are presented in order of increasing complexity as well as in terms of the difficulty of developing the needed new space technology required for these systems.

Geoplatform Concept 1

This design would be a constellation of antennas placed far apart in space (up to six antenna diameters) via a tether system so as to minimize interference, but sufficiently closely linked so that multiple antennas could be fed power from a centralized unit. This concept utilizes a number of conventional antennas of a low-mass design. These antennas are then formed into a large and stable constellation via passive tethers.

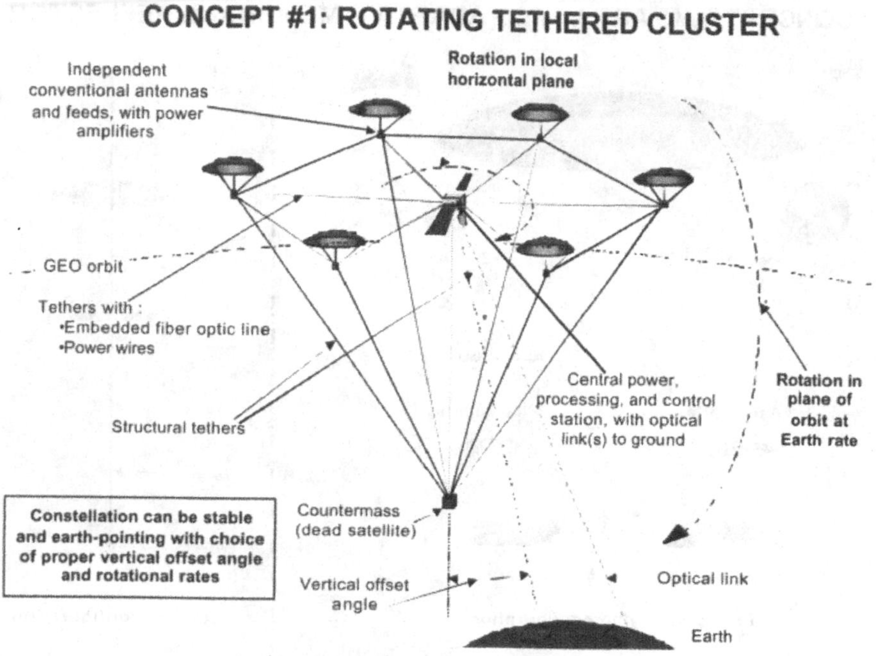

CONCEPT #1: ROTATING TETHERED CLUSTER

Independent conventional antennas and feeds, with power amplifiers

Rotation in local horizontal plane

GEO orbit

Tethers with :
•Embedded fiber optic line
•Power wires

Structural tethers

Central power, processing, and control station, with optical link(s) to ground

Rotation in plane of orbit at Earth rate

Constellation can be stable and earth-pointing with choice of proper vertical offset angle and rotational rates

Countermass (dead satellite)

Vertical offset angle

Optical link

Earth

Copyright permission granted by Joseph N. Pelton and Ivan Bekey

FIG. 14.6. Geoplatform Concept 1: A large-scale tethered platform system.

Although these antennas would be deployed only some 10 years into the future, these conventional parabolic antennas could still be quite advanced in design. The design architecture envisions that the antennas would be passively formed into a large constellation, allowing them to be separated sufficiently (several antenna diameters) so as to prevent interference while also enabling their operation either independently or in a coordinated fashion. Processing latency would remain a key issue to address in such a design. This approach is illustrated in Fig. 14.6.

Geoplatform Concept 2

This second concept of a future geoplatform would be envisioned as a free-flying antenna with feed and power systems that are not physically interconnected. This approach depends on the development of the necessary piezo-electric materials and free-flying radiators, which would allow the large-scale phased-array antennas to maintain their shape. This approach would use parabolic-type antenna systems, as illustrated in Fig. 14.7, whereas the same approach that uses more advanced-phase array antennas is illustrated in Fig. 14.8.

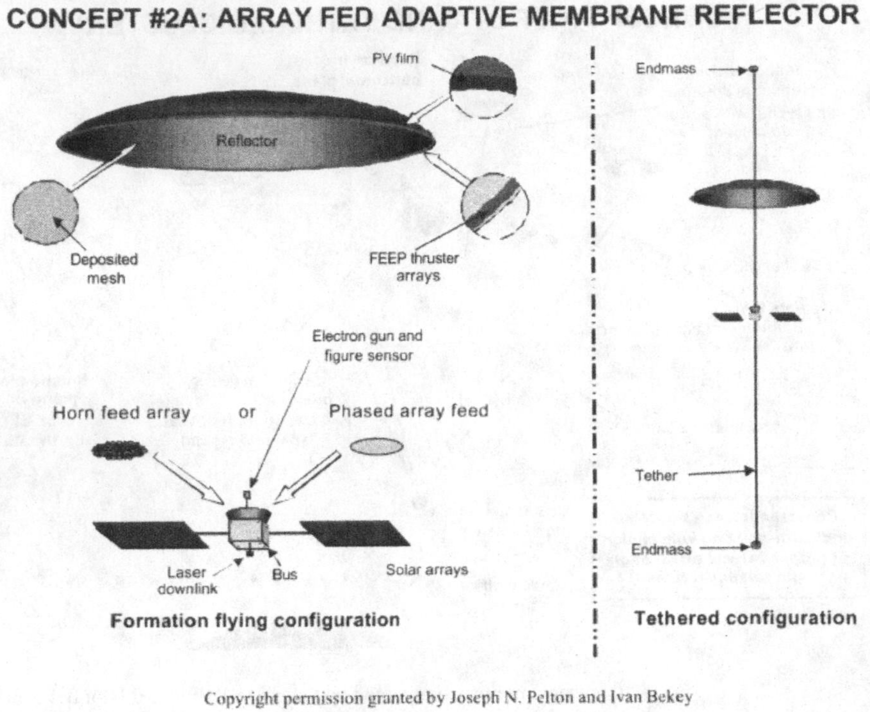

CONCEPT #2A: ARRAY FED ADAPTIVE MEMBRANE REFLECTOR

Copyright permission granted by Joseph N. Pelton and Ivan Bekey

FIG. 14.7.   A large scale geoplatform using parabolic antenna formed from piezo-electric materials.

Geoplatform Concept 3

This is the most exotic and futuristic type design, and it envisions a large-scale swarm of thousands of active-phase array elements. This approach involves not only a wide range of new satellite and launch system technologies, but also a number of policy, regulatory, and legal considerations and is certainly the longest term concept for the end of the 2010 to 2020 time period. This approach is illustrated in Fig. 14.9.

NEXT STEPS FORWARD

Research in the United States, Europe, Canada, and Japan has been directed to the development of such advanced systems at the conceptual level for some years. Although none has yet entered the demonstration phase, in the past few years there have been and continue to be technology development programs to advance the

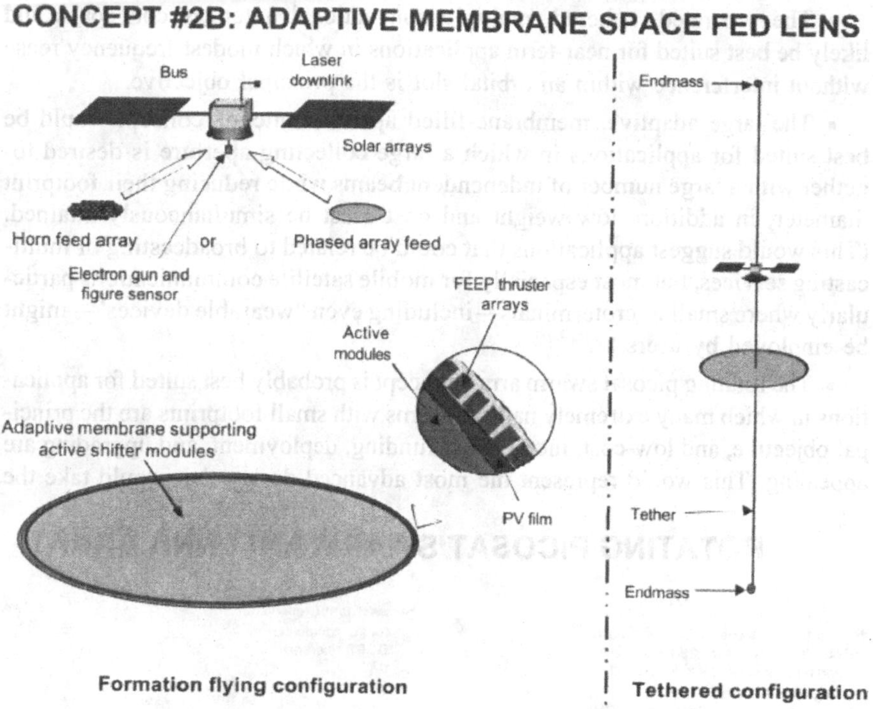

CONCEPT #2B: ADAPTIVE MEMBRANE SPACE FED LENS

Copyright permission granted by Joseph N. Pelton and Ivan Bekey

FIG. 14.8. A large-scale geoplatform using a flat antenna and phased-array components to create electronically formed beams.

key technologies to enable efficient and cost-competitive power delivery from space to the ground.

Technology programs are underway in a number of countries, led by Japan, but also being pursued in Europe, Canada, and the United States. The universal aim is to make orders of magnitude improvements in weight, cost, efficiency, and performance of phased-array, large-shaped membranes for reflection and refraction of RF and optical wavelengths, and especially the weight and cost-efficiency of solar energy conversion in space.

It seems clear that because the advances in these technologies would benefit both the communications satellite and power delivery satellite communities, they should be pursued in a coordinated fashion between the two communities, although this has not happened to date. At the least, the advanced communications satellite community should be aware and take advantage of advancements planned and attained in the space solar power area.

The three geoplatform concepts as presented earlier would each be suited for different purposes as follows:

• The separated, tethered, multiple conventional antennas concept would likely be best suited for near-term applications in which modest frequency reuse without interference within an orbital slot is the principal objective.

• The large adaptive, membrane-filled aperture antenna concept would be best suited for applications in which a large collecting aperture is desired together with a large number of independent beams while reducing their footprint diameter; in addition, low weight and cost must be simultaneously attained. (This would suggest applications that could be related to broadcasting or multicasting services, but most especially for mobile satellite communications particularly where small microterminals—including even "wearable devices"—might be employed by users.)

• The rotating picosat swarm array concept is probably best suited for applications in which many extremely narrow beams with small footprints are the principal objective, and low-cost, incremental funding, deployment, and upgrading are appealing. This would represent the most advanced design that would take the

ROTATING PICOSAT SWARM ANTENNA ARRAY

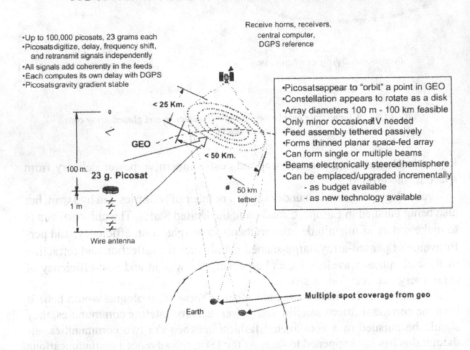

FIG. 14.9.   Very large-scale phased-array with pico satellite swarm to create ultra-high throughput satellite system for the 2020 time period.

longest to implement, but also could most likely meet the demanding economic objectives identified in Fig. 14.5.

The choice of which type of space antenna is of most interest is also affected strongly when the capability is desired because the technology development and demonstration programs required are different for the three concepts. In addition, the interface with nonantenna subsystems of the spacecraft is also different for the three concepts and could have major effects on the attractiveness and practicality of any of them. Likewise, different types of user terminals for ground, sea, or air application would be appropriate for the different geoplatform concepts. Further, Concept 3 also gives rise to the most regulatory, policy, and legal issues, but fortunately there is also the most time available to address such concerns.

In addition, it must be borne in mind that these three are only concepts. Although all have been subjected to conceptual feasibility study and each is deemed feasible, none has received even preliminary engineering analysis, let alone the disciplined detail considerations that will arise in dedicated system design studies. Thus, the specifics of each option and the choice among them cannot be made until much more preliminary engineering study is undertaken for each.

AN ALTERNATIVE CONCEPT: HYBRID OPTICAL COMMUNICATIONS WORKING IN TANDEM WITH STRATOSPHERIC PLATFORMS

An entirely different approach to future satellite communications involves the idea of having space to High-Altitude Platform Systems (HAPS) via optical links and then using RF service (probably in the Ka bands or above) to provide broadband services to cities. Such systems would be capable of high system capacities with high cost-efficiency and with a highly desirable feature of low transmission latency. The driver of such designs is the need to have larger-scale access to spectrum. This approach also reduces transmission latency, at least for localized telecommunications, multicasting, and broadcasting applications. This concept could also be coupled with the idea of establishing a high-capacity optical ring that could move traffic around the world with high efficiency and low cost.

The basic concept of HAPS linked to an optical satellite is shown in Fig. 14.10. In this type of hybrid network, it seems feasible that optical links could be used to connect with platforms sufficiently high up into the stratosphere to be above most of the atmospheric conditions that create the greatest amount of interference. Such a hybrid satellite and stratospheric platform system could allow the interconnection of dozens or even hundreds of HAPS. Ultimately, with enough satellites and HAPS, it would be possible to provide mobile and broadcast services and create a global information network capability. If the HAPS could be deployed at altitudes as high as 20 to 21 kilometers (or about 12 miles), for instance, this might be a suf-

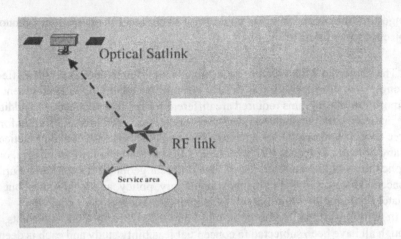

FIG. 14.10.   Hybrid satellite and stratospheric platform telecommunications networks.

ficiently high altitude to remain largely above atmospheric disturbance and rain attenuation that makes space to Earth communications via optical links today so very difficult.

Further, the HAPS would be equipped with enough on-board power to operate at broadband EHF frequencies effectively. This is particularly true because the look angle would be highly favorable and the path loss considerations minimal for the high-gain, multibeam, phased-array antennas that could be deployed on board HAPS by 2015. The detailed engineering of such a system cannot be presented in depth here, but a preliminary review of such a hybrid system is presented in Fig. 14.11.

WHY DO WE NEED ADVANCED SYSTEMS IN THE FUTURE?

There is always the question of why we need to develop advanced systems for the future. The demand for telephone, TV broadcast, and other conventional services has peaked and is now growing at only a modest rate. New Internet, multimedia, video streaming, and other IP-related services are, however, growing rapidly and seem likely to do so for some time to come.

The next decade seems likely to produce unprecedented growth in broadband mobile services led by third-generation Personal Communications Sources (PCS) and followed by so-called fourth-generation services. In addition to this service demand, there appear to be a host of new telecommunications and IT applications that are an off-shoot of new antiterrorist applications, ranging from increased telecommuting and electronic decentralization to totally new systems for monitoring aircraft, vehicles of all types, and various types of infrastructure. This new

---

Key Observations on Hybrid HAPS/Geo Satellite Network

- High-altitude platform systems would function as optical/RF transformers. (Thus, an optical signal can be sent to and from a geosynchronous satellite from a stratospheric platform.) Such a hybrid system largely overcomes problems of latency for regional and intracity communications, but interregional and international communications would still experience some 250 milliseconds of delay.
- HAPS, in comparison to geosynchronous satellites, substantially reduces problem of rain fade in Ka band and above because of high power, favorable look angle and reduced path loss (i.e., about 20 kilometers vs. 35,800 kilometers altitude).
- Optical communications satellites can be deployed in multiple locations for space diversity, and thus direct space-to-earth and earth-to-space links would be available as an emergency "fail-over" system.
- High levels of frequency reuse can be achieved by HAPS (with well over 1,000 beams and up to 300 fold frequency reuse) for such applications as broadband mobile communications. Simpler systems could be deployed for entertainment and broadcasting applications.
- HAPS could serve localized needs, and optical com satellite could serve interregional needs.
- High-altitude aircraft have now been developed, and prototypes are in operational testing, thus representing a higher state of R&D and feasibility testing than most geoplatform concepts.

FIG. 14.11.   The advantages of a hybrid satellite and HAPS network.

form of electronic monitoring will likely include water supplies and systems, energy plants including nuclear facilities, chemical plants, pharmaceutical production facilities, and other critical locations.

The past 150 years in telecommunications has seen an ongoing exponential increase in both transmission capabilities and service demand as shown in chapter 1. The increasing amount of information in the global society suggests that the demand for new, broadband, and more flexible and mobile services will continue to increase.

Beyond these historical patterns of development, it seems likely that other goals will also emerge with new demands for security, privacy, and quality of information. The new communications systems will not only demand greater throughput, but also various forms of preprocessing to screen out unwanted information. They will also target time-critical applications or create alarms or system alerts with regard to abnormal or "implicit danger signals." Thus, information transfer will increasingly merge with end user applications, including improved ways to archive and store information.

The advanced satellite technologies will more than ever need to be integrated with other media to achieve seamless connection with servers and databases around the world. Scientific, business, entertainment, educational, and health applications will all drive the demand curve; to stay relevant, satellite networks must evolve with other electronic and optical technology.

In particular, the interface to information systems in the home and office as well as at the individual user level (i.e., microterminals) is ultimately where the future development of the satellite industry will hinge. It seems likely that not only the new technology in space, but also low-cost, mobile, and compact personal user terminals are the key to the future. The availability of these terminals will ultimately define and enable a host of new applications that can now only be guessed at by today's engineers.

Prior to the Internet, we had a hard time envisioning such applications as Web mapping, video streaming, or even e-mail. The new satellite applications two decades hence are equally difficult to envision. The types of technology described in this chapter could conceivably make these new applications lower in cost, more mobile, and perhaps even more secure.

ACKNOWLEDGMENTS

The authors of this chapter wish to provide special acknowledgment to Ivan Bekey of Bekey Designs, whose concepts of advanced satellite systems are represented in Figs. 14.6, 14.7, 14.8, and 14.9 and were originally developed for a research study of advanced new satellite design concepts for the Communications Research Laboratory of Japan and carried out by the Space & Advanced Communications Research Institute (SACRI) at George Washington University.

REFERENCES

Iida, T., Pelton, J. N., and Ashford, E., (2003), *Satellite Communications in the 21st Century: Trends and Technologies*, Washington, DC: AIAA.
McAleer, Neil, (1992), *Arthur C. Clarke: The Authorized Biography*, Chicago: Contemporary Books.

BACKGROUND READINGS

Abe, George, (1993, November 15), "The Global Network," *Network Computing*, pp. 48–53.
Bekey, I., and Mayer, H., (1977, April), "1980–2000: Raising Our Sights for Advanced Space Systems, *Astronautics and Aeronautics*, New York.
Bravman, J., Fedak, J., and Giffin, G., (1999, June), "Infrastructure on Demand—Global Fiber-Optic Capacity from EHF Satellite Communications Systems," *Proceedings of the Fifth International Workshop on Satellite Communications, SS/GII*, Ottawa, Canada.
Cuccia, C. L., Lovell, R. R., and Edelson, B. I., (1987, May), "The Evolution of the Geostationary Platform Concept," *IEEE Journal on Selected Areas in Communications*, Vol. SAC-5, No. 4, pp. 601–614.
Edelson, B., and Morgan, W., (1977, September), "Orbital Antenna Farm," *Astronautics and Aeronautics*, pp. 20–27.

Iida, T., and Suzuki, Y., (2001, April 17–20), "Communications Satellite Research for the Next 30 Years," 19th ICSSC Conference, Toulouse, France, AIAA, Paper No. 233 (see also *Space Communications: An International Journal*, IOS Press, No. 3, Vol. 17, No. 4, 2001, pp. 271–277).

Landovskis, J., and Rayment, S., (1999, June 15), "Emerging Broadband Satellite Network Architecture and the Major New Developments," *Proceedings of the Fifth International Workshop on Satellite Communications in the Global Information Infrastructure (SS/GII)*, Ottawa, Canada.

Pelton, J. N., (1998, April), "The Future of Telecommunications," *Scientific American*, pp. 80–85.

Pelton, J. N., MacRae, A., et al., (1999), *Global Communications Satellite Technology and Systems*, Baltimore, MD: World Technology Evaluation Center.

Stuart, J. G., (1994), "Distributed Parameter Control of a Large, Flexible, Stiff-Membrane Phased Array Using Thin Film Piezo-Electric Actuator Patches," Unpublished doctoral dissertation, Dept. of Aerospace Engineering, University of Colorado.

PART SEVEN

SYNTHESIS AND CONCLUSIONS

CHAPTER FIFTEEN

Trends for the Future: Telepower Opportunities and Teleshock Concerns

Joseph N. Pelton
Robert J. Oslund
The George Washington University

> *If the human species is to achieve success as a species, i.e. continuous development for an eon, that is to say a billion years, the most severe test will come in the 21st century.*
>
> —Joseph N. Pelton (2001)

Modern technological progress has often been described as a "one-way gate." Once major innovations occur—regardless of whether they are helpful, harmful, or even both—they cannot simply be undone. They are almost always here to stay. Unlike the plot in "Terminator 2," we cannot send people or machines from the future back into the past to wipe out technological errors. Spandex, nuclear weapons, birth control pills, Viagra, broccoli, indoor plumbing, musical videos, and chemical and biological weapons, in one form or another, are probably all here to stay.

It is easy to identify the many ways that satellites have improved global society, human life, and commerce. We can also glibly project many ways in which satellites will continue to aid humanity in the decades ahead. Yet like virtually all technologies and industries, satellites represent a double-edged sword. In the years ahead, satellites will not only bring us new telepower, but also teleshock or negative impact on our global society.

Satellites continue to transform our world in new and powerful ways, from peacekeeping to missile targeting, from global news to porn. Satellites represent what might be called an artifact of an ultra-information society, where services dominate economic growth, and knowledge and information represent ultimate

341

power. This new way of looking at information networks and 21st-century human enterprise can be variously described as the "worldwide mind," "global consciousness," "the E-Sphere," or the "instantaneously interconnected world." Not only is today's world remarkably different from just a century ago, it will become even more different as the pace of change accelerates like a missile that blazes into the future with an infinite number of short-term stages.

TELEPOWER TRENDS DRIVEN BY SATELLITE AND RELATED TECHNOLOGY

To sum up how satellites have changed or will continue to change our world, here are our projections as to the top 10 telepower trends. In many ways, these trends will be powered by satellite systems of the 21st century, as well as by their "kissing cousins" information technology, computer networks, and broadband terrestrial telecommunications systems.

Globalism

The world is a much smaller place today than 40 years ago, when the age of satellites began. These amazing electronic relays in the sky expand the reach and power of all other electronic devices and media to the most remote corners of the earth. This extending power and global impact of satellite systems now applies to the telephone, fax, radio, TV, the Internet, the World Wide Web (WWW), electronic newspapers, digital photography, computer security systems, CD-quality music, high-definition videos, local and wide area networks, multimedia encyclopedias, and interactive digital TV. Everything from astrophysics to "I Love Lucy" from bigotry and porn to religious ceremonies at the Vatican are instantly available via satellite-linked modems even in the rainforests of Brazil, the frozen shores of Antarctica, or the middle of the Saharan desert.

Certainly fiber optic cable, coax, terrestrial wireless, and other transmission systems represent a part of the mix, but satellites currently create the largest pool of directly linked broadcast and electronic networking switched systems on a planetary scale. All the rest of our telecommunications nets are point to point and often more nationally or locally focused than global in scope.

Contemporary and future satellite systems now allow and will further enable a true anytime and anywhere broadband reach. Only satellites allow over 200 countries and territories to achieve instantaneous linkage via telephone or the Internet. Well over 100 million satellite terminals located at, in, or on offices, homes, ships, airplanes, trucks, trains, buses, and cars service a global cast of users.

These terminals are now interlinked via an amazing combination of broadcast, fixed, navigational, and mobile satellites. In a decade, that number of terminals will likely grow into the several hundreds of millions. In another two decades, that number might even exceed 1 billion and be so pervasive that we may own several satellite terminals—one of which we may "wear" within a suit coat or on our arm like a wristwatch.

Within the next 20 years, satellite user terminals will continue to shrink in size and cost and "disappear" within our clothing or briefcase or into a hand-held personal data assistant. A flip-down eye loop may even provide us access to "live via satellite" multimedia information from anywhere on the planet or even outerspace colonies. These new miniature terminals will provide an ease and convenience of communications that can only be imagined today. Certainly "wearable" satellite terminals or clip-on video-viewing units (i.e., eye-loop devices) may represent the next key satellite capability. This may seem a bit like Buck Rogers science fiction fantasies, but prototypes of microterminals that look like a clothing accessory are under development now for battlefield soldiers.

Governments and corporations that seek to control or limit the exchange of information will have a most difficult time of it when over 1 billion terminals are in use and these tiny transceivers cost the same as personal data assistants cost today. Neither totalitarian regimes nor isolationist cultures, for better or worse, will be able to remain aloof in an era of true electronic globalism. National borders will not be able to withstand the onslaught of global information distribution systems.

Both privacy and isolation will be hard to achieve in a world where mass consumer satellite services are pervasive and affordable. Within 20 years, teleeducation and telehealth services will penetrate to even the most remote regions of our planet. Even today, in remote parts of China, Indonesia, India, Malaysia, Brazil, and Nigeria, satellite teleeducation and even some forms of telehealth are routine occurrences. Global business via satellite and fiber is already commonplace, with planetary rates of electronic funds transfer projected to be near $400 trillion (U.S.) during 2003. As impressive as the growth rates in satellites services around the world has been over the past four decades, this is nothing compared to what will soon be. The exponential growth in e-business and electronic services could increase by an order of magnitude in the next 5 years. Satellites are on the front line of global growth—financially, socially, and culturally.

Telework

Fundamental shifts in patterns of work and corporate organizations are occurring in the United States and around the world. Change is now occurring at an amazing pace. As only one example, Hewlett Packard Company, together with its Compaq operations, has some 40,000 teleworkers at home or at remote work sites who are continuously drawn into flexible electronic teams formed around the world to carry out the tasks of the day. Although these efforts began through IT networking based on terrestrial cable technology, the flexibility of satellites provide the wireless networks of the future. Personnel from Boise, Idaho to Singapore, from Fort Collins, Colorado to Kuala Lumpur are available to each other at the touch of a key or multidigit call numbers via IP networks. In Japan, NEC continues to expand its network of remote telework centers to bring its workers together electronically because it can no longer afford to build new workspace in the heart of Tokyo.

Today, the Yankee Group consultants estimate there are over 15 million teleworkers in the United States. They and other market forecasters such as the Gartner Group and Nielson's expect this number to rise to above 50 million by year-end 2010. Similar patterns of growth are expected in Europe, Canada, Japan, and elsewhere around the world. Because of their flexibility of interconnection; universal coverage; increasing ability to support IP-based LANs, MANs, and WANs; and ever lower cost user microterminals, broadband satellite links are well positioned to fuel this trend.

There are also somewhere over 1 million international teleworkers (or "electronic immigrants" that serve as worldwide teleworkers). Twenty years ago, the cost of labor on an hourly basis was well below the cost of international tie-lines, but now the cost of international information networks is a small fraction of the cost of skilled personnel. We see a continued growth of "lone eagles"—people who decide to locate in beautiful and scenic areas in Colorado, Hawaii, or Switzerland. Service providers from their electronic perch can provide information or consulting services around the world via broadband Internet/IP connections. This can be easily accomplished via dedicated networks now offered through Eutelsat, Intelsat, Starband, Telesat, Panamsat, Hughes Network Services, SES Global, or Gilat/GE Spacenet services. Tomorrow they can turn to Spaceway, iPstar, Wild Blue, New Skies, Echostar, and a host of others that can provide electronic-based IP services, including multimedia links, at incredibly low prices.

Workers in many dozens of "emerging economies" are electronically linked via fiber or satellites to the United States, Europe, or Japan to provide teleservices to and from remote locations. Insurance companies in the northeast part of United States have teleworkers in the Caribbean providing word processing and data entry. Airlines hire local workers in Barbados and Jamaica to provide inventory control, while Japan hires computer software development engineers in India, Pakistan, or South Korea to provide key services at lower salary costs. Programmers from Bangalore to Lahore, from Seoul to St. Petersburg are essentially providing data and IT services around the world live via satellite. The question is whether new broadband satellite networks will beget telework and more worldwide teleservices or whether the demand for telework-based services will beget new generations of satellites.

New 21st-Century Paradigms—The Worldwide Mind and "Knowledge-Based" Organizations

The difference between the 20th and 21st centuries is only beginning to be discernible, but one of the mega-trends that is beginning to emerge is the decentralization and reconstitution of corporations, organizations, and governments in new and different ways. Satellites and broadband networking systems are the tools that ultimately allow this shift to occur. A century ago, corporations had a clear national identity, a specific "geographic center and home headquarters," and most of its employees were clustered in a specific region. This approach to corporate, organizational, or governmental organization is no longer necessary and actually is quite inefficient from the perspective of energy consumption, air pollution, real estate costs, transportation planning, or antiterrorist planning.

New knowledge-based organizations do not have to be located in a single building, state, region, or country. The realities of the 21st century are higher energy costs, diminishing oil reserves, higher pollution levels, increasing cost of urban real estate, higher urban infrastructure and office costs, and increased costs and tensions related to terrorism. In light of the spread of broadband IT and multimedia networks, falling telecommunications costs, and multigigabit/second fiber and broadband satellite networks, the urge to amass workers in gigantic high-rise centers will continue to diminish in coming decades. All of these factors and more give rise to a new 21st-century concept—the "knowledge-based" corporation, organization, and government.

This shift will change the nature of corporations. One well-known analyst of corporations, Peter Drucker (1998) suggested that for the United States and other advanced economies, an incredible shift will come. Drucker predicted—and the trend is already well established—that as we move toward the worldwide mind, key patterns of industry size and composition will shift. Today, 80% of the U.S. workforce is employed by large industries that are numbered in the few thousands. In the future, he foresees that 80% of the workforce will be employed by much smaller corporations employing only 500 workers or less (Drucker, 2002).

The Pentagon, the World Trade Towers, and 100-plus story skyscrapers of New York, Chicago, Kuala Lumpur, and Shanghai represent a 20th-century mentality. The new 21st-century paradigm—empowered by satellites and related technology—represents a move toward "distributed networking," "the worldwide mind," and smaller "knowledge-based" corporations. Thus, we will move toward a new sense of telegeography or telepresence. These new ways to develop our cities, homes, and work environments will make much greater sense than what we have done in the past—in terms of economics, environment, and security.

Even fiber optic networking that is static and inflexible does not truly meet future information needs. Fiber does not support mobility and complete global interconnectivity, and thus will ultimately be recognized as a 20th-century paradigm. Visions of the future from Star Wars to Star Trek help us to think beyond "wire technology." This does not mean that fiber will not be a key part of the future. It just means the continued evolution toward the "Pelton Merge." This is the new digital environment where fiber, coax, and computer-based MANS and WANS will be integrated with satellite and wireless technology to support complete flexibility of broadband access anytime and anywhere.

A Liquid World Economy

A few decades ago, new electronic funds transfer systems such as SWIFT, CHIPS, and so on were invented within the world banking community to move money more efficiently across countries and around the globe. The driving benefit of such systems was to reduce the float of capital that was unavailable for use while checks were being cleared through banking systems. Today, we understand that the benefits of electronic banking are far more extensive than just reducing floating cash. The entire world of banking has been revolutionized. It is not only more efficient and faster, but these systems are also more global and integrated with the world economy. Now with the Internet and the WWW, EFT systems are increasingly integrated with the new world of e-commerce and e-trade. (In time these systems can be made even more secure than they are today. In fact we need to be more concerned about the security of electronic banking systems against terrorist and criminal attack. Public keyed infrastructure (PKI) systems are really not very secure and urgently need to be upgraded.)

Since 1997, the value of electronic fund transfers (EFTs) has soared from below $50 trillion (U.S.) to near $400 trillion (U.S.) as of the end of 2003. The combined economic product of all the countries and territories of the entire world is only about $70 trillion (U.S.). These statistics alone should underscore the true importance of transnational EFT. Satellite, wireless, and cable-based EFTs represent an economic machine that is at the hub of global enterprise. Such electronic cash is therefore central to the idea of an emerging worldwide mind. Without the satellite and fiber infrastructure to support global EFT, the world economy would grind to a halt. A frontal assault on this system by cyberterrorists could be a nearly lethal blow to global economic prosperity.

Coping With Terrorism in a Post-9/11 World

The mindset of the Western or developed country has clearly changed in the wake of the terrorist attack on the World Trade Center twin towers in New York City. All of the countries of the Organization for Economic Cooperation and Development (OECD) have begun to recognize their vulnerability to terrorist attack. These can come against almost any centralized facilities, regardless of whether these are skyscrapers, sea or airports, food and water supplies, nuclear power plants and hydroelectric turbines at major dams, transportation systems, or communications systems.

After 9/11, there is a heightened awareness of the vulnerability that centralization brings to modern society. This vulnerability applies not only to man-made terrorism, but also to natural disasters such as typhoons, hurricanes, floods, and earthquakes. There is increased awareness that overcentralized facilities are the most vulnerable to attack. As a corollary, electronic systems now allow governments and corporations to distribute core assets and people in new and innovative ways to make them more secure. So far corporations have been more adept at adjusting to these new realities than governments, but here too adjustments are being made.

The great "Kobe earthquake" in Japan not only killed many people, but also wiped out transportation, fiber optic communications systems, and key infrastructure for many months. This event changed thinking about vital communications in that country. The government decided to place satellite Earth terminals throughout the country at Post Office locations to ensure that terrestrial communications in the form of fiber and coax would be backed up in times of disaster. Now over 4,000 VSAT satellite terminals are in place, and soon 10,000 of these microterminals will provide Japanese with a level of communications security in the wake of a natural disaster that far exceeds the capability of any other country. The United States, Europe, and other countries look to the Japanese model as they move to provide national telecommunications network satellites and High-Altitude Platform Systems (HAPS) to meet the consequences of terrorist attack or natural disaster. Backup data centers, distributed wireless and satellite facilities, and a new emphasis on the planning of "telecities" to replace classic mega-cities like New York City and London will likely represent a key trend of the 21st century. Unfortunately, it appears that yet other disasters must occur before the U.S. and other governments truly recognize the need to provide this form of protective backup.

Today, most communications providers and satellite system operations provide "managed network services" to offer not only enhanced security, but also protection against natural disasters and terrorist attack. Some basic issues that will be considered more carefully in the future include the following:

1. Reviewing key facilities to ensure that they are not clumped together, built in flood zones, or found in structures that are not hurricane and earthquake proof. (Verizon, for instance, put much of its wireless antenna systems for Manhattan on top of the World Trade Center, and in Florida the construction of several teleports has been allowed in flood zones.)

2. Strategic planning so that government and corporate offices will be in locations that are not only more decentralized, protected, and lower in cost, but also where their data and communications centers will be more protected.

3. Giving greater attention to disaster and terrorist risks and the implications for designing better and "smarter" airports, transportation, and energy systems. In short, there are many key facilities that need to be conceived in different ways and designed to be "smarter," more distributed, and more resilient.

Telemedicine, Telehealth, and Teleeducation

The technology that makes possible the delivery of medical, health, and education services more effectively over long distance and to remote areas must not be confused with the core objectives. The new way to look at governmental services is to seek ways to deliver more and better services to more people. A corollary is that one must use the most effective technology and system to deliver services more effectively and with the most up-to-date information. There will be more people to educate and in need of health care in the next two generations of humanity than have ever been educated or treated since the beginning of human civilization.

Satellites can and will assist in this great challenge, but the tool must never be confused with the social need. Satellites, however, allow for services to reach the most remote and isolated spaces and with the most up-to-date information. In some cases, satellites can also conserve substantial amounts of energy and provide financial savings, but this should be considered a side benefit rather than the object of telecare, telehealth, and teleeducation. In looking to the future, there are many things that satellites can do better, faster, and cheaper. The smartest civic designers will focus on better and faster, and let cheaper evolve naturally—and in terms of longer term life cycle operational costs.

Bridging the Gap

For the last half of the 20th century, there have been broad concerns expressed in terms of the economic and information gaps among the world's nations. There have also been continuing charges of neocolonialism and exploitation of emerging nations by multinational corporations. The advent of the World Trade Organization (WTO), free trade coalitions, and e-commerce has tended to increase rather than ameliorate some of those concerns. On the one hand, international trade originating from developing countries (or emerging nations) is increasing as a percentage of the world's total, but much of that trade can be traced to multinational organizations. Although teleeducation, telehealth, telemedicine, electrification, microbanking, and sustainable development programs seem to offer new hope and opportunity, the concerns remain. Satellites can help bridge the digital divide if planned and implemented in the right way.

The roughly 2 billion people in developing nations that lack educational programs, health care, electrical power, or potable water must be considered a challenge not only for the poorest countries, but for the stability of the world's economic and political systems. What seems clear is that wireless communications and satellite networks often represent the best hope for enabling sustainable development in the emerging countries of the world.

Fiber optic networks and high-end computers do not address the basic needs of many emerging economies. Satellites allow south-to-south communications and programs and facilitate teleeducation, telehealth, and growth of services in developing countries. The spread of e-commerce, the Internet, and other modern services will most likely come via wireless and satellite technology in the world's poorest or most isolated nations. Certainly satellites and low-cost VSAT technology are not a panacea, but they do represent solid ways to address basic educational, health, cultural, social, and economic needs that can be found in Africa,

Asia, South and Central America, and the Middle East. It is key to recognize that the critical pathway to success in the delivery of these services relates to the educational and health content, not the delivery system.

The Visualized World

Throughout the Medieval period of human history, virtually all knowledge and learning remained the preserve of priests and scholars in monasteries and other elite centers of learning. Only 1 person in a 1,000 had access to higher learning and scholarly knowledge. The Renaissance, the printing press, specialized economic and social systems, and the modern nation-state changed the world and created what Teilhard de Chardin called the *noosphere*. This is a condition where knowledge, learning, and wisdom spread throughout the land and over time via many communications channels, and access to information begins to transcend social class and material wealth. The birth of democratic systems; economic, technological, and scientific invention; and competition continued the trend that was extended even further with the end of the colonialist period.

Today, information available to our global society is accelerating at exponential speeds. Electronic networks and satellites interconnect even the most remote areas. The nature of learning, communications, and social interchange is being totally transformed via distance learning, international scientific collaboration (i.e., big science), access to the Internet, and the global reach of corporations and electronic commerce.

Over time knowledge and information made several transitions. First, hunter-gathers "invented" agriculture and farming, and this led to permanent human settlements. Next, the "lore" and "clan culture and knowledge" of the tribe were transferred to new generations by oral tradition, but in systematic forms of instruction. This led to refinement of language, writing, and learning and eventually led to books, plans, and diagrams. Most recently, we entered the electronic age, where written information and broadcast information, electronic memory systems and visual systems are not only preserved, but educational and learning processes allow interexchange of information and building of new information at unprecedented speeds.

The amount of information available to society increased even as proprietary information was "protected" and made "secure." The rate of speed in the circulation of information accelerated. Thus, new ways of obtaining, storing, and sharing information have evolved. Languages are transformed, "streamlined," "shortened," and, in some cases, have even faded away. Thousands of languages have been lost in the last few decades. The final trends regarding the development, sharing, protection, and refinement of information in the 21st century are still evolving and are yet to be fully understood. One thing that is now clear is that satellites and broadband fiber optic networks will make the world ever more visually oriented and less text oriented. We have yet to understand the full implications of this change. Yet just as more and more information depreciates the value of any one intellectual item, more and visual information also likely depreciates both visual and text information.

The patterns by which information is circulated in modern "electronic culture" is changing, and quickly so. This changing pattern in the distribution of information appears driven by many factors. These factors include: global reach; increasing speed and volume of the information being circulated; magnitude and educational level of the populations of modern countries; ownership patterns of media by private and public entities; interaction of "news and information" with sports; entertainment, and cultural content; and the increasingly low cost of computing and communications. Satellites impact many of these factors and certainly increase the scope, size, and speed of information distribution.

The end result is that information is not only becoming increasingly visual, but almost randomly prioritized in terms of its meaning, importance, or relevance. We see some unexpected results that stem from multichannel TV, movies, multimedia distribution, the Internet, and electronic networking systems. These include such factors as "global culture and information production," information overload, decreased vocabularies, educational systems that create more and more specialization, and fewer broadly educated generalists or systems analysts. The meaning of *visually dominant* information systems and reduced reliance on the written word will accelerate the speed of change in the world of the global consciousness.

The speed and low cost of modern transmission systems such as fiber, coax, terrestrial wireless, and broadband and broadcast satellites certainly appear to be major drivers in this transformation—not only in the most developed countries, but around the world as well. The fact that TV entertainment, sports, movies, and multimedia-based games and e-commerce now dominate the electronic channels, not only for traditional telecommunications networks, but now also on the Internet, appears to represent a fundamental shift in the world today—with much more to come tomorrow.

The "Everywhere Internet" and Digital Convergence

A key part of this transformation in the production, reception, and use of information appears to relate to two closely related phenomena. Digital convergence is merging together telecommunications services (telephone, fax, telex, multimedia) with electronically networked (or IP) services that now include e-mail, the WWW, IP-based voice, multimedia, and video on a real-time basis. Further, the user terminals of the future will increasingly follow the "digital convergence" route. This will be in the form of the so-called "I-Phone," which provides mobile voice, data, e-mail, and Web access in the form of third-generation PCS services. Now "Wi Fi" broadband wireless services (standard 802.11g) are growing at exponential speeds, and the major new allocation of rf spectrum to this service at the 2003 World Radio Conference will only speed this trend. Closely paralleling this trend will be the development of improved software so that new "digital user terminals" will also combine all the other desired information services such as computer functions (i.e., computing, spreadsheets, graphic presentations, Web mapping, emergency alarm and rescue, etc.) and ultimately GPS space navigation services as well.

In this new world, the line among communications, computing services, navigation, emergency services, news, sports and entertainment services, and e-commerce operations will largely dissolve and become irrelevant except in terms of how these activities are paid for through subscription, per minute access, advertisement, or other mechanisms. The entire business and economic models for telecommunications industries, computer networking, e-commerce, emergency, sports, news, and entertainment industries could well be thrown into a tail spin. Already the digital video recorder (DVR) industry and third-generation PCS user terminal devices are merging markets and confusing business models as never before. With a DVR one can thus bypass cable-TV set-top boxes, record, and watch TV shows suited to your own schedule and often with commercials zapped out.

With a "smart" third-generation user terminal, one does not need a computer or TV to access the Web, get and send e-mail, watch TV, access your schedule or address book, or many other functions. Because of their broadband capabilities, mobile access, and diversity and flexibility of network design, satellites can help make all of these digitally converged services available to users with increased speed and ease of operation.

Satellites and the Global Environment

In many cases, if one steps back and looks at the digital satellite revolution, one finds that it serves to make more information available faster, more flexibly, and at more access points from more varied types of media and applications. The question thus becomes—how does it make the lives of humans better and more sustainable? In terms of political strife, warfare, and terrorism, these tools can also serve to provide greater security and homeland defense. The most important application of satellites may be to help us monitor, protect, and rescue our global environment.

The various activities such as "Mission to Planet Earth"; the Earth Observation Systems; and commercial remote systems such as Spot, Ikonos, Earth Imaging, and so on now serve to provide us with increased data about patterns of air, water, land, polar, glacial, and wetland pollution. Increasingly, sophisticated, high-resolution, multispectral sensors; side-looking and synthetic aperture radar; infrared, thermal mapping systems; and other devices allow us to collect data from all over the world. Communications satellite technology, however, is key to transferring these data from remote sensing satellites to data analysis centers and from there to end users wherever they may be.

Satellites now are key to linking scientists together to carry out big science projects and support super-computer processing to help protect against earthquake, tsunami, typhoons, hurricanes, volcanoes, and forest fires. The key is not only to do this effectively and at low cost, but to provide the coverage and bandwidth to be able to do it quickly and accurately, and to allow the process to be increasingly decentralized so that the right person or organization can obtain the right information at the right time.

The prior list of mega-trends driven by satellite communications (or at least seriously impacted by them) could be continued at some length. Yet the list is indicative of the importance and significance of modern communications in our lives. Not until a "Galaxy Satellite" that supported paging for a large number of U.S. doctors failed (as happened in 1998) did people realize how central to their lives satellites had become—and also how hidden they are in hundreds of important applications.

The previous "top 10 list" highlights developments that are generally perceived to be "positive" or "telepower" trends. Not all of the trends that are driven by satellites can be considered to be on the plus side. The power of satellites can cut two ways—for positive or negative results. The areas of concern, or "teleshock" factors, must also be considered.

TELESHOCK EFFECTS OF COMMUNICATIONS SATELLITES

The ambiguity of technology and its impact can be seen almost everywhere. Most would start with the assumption, as presented in the earlier chapter on the benefits of satellite communications, that more and better information at lower cost would be a positive. However, there are limits. The torrential outpouring of information that now comes to us electronically has given increased concerns about information overload and the overall depreciation of the value of news and information. The huge amount of data on the Internet seems to be a good, but what about information that is out of date, passé, wrong, pornographic, criminal in intent or purpose, or supports the work of terrorists? Clearly almost every aspect of communications satellite technology and application has a plus and minus to it.

The Internet and Computers Versus the Loss of Language and Culture

The dominant language in the world today is English. This is because of the huge number of movies, TV shows, sports programming, computer software, and Internet Web sites that are programmed in English. The government of Iceland is desperately seeking to convince Microsoft to produce an Icelandic version of Word because it fears the death of their many thousands of years old language if it is not put onto local computers. Satellite distribution of media around the world is creating the predominance of a few world languages such as English, Chinese (Mandarin), French, Spanish, Japanese, Arabic, and so on.

Within another half century, thousands of the world's languages will have been rendered obsolete. Of the 100 most visited Web sites in the world, 40% of them are in California and 96% of them are in the United States. Satellite-based ISPs and DTH satellite TV are most assuredly changing the linguistic map of the world and at an increasing pace. It is not possible to stop change and innovation and create an "anthropological zoo," where dead languages are kept on display like trained seals. Nevertheless, the power of satellites, media, and the Internet to carry out a form of linguistic colonialism can no longer simply be ignored. It is an important global cultural issue.

Teleterrorism, Telecrime, Telepornography, and Telewar

Freedom and democratic governance ultimately must rest on free speech. Nevertheless, the hazards of global freedom of speech via satellite and the Internet must also be recognized. One can now, through the global reach of satellites and fiber-interconnected networks, reach almost any sort or type of information. The Internet is not a network, but an interconnected maze of some 100,000 networks, many of them programmed by a single individual or small group. This programmer may well be a terrorist, extortionist, "electronic bank robber," pornographer, distributor of computer viruses, neofascist, racist, or some other purveyor of hate, intolerance, or criminal behavior. Satellites are not the cause of this hateful or harmful information, but it is most definitely an amplifier that allows potential access to this information to a global population that ultimately may represent billions of people.

Super-Speed Living and Information Overload

As the 21st century begins, humans are working fewer hours. This ranges from nearly 55 hours a week in South Korea and Taiwan to 35 to 32 hours a week in France. Meanwhile our computer networks, banking systems, e-commerce networks, stock markets, travel agencies, transportation-control centers, security systems, and much, much more are moving to 168-hour work weeks or more familiarly to "24/7" operations. The combined forces of globalism and satellite- and fiber-linked computer systems are driving us toward continuous operations that never shut down.

These advanced satellite and fiber systems are creating far more information than individuals can read. Our most advanced telecommunications systems are now transmitting data in the 100 gigabit/second to the terabit/second range. A terabit is a trillion bits of information. This is, in the American system, 1,000,000,000,000 bits of information. This would mean that a single communications device, operating at a terabit/second, could transmit enough information in about 36 hours time to be equivalent to 6 billion books or the equivalent of a book for every man, woman, and child on the planet. (This assumes that a 300-page book represents 2.5 megabits of information with about 8,000 bytes or 64,000 bits of information to a page.) Such a device, of course, is just one of hundreds or even thousands of such systems. A single Intelsat-9 satellite alone can move the equivalent of the Library of Congress in a matter of minutes. There are already five of these new satellite networks deployed around the world. When one considers that there are nearly 300 geosynchronous and about 200 low or medium Earth orbit communications satellites now deployed and operational around the world, one begins to understand the image of an emerging worldwide mind.

Humans can only process about 56,000 bits of information in a second at top speed. If we were required to think at such speeds for prolonged periods of time, our brains not only would begin to hurt, but we would experience other symptoms as well. We would become apathetic, distracted, in need of sleep, and then seek to become "couch potatoes." As noted earlier, the Lazarfeld–Merton experiments with white rats showed extreme sensory stimulus ends with inability to eat, loss of sex drive, and death.

Within a decade, we may well literally be paying more for the organization, sorting, culling, and selection of information that we receive than we will be paying to plug into information networks. With features like caller ID and antispam programs, we are already well along this road.

Already over 100 million people around the world can select between 100 and 300 different TV channels transmitted 24 hours a day by DTH satellite TV, and the number of available channels keeps expanding. Studies by sociologists over the last 50 years, going back to the Lazarfeld and Merton experiments, confirm that information overload can lead to a host of mental and physical dysfunctions. Satellite distribution is widening the sphere of information overload, super speed-living, and attempts to lead 24/7 lives. Recent studies released by a U.S. medical research institute indicate that sleep deprivation in the United States has increased by over 15% in the last 5 years. An increasing percentage of drug use is not to seek "highs," but to remain awake and alert longer. In the recent Iraq war, pilots flying long missions were routinely given drugs to help them fly 24-hour missions.

Automation and Telework Versus Teleisolation and Culture Shock

When John Kenneth Galbraith wrote about growing affluence in the most economically advanced nations back in the 1960s, he predicted that automation and advanced communications networks such as computers and satellites would provide us with new levels of wealth and leisure time to enjoy family, recreation, and culture never before encountered in human history. Dr. Galbraith's prediction did not come true. We are indeed seeing productivity going up, but our economic systems are geared toward more and more throughput and increased levels of consumption. As a result, professional people are working longer hours, and employment among women in the United States and Europe continues to increase.

In the United States, in particular, married couples—both working-class and especially professionals—are often finding they are working 10-hour days or longer, especially when long commutes are added to the process. (The officially recorded hours worked as recorded by government and industry statisticians and the actual hours put in by workers at home, on the road, and in home offices are frequently well out of sync with one another. This allows productivity numbers to increase even when this may actually not be the case.)

The advent of telework and telecomputing was once believed to be a panacea that would allow working people to spend less time on the job, more time with their children, and less stressful lives. This may ultimately prove to be true. The reality of telework today, however, has indicated that new concerns must be recognized and faced. Especially in the U.S. model, where people are working from their homes, there are new concerns about unregulated child labor, a lack of control of hazards in the home workplace, and new forms of isolation and even "piece rate" electronic sweat shops.

There can be no doubt that telecommuting will help conserve energy, reduce air pollution, increase industry productivity, and reduce costs, but there must be doubt that it can and will provide increases in "quality of life" unless socioeconomic protective measures are taken. This can be particularly true with regard to international teleworking via satellite links where child labor and other abuses are not curtailed or monitored.

A LITERARY VERSUS A VISUAL WORLD

When can more information be less information? Clearly this can be the case when information shifts from word-based systems to multimedia- and visual-based networking. The number of pixels in a TV screen that are flashed before the viewer many times a second is 10,000 times greater than the bytes of information in a written document, but the transferred knowledge can be many times less. As we move to high-definition and three-dimensional video systems, the amount of "wasted" or "discarded information" in a single video frame exceeds 99.999%. Recent studies confirm that those who live in a video and computer game world suffer a loss of vocabulary and decreases in certain critical thinking skills. There is a serious question as to whether satellites that provide us only broader band services and faster throughput rates are sufficient. In the future, do we need to develop satellite systems that also serve broader social and educational goals? Would satellite- and Internet-based educational programs actually be more effective if they were more text than multimedia oriented?

Teleeducation Versus Electronic "Know-Nothings"

Certainly satellite-based teleeducation and other social services offer great hope of reaching the billions of people in the so-called Third World without health care or educational systems. Low-cost microterminals can provide access to previously underserved areas, and educational and medical software can allow schooling and health care instruction in remote locations in Africa, Asia, South America, and other areas that lack basic social services. (Indeed there are significant pockets of need to be found in countries such as the United States, Canada, Australia, and Europe as well.) Ironically, as new electronic tools and software are being developed to address the social needs of emerging economies, the academic achievement levels in the most developed cultures are currently dropping. Functional illiteracy in the United States is estimated to be around 17%, and even in Europe and Japan it is now rising to around 5%. Not long ago, functional illiteracy was almost unknown in Japan, South Korea, and many parts of Europe, but this is no longer true. Although there are many hypotheses as to why this is happening, TV, computer games, "de-skilling" of dead-end service jobs, and school dropouts are recurring themes in possible explanations.

Economic Growth-Oriented Telesociety
Versus Survival-Oriented Values

There is increasing concern that the current economic model of modern Western industrial societies may well ultimately become obsolete. Critics suggest that there are seeds of defeat embedded in economic systems that depend on growth values rather than survival values. They suggest that sustainable development is incompatible with continued economic expansion, ever-increasing growth of population, higher levels of consumption, and modest or no reward for environmental protection or recycling. They also note the problems inherent in the ever-widening separation of elements of consumption and production in modern global economic systems (with a corresponding drop in pride and integrity of creation and quality of service). The local economy may have had its faults, but it had built-in elements of self-correction. If you traded your wool for a neighbor's food produce and the produce was bad, you could easily correct the problem.

If one actually took a longer term view of survival challenges and needs for humanity in a 21st-century economy, there would likely be a number of changes made in our approach to transportation, energy, housing, and most certainly communications and information networking. Satellite communications certainly do not create these problems, but they greatly expand the sphere of the difficulty and amplify the degree of impact.

FINAL THOUGHTS

One could continue to expand this discussion endlessly. We can find more and more examples of the positive and negative impact of modern satellite systems. We can certainly speculate about how new, faster, and cheaper satellite technologies will continue to impact 21st-century patterns of human living, working, war making, politics, and recreation. Undoubtedly, some of these innovations will make our life better and easier. Others will place us in greater peril. In some cases, new technology will bring increased telepower accompanied by greater risk of teleshock outcomes as well. It is hard not to acknowledge that satellite networks have changed almost all aspects of our lives from how we get news or are entertained to how we buy things or how we work. Satellites, for better or worse, are key change agents in 21st-century life. As new satellite and fiber technology explodes in terms of increased digital throughput and still lower telecommunications and information costs, these impacts will only increase in magnitude.

The true question that we all need to ask ourselves—and our economic, social, and political leaders—is not about our new information technology, but about our long-term and short-term values. Often our short-term calculations are at odds with the interests of our children and grandchildren. Questions that need asking include:

1. Is it enough to ask of satellite system and fiber operators that they develop networks that are simply broadband, faster, and lower in cost?

2. How can we create not only better electronic and photonic machinery, but also actually create improved educational, medical, social, and cultural systems that serve the needs of our grandchildren and generations to come? How can we create economic and political systems that are wise enough to look further into the future?

3. As we move from national and regional to international and global economic, social, and cultural networks and systems, how can we use new technology not only to transact business faster and at higher volume, but also to improve the human condition? How can we improve "human value" in terms of security, environment, world peace, prosperity, and sustainable livelihood for ourselves and for our progeny?

4. Can we find ways—though economic, legislative, regulative, or judicial systems—to control the use of digital information systems to produce a higher quality of life, protect our privacy, offer greater protection of our environment, and sustain human life and the ultimate survival of *homo sapiens* on our planet?

We know we can build and deploy faster, higher volume, and more interconnected satellite systems, and over the last 40 years we have proved it. The challenge of the next 40 years is to design satellite and communications systems that increase the spirit and value of human life.

REFERENCES

Drucker, Peter, (1998, October 5), Management's New Paradigms, *Forbes*, pp. 152–177.
Pelton, Joseph N., (2001, February), Arthur C. Clarke Lecture, Washington, DC.

BACKGROUND READINGS

Iida, T., and Suzuki, Y. (2001, April 17–20), "Communications Satellite Research for the Next 30 Years," 19th ICSSC Conference, Toulouse, France, AIAA, Paper No. 2001-233.
Landovskis, J., and Rayment, S., (1999, June 15), "Emerging Broadband Satellite Network Architecture and the Major New Developments," Proceedings of the Fifth International Workshop on Satellite Communications in the Global Information Infrastructure (GII), Ottawa, Canada.
Pelton, Joseph N., (1992), *Future View: Communications, Technology and Society in the 21st Century*, Boulder, CO: Baylin Publications.
Pelton, Joseph N., (1998, April), "The Future of Telecommunications," *Scientific American*, pp. 1–85.
Pelton, Joseph N., (2000), *E-Sphere: The Rise of the World Wide Mind*, Westport, CT: Quorum Books.

Glossary

AEEC—Airline Electronics Engineering Committee

BSS—Broadcast Satellite Service

CCIR—Consultative Committee on Radio (established by the ITU and now known as ITU–R)

CCIT—Consultative Committee on Telegraph (established by the ITU and now known as ITU–T)

CNES—the French National Space Agency

CNN—the Atlanta-based Cable News Network

COFDM—Coded Orthogonal Frequency Division Modulation

Comsat—or COMSAT—Communications Satellite Corporation that was acquired by Lockheed Martin Corporation and then disbanded

C-band/Ku-band/Ka-band/L-band—the frequencies in which satellite transmissions operate

DASA—the German National Space Agency

DBS—Direct Broadcast Satellite

DOS—Disk Operating System

DTH—Direct-to-Home satellite service

DVD—Digital Video Recorder

EADS—European Aerospace and Defense Systems

EFT—Electronic Funds Transfer

EIRP—Effective Isotropic Radiated Power

359

ELV and RLV—Expendable Launch Vehicle and Reusable Launch Vehicle

ESA—European Space Agency

E-Sphere—Concept of a electronic interconnected world that is also referred to as "the World Wide Mind," "the Global Brain," "One World," or the "Global Village"

Eutelsat—or EUTELSAT—European Telecommunications Satellite Organization

FSS—Fixed Satellite Service

GMDSS—Global Maritime Distress and Safety System

GMPCS—Global Mobile Personal Communications Services.

HAPS—High Altitude Platform System

IBRD—International Bank for Reconstruction and Development (World Bank)

IGO—Inter-Governmental Organization

IRC—International Record Carriers (these U.S. record carriers included: ITT, Western Union International, and RCA)

IMCO/IMO—Intergovernmental Maritime Consultative Organization

IMF—International Monetary Fund

Inmarsat—or INMARSAT—International Maritime Satellite Organization

Intelsat—or INTELSAT—International Telecommunications Satellite Organization

IP—Internet Protocol

ISP—Internet Service Provider

ITU—International Telecommunications Union

LAN/MAN/WAN—Local, Metropolitan, and Wide Area Network

LEO/MEO/GEO/HEO/NGSO—Low-Earth Orbit, Medium-Earth Orbit, Geostationary, or Geosynchronous Earth Orbit, Highly Elliptical Orbit, and Nongeostationary Orbits

LMSS—Land Mobile Satellite Service

MSS—Mobile Satellite Service

MARECS—Maritime Experimental Communications System

MARISAT—Maritime Satellite System

MPEG—Motion Picture Expert Group

Microwave—Radio Frequencies in the 3 GHz to 30 GHz range that include the C, Ku, and Ka bands that are used for satellite communications

NACA—National Civil Aeronautics Agency (predecessor organization to NASA)

NASA—National Aeronautical and Space Administration

OECD—Organization for Economic Cooperation and Development

PTT—Governmental Postal, Telegraph, and Telecommunications administration

Panamsat—or PanAmSat—Pan American Satellite Corporation, Inc.

Pelton Merge—The concept that digital communications and IT systems will be increasingly be "seamlessly interconnected" so that fiber optic cables, coax, terrestrial wireless, cellular, and satellite can be linked without difficult via compatible protocols

RF—Radio Frequency (a wide range of electromagnetic frequencies that are used for telecommunications and other practical and scientific purposes)

RLV—Reusable Launch Vehicle

SoHo—Small Office/Home Office

TDM/QPSK—Time Division Modulation/Quadraphase Shift Keyed

TVRO—Television Receive Only satellite terminal

TWTA—Traveling Wave Tube Amplifier

UNESCO—United Nations Educational, Scientific, and Cultural Organization

USAT—Ultra Small Aperture Terminal

VSAT—Very Small Aperture Terminal

WARC—World Administrative Radio Conference

Biographies of the Authors

JOSEPH N. PELTON (EDITOR)

Dr. Joseph N. Pelton is director of the Space & Advanced Communications Research Institute (SACRI) at the George Washington University. He also serves as director of the Accelerated M.S. Program in Telecommunications and Computers at the George Washington University Virginia Campus. Further, on a volunteer basis, he is the founder and executive director of the Arthur C. Clarke Institute for Telecommunications and Information (CITI), which works in partnership with telecommunications research institutes and foundations in Europe, North America, and Asia. He is the author of 22 books in the fields of telecommunications, telecommunications and space policy, and satellite communications, including Pulitzer Prize-nominated *Global Talk*.

He has served as chairman of the board (1992–1995) and vice president of Academic Programs and Dean (1996–1997) of the International Space University of Strasbourg, France. This experimental international academic institution specializes in graduate interdisciplinary studies and hosts study programs at leading universities around the world, in addition to its master's program conducted at its main campus in France. Dr. Pelton has also been director of the Interdisciplinary Telecommunications Program at the University of Colorado at Boulder, director of Strategic Policy at Intelsat, and founding president of the Society of Satellite Professionals International in 1983. He has received the Arthur C. Clarke Award, the SSPI Hall of Fame Award, and the H. Rex Lee Award from the Public Service Satellite Consortium for leading Project Share at Intelsat. He is a full member of the International Academy of Astronautics.

Dr. Pelton was the 1997–1998 chairman of a NASA and the National Science Foundation Panel of Experts on Satellite Communications, and from 1998 to 2001 on the External Evaluation Committee for the Japanese Space Agency (NASDA). Dr. Pelton holds BS, MA, and PhD degrees from the University of Tulsa, New York University, and Georgetown University, respectively.

ROBERT (JACK) OSLUND (EDITOR)

Dr. Jack Oslund is an Adjunct Professor on the faculty of the George Washington University's Graduate Telecommunication Program, and Coordinator of the interdisciplinary Graduate Certificate Program in "Telecommunication and National Security." Dr. Oslund retired in 2000 from the former Comsat Corporation, where, since 1974, he served in a variety of management positions involving the U.S. Signatory roles in Intelsat and Inmarsat and was Director of Corporate Regulatory Relations. He also served as a private-sector advisor on U.S. delegations to the International Maritime Organization, the International Telecommunication Union, and the Inmarsat Assembly. During the latter part of his career at Comsat, he was selected as five-term Chairman of the Legislative and Regulatory Task Force of the President's National Security Telecommunications Advisory Committee (NSTAC). He has testified on Critical Infrastructure Protection issues before Congress, and was a contributor to the Congressional Joint Economic Committee Report on *Security in the Information Age: New Challenges, New Strategies*. He also has contributed chapters to books and articles to journals and trade publications on satellite and international communications issues. Prior to joining Comsat, Dr. Oslund was an International Staff Officer in the White House Office of Telecommunications Policy, a Senior Associate and Executive Director of the Communications Committee at the Chamber of Commerce of the United States, and a faculty member at the Joint Military Intelligence College (formerly the Defense Intelligence College). He also was an officer in the U.S. Marine Corps. In addition, he has been an Adjunct Professor in the International Communications Program of The American University's School of International Service, and a lecturer at the Department of State's Foreign Affairs Service Training Center. He received his PhD in International Studies from American University's School of International Service.

PETER MARSHALL (EDITOR)

Peter Marshall's career bridged the two disciplines of TV news and satellite communications. He started as journalist with BBC News in the 1960s. From the 1970s to the mid-1980s, he was a senior executive with Visnews of London, one of the first companies to pioneer and develop the use of global satellite systems for the coverage and distribution of TV news. He moved to the United States in 1986 to create the Broadcast Services Division of Intelsat, and 3 years later he returned to the private sector to become the president of Keystone Communications. This organization, now known as Globecast (a subsidiary of France Telecom), became the largest provider of global satellite broadcast services, with some 50 transponders in service worldwide.

Peter Marshall, now semiretired and living in the United Kingdom, was the president of the Society of Satellite Professionals International (SSPI) from 1988 to 1993, and he was elected to the Society's "Satellite Hall of Fame" in 2003. He continues to serve as a board member of the Arthur C. Clarke Foundation of the United States.

TAKASHI IIDA

Dr. Takashi Iida received his BE, ME, and Dr. Eng. degrees from the University of Tokyo in 1966, 1968, and 1971, respectively, all in electronics engineering. He joined Radio Research Laboratories (RRL), Ministry of Posts and Telecommunications (MPT), in 1971. From 1975 to 1976, he was a visiting researcher at Communications Research Centre, Department of Communications, Ottawa, Canada. In 1987, he was transferred to National Space Development Agency of Japan (NASDA). He was senior engineer of Satellite Program Department, NASDA, and involved in the communication and broadcasting satellite projects. He was appointed director of Space Communications Division, Communications Research Laboratory (CRL) (former RRL), MPT, in 1989. He was visiting professor of the Interdisciplinary Telecommunications Program, University of Colorado at Boulder, from 1991 to 1992.

He was appointed director general of CRL in 1998. He was subsequently appointed president of CRL, Independent Administrative Institution and continues to hold that position at the present time. He was general chair of the 21st AIAA ICSSC held in Yokohama, Japan, on April 15–19, 2003. He is editor and author of a textbook entitled "Satellite Communications-System and Its Design Technology," Ohmsha and IOS Press, 2000, and co-author of "Communications Satellite Systems," Encyclopedia of Physical Science and Technology, 3rd Edition, Academic Press, 2002. He received the AIAA Aerospace Communications Award for 2002. Dr. Iida is Associate Fellow of AIAA, Member of IAA, Fellow of IEEE, and Fellow of the Institute of Electronics, Information, and Communication Engineers of Japan (IEICE).

LOUIS IPPOLITO

Dr. Ippolito is vice president and chief scientist for ITT Industries, Advanced Engineering and Sciences in Ashburn, Virginia. His current responsibilities include telecommunications systems development and analysis for a broad range of government and commercial programs. He is the author of over 60 publications, including the books *Radiowave Propagation in Satellite Communications Systems* and the *NASA Propagation Effects Handbook for Satellite Systems Design*. Dr. Ippolito is a contributing author for The Communications Handbook, J. D. Gibson, Editor. He teaches graduate courses and continuing engineering courses for the George Washington University, and he is also an adjunct professor in Telecommunications Management at the University of Maryland University College. Dr. Ippolito is a Fellow of the IEEE, Senior Member of the AIAA, and a Member of the International Radio Science Union, and he has participated as a U.S. Delegate to International Telecommunications Union Radio Conferences.

NAOTO KADOWAKI

Naoto Kadowaki received his BS in communications engineering and his master's degree in information engineering, both from the University of Tohoku, Sendai, Japan, in 1982 and 1984, respectively. From April 1984 to March 1986, he was with Mitsubishi Electric Corporation. He joined Communications Research Laboratory of Ministry of Posts and Telecommunications in 1986. His research interests are high data rate satellite communications systems, mobile and personal satellite communications, computer networks, and communications protocols. From July 1990 to June 1991, he was a visiting researcher to AUSSAT that was reformed as OPTUS Communications, Sydney, Australia. He is currently the manager of Gigabit Satellite Communications section, Space Communications Division of Communications Research Laboratory. He is a member of the IEEE and IEICE of Japan. He is also an active member of the AIAA, the Japan Forum, and the Japan U.S. Science Technology and Space Applications Program.

NORM LERNER

Dr. Lerner is the founder and president of TRANSCOMM since 1970, and has a distinguished professional background in engineering, management science, and financial economics. He has been responsible for the management of TRANSCOMM's major domestic and international telecom, energy, and postal projects. The author of many publications, Dr. Lerner has a BSEE from MIT, an MBA from Columbia University, a PhD in Mathematical Economics from American University, and is a licensed professional engineer.

He has been an active participant in the Presidential Task Force on Communications Policy, FCC Cable Television Advisory Group, NASA Study Group for Satellite Communications in Underdeveloped Nations, Congressional Office of Technology Assessment Working Groups for Common Carriers and International

Communications, FCC Advisory Group for the Landmobile World Administrative Radio Conference, and the U.S. Telecommunications Protocol Mission to the People's Republic of China.

Dr. Lerner has contributed to numerous economic and policy studies and has provided expert witness testimony in over 40 federal and state regulatory proceedings concerning telecom, postal, and energy issues. He has been involved in numerous market research, regulatory, and financial studies for private clients, multilateral organizations, and federal government policymaking agencies. Domestic and international telecom analyses have emphasized domestic satellite, wireless, and cable/wireline facilities and services.

Dr. Lerner has also participated in many foreign telecommunications projects for both governmental and industrial clients, particularly in Mexico, Central and South America, Central Europe, and Asia. Activities have included privatization/liberalization and PTT organizational and financial assessments, regulatory policy and structure analyses, market/demand studies, cost and pricing analyses, new facilities and services development, and financial planning studies.

Prior to forming TRANSCOMM, Dr. Lerner was a consultant to the executive office of the president of the United States, Office of Telecommunications Policy. Previous management and engineering positions were with Computer Sciences Corporation, MITRE Corporation, International Telephone and Telegraph, and RCA Corporation. He was also adjunct professor of Management Science and Technology at George Washington University and the University of Maryland. He is also a guest lecturer at the Inter-American Defense College.

HAMID MOWLANA

Hamid Mowlana is professor and founding director of the International Communication Program at the School of International Service, American University, Washington, DC. He has been on the American University faculty since 1968. He served as the president of the International Association for Media and Communication Research (IAMCR) from 1994 to 1998, and he is now an honorary president of the association. He has been a visiting professor at a number of universities in Europe, Asia, Latin America, and Africa, and he has worked with UNESCO in Paris.

Professor Mowlana is the author and editor of numerous books and publications on international communication and international relations, and he serves on the editorial board of various scholarly journals. He is the recipient of various national and international awards and was name Distinguished Scholar by the International Studies Association (ISA) and as Scholar/Teacher of the Year of the School of International Service. Professor Mowlana received his PhD at Northwestern University.

ERIC J. NOVOTNY

Dr. Eric J. Novotny is presently a vice president with International Launch Services, a joint venture of Lockheed Martin Corporation and the Khrunichev State Space Research Center. He was formerly vice president of Europe, Hughes Electronics, Brussels Belgium, and held a variety of positions with COMSAT Corporation and as an officer in the U.S. Air Force. Novotny was also adjunct professor of Management Science at Vesalius College, Brussels. He received his MA and PhD in government from Georgetown University and his diploma in management studies from Oxford. The opinions in this book are those of the author and not necessarily those of any organization with which he is affiliated.

D. K. SACHDEV

D. K. Sachdev is the founder and president of SpaceTel Consultancy based in Vienna, Virginia, which provides business strategy and engineering support for satellite and wireless systems for broadband, telecommunications, and broadcasting services.

D. K. Sachdev teaches courses in satellite communications as an adjunct professor at George Mason University, Virginia, and he is also a visiting scholar to the Space & Advanced Communications Research Institute (SACRI) at the George Washington University. He has over 50 technical and research publications to his name.

He is also the cofounder and president of Nirvano Technologies, Virginia, a company planning to provide advanced wireless-based health surveillance systems with MEMS and nano sensors.

He also served as senior vice president of Engineering Operations, WorldSpace in Washington, DC. In this position, he led the development and subsequent deployment of complete digital radio systems in Africa and Asia via direct access radio satellite networks. He also led the initial engineering development of the XM radio system.

At Intelsat, he served as the head of managed technology, R&D, and system development programs, and in this capacity he designed and developed 16 large satellites that today form the backbone of the Intelsat and New Skies international networks.

ROBINDER N. SACHDEV

The cofounder and corporate development officer of Spectrum Access, a leading edge provider of broadband wireless applications, Mr. Sachdev focuses on the strategy, technology research, and finance needs of the company. Mr. Sachdev is a strategy and communications expert with several innovative ventures behind him. Starting his career as an education specialist, teaching systems analysis and design to top corporate officials and bureaucrats, he went on to cofound businesses in software consulting, electronic publishing, and international trade. He has been an assistant to Mr. Rajiv Gandhi, then prime minister of India, and he devised software and communication strategies for two general elections in India.

Mr. Sachdev has authored articles on technology and strategy, and has worked to translate strategic planning into viable projects—with institutional, public, and private partners, including the International Finance Corporation and the private equity markets. Recipient of several scholastic achievement awards, Mr. Sachdev is also an adjunct faculty in International Communications at American University, Washington, DC, and he founded the *Intercultural Management Quarterly*, where he serves on its editorial board. Mr. Sachdev holds MBA and MA degrees in International Communications, both from American University, Washington, DC, and a bachelors of technology degree in electrical engineering from G B Pant University, India.

INDU B. SINGH

Dr. Indu Singh is the founder of Via AsiaNet and serves as chairman of this company. Via AsiaNet owns a telecom software company, TBL Software, in India. TBL is a joint venture between Via AsiaNet and the Indian Government Ministry of Communication. Via AsiaNet is currently implementing nationwide VOIP and Pre Paid Calling Cards and international long-distance services in India. From 2001 to 2002, Dr. Singh also served as president and CEO of Transat Holdings, a satellite-based ISP provider in South America.

Prior to serving as president of Transat Holdings and creating Via AsiaNet, Dr. Singh served as founder, president, and CEO of JMS North America, Inc. (formerly James Martin Strategy, Inc.), a wireless engineering company based in Northern Virginia. JMS provided wireless engineering and network implementation support to leading telephone companies in North America and Europe as well as consulting support in Asia.

Dr. Singh holds BS, MS, and doctoral degrees in telecommunications. Earlier in his career, Dr. Singh was an associate professor of Telecommunications at Rutgers University in New Jersey.

DAN SWEARINGEN

Dan Swearingen is a communications systems engineering consultant. He served in a variety of technical and management positions during his 29 years with COMSAT Corporation, including engineering vice president for COMSAT Mobile Communications. In the 1970s, he led the systems engineering group that defined the civilian MARISAT communications system architecture (later known as INMARSAT-A). From 1980 to 1981, he served in London at the newly created International Maritime Satellite Organization (INMARSAT) preparing for the beginning of INMARSAT operations in early 1982. After his return to COMSAT, he continued to support the evolutionary planning for the INMARSAT system through his contributions to the INMARSAT Advisory Committee on Technical and Operational Matters (ACTOM).

ROBERT N. WOLD

Robert Wold is a pioneer in the use of U.S. domestic satellites for broadcasting. His role in satellite communications broadcasting started in the late 1970s in the domestic markets, but his LA-based company, Wold Communications, expanded into international markets in the 1980s. This was accomplished in partnership with Visnews of London, then headed by Peter Marshall. Together they developed transatlantic and transpacific satellite news services and worked with the Japanese TV networks to create the Japanese International Satellite Organization (JISO). Wold Communications became part of Keystone Communications in 1989.

Robert Wold was a founder and chairman of the Society of Satellite Professionals International (SSPI) and is a member of the SSPI "Hall of Fame." Since retiring from Wold Communications, Robert Wold has become a leading writer and historian on the satellite industry.

Author Index

Subject Index

Printed in the United States
by Baker & Taylor Publisher Services